BIOSCIENCES

INFORMATION SOURCES AND SERVICES

BIOSCIENCES

INFORMATION SOURCES AND SERVICES

Y. Alston
J. Coombs

M
stockton
press

Published in the United States and Canada by
STOCKTON PRESS, 1992
257 Park Avenue South, New York, N.Y. 10010, USA

ISBN 1-56159-047-9

First published in the United Kingdom by
MACMILLAN PUBLISHERS LTD, 1992
Distributed by Globe Book Services Ltd
Brunel Road, Houndmills,
Basingstoke, Hants RG21 2XS, England

ISBN 0-333-57639-X

A catalogue record for this book
is available from The British Library.

While every care has been taken in compiling the information contained in this publication, the publishers and editor accept no responsibility for any errors or omissions.

Typeset and printed in Great Britain

CONTENTS

PREFACE

This book has its origin in the Macmillan publication *The International Biotechnology Directory*, first published in the early 1980s in response to the growing need for a reference book covering information sources, organizations and companies involved in biotechnology. Since then there has been an explosion of scientific, commercial and industrial activity in the biosciences; not only in respect of biotechnology, but also covering medicine, food and the environment. Many of these developments have been linked with the parallel expansion of the microelectronics industry linked to the PC (personal computer), data handling and information technology; including desktop publishing, on-line searching, electronic mail, CD ROMs, computer imaging and the ability to reproduce complex chemical formulae and/or reactions on screen.

This explosion of information in the biosciences generates many problems for both those seeking facts and those handling experimental data. Hence this book which serves to bridge the information gap.

The contents of this book have been extended well beyond that of the original information services section of the *Directory*. First, activities, products and developments in *BIOSCIENCES: Information & Services* are described in more detail than was possible in the *Directory*. Second, the scope has been extended to cover a wider range of different types of material and a broader range of scientific, commercial and industrial disciplines. Again, reflecting the advances in software and equipment now available to handle the information explosion, the copy has been prepared from databases developed in-house by Katy Hall and Michael Spooner - to whom our thanks are expressed.

Y R Alston and J Coombs

INTRODUCTION

This book covers sources of scientific and technical information of interest to those working in the biosciences. The term, as used, covers the more fundamental aspects of biology as well as those areas of application which depend on scientific principles rather than the empirical use of living organisms in a traditional manner. Although aimed at both academic and industrial users this book does not cover commercial or marketing information - a guide to which can be found in the Macmillan *Directory of Business Information Sources*.

In general medical and veterinary aspects which relate to patient care are not included. However, information sources which cover new developments in techniques, such as genetic engineering and hybridoma technology, which can lead to commercial products with applications in diagnostics or therapy are included. Agriculture, horticulture and forestry are dealt with in the same way. General contact addresses are given to relevant international and national organizations, whilst the more detailed information sources concentrate on those describing investigations or applications involving reactions at the molecular level.

International Organizations, Government Departments, Scientific Societies and Trade Associations are listed in Part II (ORGANIZATIONS).

Part III (PUBLICATIONS) covers established scientific journals, newsletters and some trade publications as well as abstracting services and directories.

In fact the development of the capability to access many of the abstracting services through computer links (modems) leads to an overlap with Part IV (SERVICES) which covers such databases, as well as others which offer access to protein or nucleic acid sequences, the contents of culture collections or availability of equipment. This is supplemented by lists of hosts and operators of such databases as well as suppliers of software and systems for capture and handling of the large amount of analytical data associated with developments in molecular biology.

Commercial organizations offering consultancy, especially those offering services related to the provision, interpretation and use of scientific and technical information, are also included in this section. This part concludes with a list of culture collections, including those which offer the opportunity for deposit of material for the purpose of patent registration.

The scope of the contents of this book is broad, reflecting the increased impact of the biosciences in all walks of life. In order to facilitate the rapid identification of the required source of information a complete list of all entries can be found in the main index. In addition a further index is provided for each section. The level of discrimination in these indexes reflects the appropriate degree of specialization of the type of entry.

For example, organizations, societies, associations and publications are listed under generic headings ranging from 'Agriculture' to 'Waste Treatment'. For the databases and services section a further level of selection has been used, with specific identifiers covering aspects such as sequences, cell lines or hybridomas. Additional information is provided with each entry as appropriate. However, to avoid repetition or redundant information, where the title of an organisation or publication is self-evident such notes are avoided.

Obviously the content of this book reflects its origin in the *Biotechnology Directory* and it is anticipated that the readers will have the same interests, the two volumes being used in a complementary manner.

Although, since the *Directory* was first published some ten years ago the number of biotechnology firms and their products has expanded enormously, this growth has not necessarily been in the directions anticipated. Increase in biological knowledge has been associated in particular with those aspects related to molecular biology, the understanding of inheritance and the expression of such information as proteins. The ability to perform *in vitro* genetic manipulation (otherwise known as recombinant DNA technology, or genetic engineering) has promised much and major markets have evolved covering diagnostics and test systems on the one hand and therapeutic peptides and proteins on the other. However, these are just the most visible tips of a multi-billion dollar, worldwide industry, where the word industry is used in its widest sense to encompass research, commercial exploitation and information. In fact it is the information barrier which is probably the hardest to surmount. It is hoped that this book will assist in this respect acting as an initial guide to this complex meeting ground between academia and commerce as well as the resulting ramifications through government and society.

PART I

DEVELOPMENT OF INFORMATION AND TECHNOLOGY

THE EXPANSION OF THE BIOSCIENCES

Initial concepts and ideas suggested a biological revolution that would rival the expansion in micro-electronics and IT (information technology) and which would bear fruit in the 1980s. In the event this has proved too optimistic. In part the new technologies have suffered at the hands of the general worldwide recession commencing in the late 1980s and still deepening. Even allowing for this debilitating effect of national economies the commercial perception of biotechnology as the *opportunity of the 1980s* has failed to live up to expectations. In part this reflects a lack of appreciation of the complexity of biological systems and the time required to harness the novel techniques and in part a failure to appreciate fully the impact of such radical concepts on the public and the concerns they generate.

These concerns arise in particular in the areas of human genetics, the production of transgenic animals and to a lesser extent with transgenic plants - especially where the products may be used as food. Worries about the release of genetic manipulated organisms (GMOs) into the environment represent one facet of increasing concern about the environment in general. This may be local (typified by responses of the NIMBY or Not in My Back Yard type), covering waste disposal, factory effluents, new roads or indeed anything else seen to disturb the peace and ecological balance of living organisms. Such concern may also be global, with concepts of acid rain, the hole in the ozone layer, atmospheric warming due to increased release of so-called greenhouse gases, the survival of the rain forests, biodiversity and sustainable agriculture at the forefront of debate.

Within this framework the concept of biotechnology has become a focus for debate over what can be described as non-technical barriers to exploitation - especially in Europe. European and related national legislation has singled out this branch of technology from all others, approaching it with the trepidation some ascribe to the concept of nuclear power - or the fear of the global catastrophe. Elsewhere, in the USA, the approach has been more rational with legislation tending to move towards a case by case analysis of processes or products in terms of their safety and efficacy irrespective of the means of synthesis or manufacture.

Further problems are associated with the north/south divide, where the concern is that genetic material isolated from tropical regions of the less developed nations may be exploited only by those with the knowledge and funding to do so, located in the developed countries (North America, the more northern regions of Europe and Japan in particular). At the same time global economies based on agriculture and international trade in food and other biologically-based commodities are under threat from surpluses (as thousands starve in other countries) inducing schemes for set-aside of good agricultural land, whilst applications of genetic engineering promise still higher yields. Hence, on a global basis the perceived problems of the 1960s and 1970s, a need to establish security of supply in respect of food and energy, have changed in the 1990s to those indicated above - largely related to the environment and the need to improve human health, nutrition and general life style on a global basis.

The achievement of such objectives is in turn highly dependent on the rational application of discoveries and technical developments in microelectronics and biology.

In fact the two major areas of industrial development and environmental concern are linked through and dependent on what can be grouped together as the biosciences. These continue to generate an ever-increasing deluge of complex data, information and legislation. This flood can be expected to increase by a further order of magnitude with ambitious plans to provide a complete sequence representing the human genome, as well as other representative organisms from the plant, animal and microbial kingdoms.

In the past the life sciences were divided neatly according to kingdom. The overall science would, in the last century, have been clearly recognized as biology, becoming more specialized as botany and zoology and dividing further into taxonomy, morphology, anatomy, physiology and so on. At the same time further specialisation developed covering various parts of the microbial, plant and animal kingdoms resulting in disciplines such as mycology, entomology, parasitology and many others whilst greater understanding of the activity of cells was provided by the application of increased chemical and physical knowledge, as biochemistry and biophysics.

The realisation that like bred like, that characteristics could be inherited and that by applying such patterns of inheritance, plants and animals useful to man could be improved, led through genetics, to animal and plant breeding as an input to farming, horticulture and forestry. At the same time increased emphasis on the science of life attracted the application of radioactive tracer technology, the electron microscope, spectroscopy and various separation technologies (chromatography, electrophoresis) enabling the living cell to be taken apart - both mechanistically and literally. This in turn resulted in the well documented discovery of the structure and function of nucleic acids embodied in the genetic code and the subsequent commercial potential of genetic engineering.

These changes in scientific emphasis have been reflected in the academic world, with university departments, journals and societies moving away from the classical areas of descriptive biology towards the molecular aspects. For instance the term botany has lost favour in preference to the alternatives of more specialized (eg plant sciences, plant biochemistry) or applied (eg agricultural botany) branches of the science. The same is true for zoology and indeed in many universities the old distinctions have been removed with the establishments of departments or schools of biology, again with the emphasis on biochemical, physiological and genetic aspects.

Since many journals are almost entirely dependent on academia as a source of both contributions and subscriptions, the contents of journals also reflect these changes. Hence, the emphasis of the dozen or so plant journals listed here is on plant molecular sciences, protein metabolism and physiology overlapping with those journals devoted to such topics on a generic basis covering microbial, animal and plant sciences. At the same time there is still a role for applied journals. These can again be recognized on the basis of either the sector (agriculture, horticulture, forestry), the product or the specific area of concern (plant breeding, disease), with national journals (eg Canadian, Japanese, etc) reflecting local science interests.

At the same time physiology and biochemistry have also tended to become more specialized, reflecting an emphasis either on microbiological activities, on applied aspects (as biotechnology) or on cellular aspects ranging from molecular biology to membrane science. In parallel there is a further divergence into numerous aspects of biotechnology, including an emphasis on both the scientific issues and the commercial aspects as discussed in the next section. As far as the needs of these industries are concerned they can be summarized as shown in figure 1, with the emphasis on identifiable chemical entities. These can be divided into those interacting with life forms in a defined manner, as either pharmaceuticals or pesticides, and those which may be used as components of manufactured goods, where the emphasis is on manufactured foods and fine chemicals.

The problems in handling the diversity of information and source material (animals, plants, microbes and cell-lines) are compounded by problems of identification and agreement on nomenclature, dealt with by international organizations such as IUPAC, CODATA and COBIOTECH. These concerns are parallelled by the need to document accurately the ever-increasing range of plant and animal cell lines as well as establish the true relationships between various species, varieties and strains of microorganisms, plants and animals as new information becomes available through the use of molecular techniques, rather than conventional comparative descriptions, to facilitate commercial development. For, although much of the basis for the expansion of the biosciences has come from the activities of pure scientists in academia the most dramatic impact has been, and will continue to be, through the industrial application of this knowledge.

In particular a Commission on the Terminology and Nomenclature of Biology has been established by CODATA. The major purpose of this commission is to focus the attention of those working in the biological sciences and in particular those interested in bioinformatics on such problems. The participants include the American Type Culture Collection (ATCC) which has both the largest collection of organisms and an internal conviction that database integration and standards for data exchange are absolute requisites for the future. As yet such consistent use of standardized nomenclature and terminology by all database developers does not exist - indeed it is unlikely that all existing working committees within all of the relevant sub-disciplines of biology who are working on standards for terminology and nomenclature have yet been identified - such is the size of the problem. Once this has been done the sanctioned lists of terminology which they have produced will become available in electronic form. Various international bodies are supporting the activities of this commission and CODATA is optimistic that the work of database users will be greatly facilitated by having a uniform and accepted information resource or biological terminology with which to work.

Figure 1. Areas of bioscience information priorities.

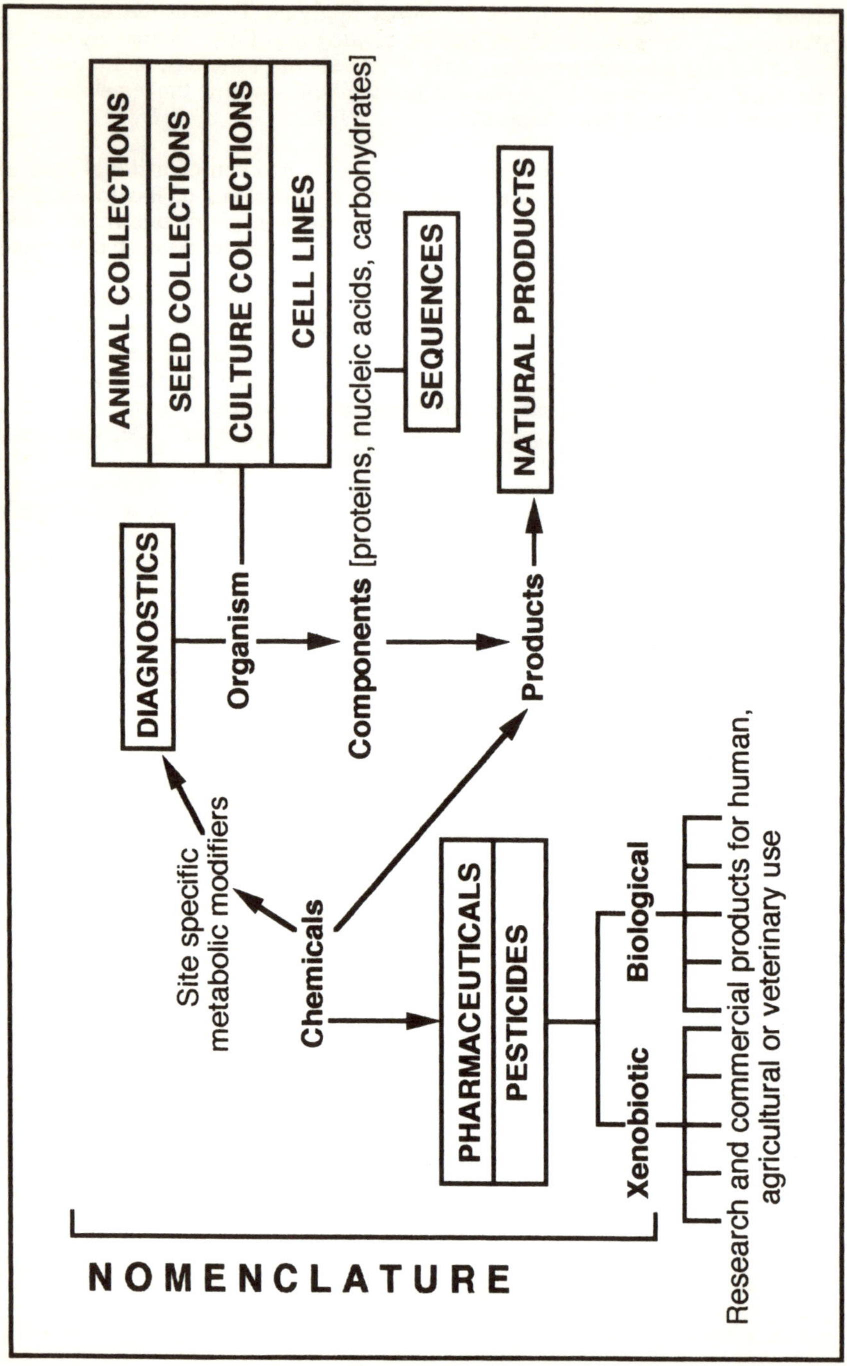

THE EXPANSION OF BIOSCIENCE DATA

The industrial application of life forms extends backwards in time through the involvement of bacteria, yeasts and moulds in fermented foods and beverages; complemented over the past 50 years or so by the growth of the pharmaceutical (antibiotics) and speciality chemicals (amino acids, organic acids, polysaccharides) industries. This activity parallelled an increase in scientific inputs into agriculture, horticulture and forestry which can be encompassed under the general headings of propagation material (improved breeds or varieties of both animals and plants), improved nutrition for animals or fertilizers for plants, the application of growth promoters (again for both plants and animals) and improved control of disease and pests (veterinary and plant protection products).

The more conventional fermentation processes are largely encompassed within the classic definition of biotechnology, strongly pushed in Europe through the European Federation of Biotechnology, which defined the concept of biotechnology as 'an industrial process in which biological organisms, enzymes or organelles were used as catalysts to convert a raw material, generally assumed to be of biological origin, into some product of commercial value'.

In the early 1980s many branches of biotechnology, seen as an extension of the fermentation industry, were promoted in anticipation of growth which has not in general been realized - mainly as a result of high feedstock prices and the inability to compete with low price fossil fuels but also due to the lack of distinct products (other than food, beverages and some pharmaceuticals) which can better be produced by fermentation. Hence, anticipated areas of biotechnological growth (energy, biodegradable plastics, organic acids, oil recovery, mineral extraction and so on), have not really come to fruition.

In contrast the discoveries in molecular biology and related research and business activities have continued to receive ever-growing attention from scientists, industrialists, politicians, legislators (government), educationalists and the religious or theological sectors alike. This sector has thus generated the greatest increase in information and at the same time the widest range of problems.

Activities centre on an understanding of the genetic code (as detailed below) and the application of this to forensics, diagnostics, new product development (diagnostics and therapeutics) and an understanding of the genetic makeup in terms of sequence analysis of the genes from a vast array of organisms. This activity has been complemented by developments in cell fusion and gene transfer which have developed in two directions.

The first is the production of fused cells (hybridomas) capable of producing unique proteins with specific immunological properties (monoclonal antibodies) of use in diagnostics and therapeutics. The second area of activity has been in the production of transformed lines, cells or organisms containing new genetic information as well as the tools (restriction enzymes and vectors) used in this area of science. At the same time, the emphasis on microorganisms as industrial catalysts has focused increasing attention on culture collections.

Attention has become focused on the ability to determine, manipulate and express nucleic acid sequences as new peptides or proteins; to achieve a better understanding of how to unravel the genetic basis of many plant, animal and human characteristics as well as subsequently manipulate them. In the short term a major business oportunity has developed, depending on a wide range of enzyme, immunological or nucleic acid-based procedures as tests for a wide range of analytes of interest to the medical, agricultural, food and environmental sectors. At the same time the possibilities of introducing new genetic material into both plants and animals as well as simpler forms of life ranging from viruses to fungi, and moreover manipulating the structures of their proteins by protein engineering, have opened up new concepts for production of pharmaceuticals, long-life tomatoes, high-lysine grain feeds, disease- and pest-resistant organisms, new vaccines and above all the longer-term possibility of establishing the blueprint of life itself, in terms of the base sequence of DNA or the human genome.

Already considerable progress has been made, not only with humans but with archetypic plant systems and yeasts (where the sequence of one chromosome is already available). The question of whether such sequences should be patented is only one facet of an increasing debate as to whether life itself is patentable.

The problems with the information explosion arising from this activity liein the logical simplicity of the way in which most of the information is stored in living cells and subsequently decoded and used to express a product (protein) in terms of a sequence of amino acids. The investigation of protein or nucleic acid sequences and the need to handle, store and compare the data produced can be broken down into a number of specific areas of activity which include:

1. enumeration and comparison of the results of the preliminary analysis, including data capture and processing. Often this involves the scanning of series of bands generated on plates of gel which can be visualized by staining or incorporation of radioactive markers.

2. encoding of the sequences in terms of the genetic code, based on a series of three-letter sequences (codons) forming what has been termed the triplet code, in which each common protein amino acid has one or (usually) more sequences of base pairs which determine the position of each amino acid in any specific protein. The main bases are depicted by the initial letter of the respective bases, which are either purine- or pyrimidine- based nucleotides. The most common, indeed almost universal, being adenine (A), guanine (G), cytosine (C) and thymine (T). The basis of the code is outlined in table 1 which shows the corresponding RNA codes in which uracil (U) replaces thymine (T). Sequences are often measured in terms of kbp, or kilobase pairs, reflecting the size and extent of information required to encode a protein. With essentially four letters only to encode such information the ability to recognize similarities or distinguish differences would be a task beyond even the most astute and retentive human mind. Hence, the computer has come to play an ever-increasing role as a means of storing, retrieving, comparing and collating information which then has to be extended further to cover the sequences and structures of the encoded protein generated on transcription.

Table 1. The genetic code based on nucleotide triplets. The three-letter amino acid abbreviations are the same as those shown in table 2.

1st base	2nd base				3rd base
	U	C	A	G	
U	Phe	Ser	Tyr	Cys	U
	Phe	Ser	Tyr	Cys	C
	Leu	Ser	STOP	STOP	A
	Leu	Ser	STOP	Trp	G
C	Leu	Pro	His	Arg	U
	Leu	Pro	His	Arg	C
	Leu	Pro	Gln	Arg	A
	Leu	Pro	Gln	Arg	G
A	Ile	Thr	Asn	Ser	U
	Ile	Thr	Asn	Ser	C
	Met	Thr	Lys	Arg	A
	Met	Thr	Lys	Arg	G
G	Val	Ala	Asp	Gly	U
	Val	Ala	Asp	Gly	C
	Val	Ala	Glu	Gly	A
	Val	Ala	Glu	Gly	G

The handling of protein sequences represents a further order of difficulty since most proteins can contain any of at least 20 different amino acids to which may be added various unusual moieties. Again the traditional nomenclature was based on an (arbitrary) three-letter code based on the amino acid name (eg gly for glycine or glu for glutamic acid). However, in order to reduce the amount of information which has to be stored alternative conventions have been adopted, as illustrated in Table 2.

The problems of handling such sequence information has been addressed by GenBank (Genetic Sequence Data bank) and similar organizations as detailed below, largely funded by the US National Institutes of Health. The purpose is to collect and distribute nucleic acid sequence data and related information. Collaborators at the Los Alamos National Laboratory gather and annotate the data whilst the European Molecular Biology Laboratory collects and distributes their data, principally in Europe. The information is updated every few weeks, whilst software information, electronic mail, on-line help and a telephone hotline are available. The technical challenge to be faced is making such databanks available to the scientific community worldwide, whilst developing the computing tools to allow the information to be used effectively as the total amount of data stored runs into millions of base pairs.

Table 2. Single-letter and three-letter codes for common amino acids.

Symbol		Amino acid	Molecular weight
One-letter	Three-letter		
A	Ala	Alanine	89
C	Cys	Cysteine	121
D	Asp	Aspartic acid	133
E	Glu	Glutamic acid	147
F	Phe	Phenylalanine	165
G	Gly	Glycine	75
H	His	Histidine	155
I	Ile	Isoleucine	131
K	Lys	Lysine	146
L	Leu	Leucine	131
M	Met	Methionine	149
N	Asn	Asparagine	132
P	Pro	Proline	115
Q	Gln	Glutamine	146
R	Arg	Arginine	174
S	Ser	Serine	105
T	Thr	Threonine	119
V	Val	Valine	117
W	Trp	Tryptophan	204
Y	Tyr	Tyrosine	181

In the same way the ability to introduce changes in the genetic make-up of any organisms may involve the use of a particular class of proteins known as restriction enzymes which have specific properties in cleaving nucleic acids in a specified way at specific sites, whilst the gene (piece of nucleic acid of interest) will be incorporated into a suitable virus-like particle or cloning vector. Again both aspects lead to a need for an agreed terminology as well as databases of such materials. There is also a need to identify and then distinguish between transformed genes and their expression products, all of which has led to further need for shorthand conventions, such as developed for restriction enzymes. These all have the general name endonuclease R, followed by a system name which is derived using the following rules. The species name is identified by the first letter of the genus name and the first two letters of the specific name to form a three letter abbreviation used in italics, for example an enzyme from *Escherichia coli* becomes *Eco*. Any strain, type, virus or plasmid associated with the specific enzyme is indicated by a letter, whilst variations are indicated by roman numerals. The site at which such enzymes attack nucleic acids can also be described using the same nomenclature, but in upright, rather than italic script.

Unfortunately the same rigorous treatment has not been applied to culture collections, many of which have been built up over the years on an *ad hoc* basis. There is a need for data sources containing phenotypic strain descriptions and to this end the Microbial Strain Data Network has been set up as a locator service for clone repositories having defined combinations of attributes. In the same way the Microbial Culture Information Service provides an on-line database giving the source of supply of named organisms which may be searched in terms of names, appearance, substrates, industrial properties, literature citations and so on.

A further development as far as culture collections are concerned has been their involvement in patenting. Business cannot survive without protecting such information as provides it with a competitive advantage over competitors. At the same time there is a need to protect invention as enshrined in the patent law. Due to the range of inventions an equal range of patent literature exists, covering both various topics and various patent agencies which include national and regional (European) requirements. The basic principle of all patent systems is the same - the inventor of a new product or process publicly discloses the details of their invention and in return is granted a legally enforceable right to exclude others from exploiting it for a prescribed time. Disclosure of an invention is effected by means of a patent specification, which is published by the relevant patent offices. To be disclosed adequately the invention must be described in sufficient detail to allow a skilled person to reproduce it. This is done by a written description illustrated by drawings and calculations as appropriate.

Although the patenting of inventions is established, aspects relating to living organisms are more complex. Obviously it is not the purpose of this book to discuss the issues surrounding the question of patenting of genes, sequences, organisms and so forth. However, the access to such information is obviously of importance to anyone thinking of using a particular organism or following a particular line of business. A fundamental requirement of patentability is that all details of the invention must be disclosed fully to the public. Hence, it has been argued that where a patent covers a life form, for instance a microorganism, such disclosure is incomplete unless cultures of the organism are also available to the public. For this reason a system has been set up which allows such organisms to be deposited with a suitable culture collection and maintained for the life of any patent relating to it. The mechanism of deposition is regulated by an international agreement (the Budapest Treaty), although further details vary according to national law. Such laws may not permit such patents anyway, or may specify when, to whom and under what conditions deposited organisms can be released.

The main reason for having to deposit a sample is that in many cases the invention could not be described as it may represent a wild-type organism which has been improved by mutation or genetic engineering. The actual treaty, *Budapest Treaty on the International Recognition of the Deposit of Microorganisms for the purpose of Patent Procedure*, was concluded in 1977 and came into force towards the end of 1980. Under the terms of this treaty certain culture collections are recognized as IDAs (International Depository Authorities) and a deposit made to any one of these is recognized as valid by other treaty members. Any culture collection facility can become an IDA, so long as it has the capabilities to discharge the function efficiently. Most of the recognized collections are listed in the section on culture collections in part IV. However, it should be noted that the types of organism which they hold differ depending on their interests, facilities and capabilities - especially in respect of potential pathogens. It should also be noted that some collections extend beyond the boundaries of classic microbiology to include viruses, phage, protozoa and cell-line cultures. Whilst the most comprehensive facility is offered by the American Type Culture Collection in the USA, in fact the UK has the greatest and most diverse range of facilities extending beyond the usual bacteria, yeasts and moulds to cover algae and protozoa, pathogenic bacteria and cell lines, as well as industrial organisms.

Some examples of the directions in which advanced computer techniques are being used to establish on-line databases are given below, covering the major areas where such data are increasingly becoming necessary.

The World Data Centre on Collections of Microorganisms, located at RIKEN, Saitama, in Japan, has developed a database of culture collection information which is available on-line through the international packet switching service (PSS). The centre is part of the World Federation for Culture Collections (WFCC). The database lists culture collection information on over 300 collections worldwide. The kind of information listed includes contact information for the collection, sponsors, number of staff, main interests and functions, if any sub-collections exist, availability of cultures, whether an identification service is provided and if a printed catalogue is available. Additionally, the species maintained by each collection are listed. The World Federation of Culture Collections is a commission of the International Union of Biological Sciences.

The *Microbial Strain Data Network (MSDN)* is an on-line referral database to centres or laboratories having information on the properties of microorganisms and cell lines used by others. The MSDN Central Directory can be used to locate the appropriate laboratories. Most of the tested properties are described as precise statements for inclusion in an on-line thesaurus within the directory. The MSDN Secretariat will make searches of the directory for interested persons without access to computer facilities. It was formed by the Committee on Data for Science and Technology (CODATA), WFCC and the International Union of Microbiological Societies (IUMS). Access to the MSDN network is possible by direct dial-up to the Telecom Gold computer in London, UK, or the Dialcom computer in Washington DC, US, or through the international and/or national packet switching service.

MSDN has a number of on-line services. These include the provision of databases, provision of facilities for database producers wanting international access, electronic mail, telex, fax, bulletin board, computer conferencing, software distribution (microbiological shareware programmes), training and consultancy. Currently several hundred electronic mailboxes have been issued to scientific users from over 30 countries.

MSDN provides an integrated 'one-stop-shopping' service through a system of gateways to other databases and information centres and eases the difficulties in arranging access to the wide range of data available on microorganisms and cultured cells. It is accessible to academic and industrial organizations throughout the world. The directory includes descriptions and references to sources of bacteria, yeasts, actinomycetes, filamentous fungi, algae, as well as vectors, hybridomas and animal cell lines. The following catalogues are available through MSDN: American Type Culture Collection (ATCC) catalogues of cell lines, protozoa and algae, human gene probes/chromosome libraries, and bacterial, phage and media; CAB International Mycological Institute Culture Collection catalogue, UK; Deutsche Sammlung von Mikroorganismen und Zellkulturen (DSM) catalogue, West Germany; European Collection of Animal Cell Lines (ECACC) catalogue, UK; National Collection of Yeast Cultures and National Collection of Food Bacterial catalogues, UK.

Other major sources of strain information that can be found in MSDN include the Microbial Culture Information System and Microbial Information Network Europe (MINE).

The *Microbial Culture Information Service (MiCIS)* is another bibliographic life-science reference database which contains strain data on approximately 30,000 microorganisms held in national and private culture collections in the UK. It includes name and synonyms, source, hazards, morphology, enzymes, culture conditions, metabolites, sensitivities, maintenance requirements, industrial applications and properties, alternative strain designations, patent status, and literature references. MiCIS currently offers an integrated service on behalf of all the collections of microorganisms in the UK. It provides detailed information about the strains held, and can be searched using the name of the strain and the field of information required (eg physiology, special features). Searches can also be made by depositors, habitat, references and so on. The contributing collections are: National Collection of Type Cultures (NCTC); National Collection of Industrial and Marine Bacterial (NCIMB); National Collection of Yeast Cultures (NCYC); CAB International Mycological Institute (incorporating the National Collection of Fungal Cultures); National Collection of Food Bacterial (NCPF); National Collection of Pathogenic Fungi (NCPF); and National Collection of Plant Pathogenic Bacteria (NCPPB). Previously hosted by the Laboratory of the Government Chemist (LGC), part of the Department of Trade and Industry (DTI) in the UK, it is now maintained at the Information Centre for European Culture Collections (ICECC), Braunschweig in Germany.

A different type of service is offered by products such as *COMPASS* which consists of a probabilistic system for yeast identification that allows the user to enter the results of up to 46 morphological and physiological tests and receive the probability of match with any of over 450 species. In contrast the *Taxonomic Reference File* (TRF) developed by the Bio-Sciences Information Service (BIOSIS) includes a 12,000 entry database of bacterial names and a bibliographic field of more than 13,500 bacterial taxonomy references published since 1969. Several related files containing information related to microorganisms are also available on TRF, including Staphylococcus Genetic Stock Center, Listing of Molecular Biology Databases (LiMB) and several amino acid and nucleotide sequence databases.

A number of databases provide access to and describe nucleotide and protein sequences and usually include references to the sources describing them. Sequence analysis software packages allow researchers to compare unknown sequences from their work with those that are published. Many of the databases and corresponding software are available in a variety of formats, including CD-ROM, floppy disks, on-line via a number of networks, or in hard copy. A comprehensive listing of molecular biology databases is provided in LiMB. The CASSEQ Biopolymer Sequence File (Chemical Abstracts) Service and their fee-based services can provide comprehensive sequence database searches. CASSEQ, the only commercially available source of patent sequence data, covers protein and nucleotide information in patents from 1981 to the present.

The main databanks carrying the basic sequence data are GenBank (USA), NBRF (National Biomedical Research Foundation, Washington), EMBL (European Molecular Biology Laboratory, Heidelberg) and DDBJ (DNA Databank of Japan). These are only useful in combination with suitable software although they can be accessed with many general software packages. Details are published in the Software Directory but much more up-to-date information can be obtained on-line from LiMB which lists all molecular biology databases. IntelliGenetics Inc has software which is available in a comprehensive series of interactive programs on BIONET. For UK users, the SEQNET node at Daresbury offers an extensive, free (for non-profit organizations) service via JANET and other networks. Its databases include the main nucleic acid, protein and vector databases plus software packages. For human genes, the Human Genome Mapping Library (HGML) at Yale has databases carrying information on mapped genes, RFLPs, probes, addresses of people from whom they may be obtained, and references and abstracts from relevant literature. This is on-line, but they also publish free directories regularly.

The following range of sequence databanks can be accessed by MEDLINE and BIOSIS: DDBJ, DNA Data Bank of Japan; CarBank, Carbohydrate Structure Database; EMBL, European Molecular Biology Laboratory (neucleotide sequences); HDB, Hybridoma Data Bank; GenBank, GenBank Nucleic Acid Sequence Database; HGML, Howard Hughes Medical Institute Human Gene Mapping Library; MIM, Microbial Strains Database; PDB, Protein Data Bank (Brookhaven Crystallographic Database); PIR, Protein Identification Resource (amino acid sequences); PRFSEQDB, Protein Research Foundation (amino acid sequences, Japan); PDS, Protein Sequence Database (Kyoto); SWISSPROT,Protein Sequence Database (translated EMBL).

A major source is *GENEBANK* - a bibliographic reference database containing more than 27,000 citations to journal articles and technical reports in genetic research, as well as descriptions of DNA and RNA sequences with 50 or more nucleotide bases - including more than 22,000 reported sequences, totalling over 26 million bases. Each record contains 16 data items, including bibliographic data, sequence annotations (including significant features in each sequence), source organisms and starting point of sequence described. Other sources include the *EMBL Nucleotide Sequence Databank* which includes over 25,000 nucleotide sequences covering approximately 27 million bases. Information for each sequence includes title, species, nucleotide composition, length, promoter and gene regions, topology, nucleic acid sequence and literature references. Sequences are from eukaryote, prokaryote, virus, bacteriophage genetic DNA or RNA, mRNA, and functional RNA molecules. In comparison the *Protein Sequence Database* contains descriptions of over 6,700 partial and whole protein sequences representing over 1.7 million amino acids that were isolated or inferred from the gene sequences. Descriptions include function of protein, taxonomy, sequence features of biological interest, how the sequence was experimentally determined, unambiguously determined residues within the sequence and citations to relevant literature.

The *Hybridoma Data Bank* consists of a database of commercially available cloned cell lines and their immunoreactive products. It is sponsored by the International Union of Immunological Societies (IUIS) and the Committee for Data on Science and Technology (CODATA). It lists information about reactants, cross-reactants, distributor of cell line or product, immunocyte donor, immortal partner, type and classification of product, availability restrictions, special conditions, and literature references. Hybridomas are selected for their ability to secrete immunoreactive substances. The most common of these substances are monoclonal antibodies which react with defined specificity to particular structural components of proteins and carbohydrates. Monoclonal antibodies are used in such diverse disciplines as microbiology, immunology, biochemistry, agriculture, environmental science, botany, and clinical medicine. Because of the wide scope of interest in these products of cloned cell lines, there is a growing demand to access data generated by laboratories which create and test these immunoreagents.

HDB is now an international network with four nodes to collect and disseminate data. The four major nodes are located at ATCC, the Institute of Physical and Chemical Research (RIKEN) in Saitama, Japan, the Centre European de Recherches, Documentaires sur les Immunoclones (CERDIC) in Nice, France, and the National Institute of Immunology (NII) in New Delhi, India. There is also a collecting node at the European Collection of Animal Cell Cultures at Porton Down, UK, and a disseminating Node in Ottawa, Canada, which provides on-line access to the HDB through the National Research Council of Canada's CISTI system.

Further databanks are available in the USA (see part IV) covering both enzymes and vectors. In particular details of restriction enzymes are available in a bibliographic reference database which contains descriptions of some 1,000 restriction enzymes, including details of the enzyme, recognition sequence and site of cleavage. A similar vector databank contains the nucleic acid sequences of several hundred frequently used cloning vectors as well as relevant restriction maps.

Plant genetic resource databanks are also available, for instance through IBPGR - the International Board for Plant Genetic Resources. IBPGR is a centre of the Consultative Group on International Agricultural Research (CGIAR), sponsored by the Food and Agriculture Organization, United Nations and World Bank. It coordinates an international network of crop genetic resources. The seed collections have developed over a decade, and a network of active collections is now being created. IBPGR has published five directories and catalogues of significant germplasm collections. They have a database of gene banks and their standards, and they are currently completing the compilation of databases of information from the genetic resource centres. Work is being concentrated on 10% of the core collections because there is such an abundance of data. The databases cannot be directly accessed; requests should be addressed in writing to IBPGR.

The *International Union for the Conservation of Nature and Natural Resources* (IUCN) also maintains a database of rare plants in botanic gardens worldwide with details of who has what, its wild origin and availability of material. The collections covered include 600 seed banks.

For industry, details of equipment and services are also available through the efforts of DECHEMA (*Deutsche Gesellschaft fuer Chemisches Apparatewesen, Chemische Technik und Biotechnologie e.V.*, supplied by FIZ Chemie GmbH to STN International, Germany, who act as host for the following series of databases:

BIOQUIP (Biotechnology Equipment Suppliers Data Bank). A reference database which contains descriptions of about 2,500 biotechnology industry products available in Germany from about 400 manufacturers. Covers equipment for mechanical and thermal processes, packaging and storage, measurement and control; chemicals and instruments for laboratory analysis, safety devices; and pumps, compressors, valves and gaskets. Includes product description, name and address of manufacturer and information on distributors.

DEQUIP (DECHEMA Equipment Suppliers Databank). A similar database contains descriptions of about 50,000 products from some 2,800 companies worldwide. Covers product categories for computer-aided engineering hardware and software; general and specialized analysis and testing equipment; laboratory instruments and chemicals; pumps, compressors, valves and fittings; process instrumentation, control, and automation equipment, and equipment for use in mechanical and thermal processes, nuclear and radiochemical applications, materials handling and storage, and safety engineering. It includes manufacturer's name, address and products.

THE EXPANSION OF BIOSCIENCE INFORMATION

The expansion of information *per se* in the life sciences can be matched to the changes in science and industry detailed previously and then coupled to the evolution of the media through which such information is stored, manipulated and disseminated. Although computer and electronic data storage are becoming ever more important the printed record remains the mainstay of the biosciences - whether the information is academic, commercial, industrial or government based (figure 2).

Figure 2. The written word.

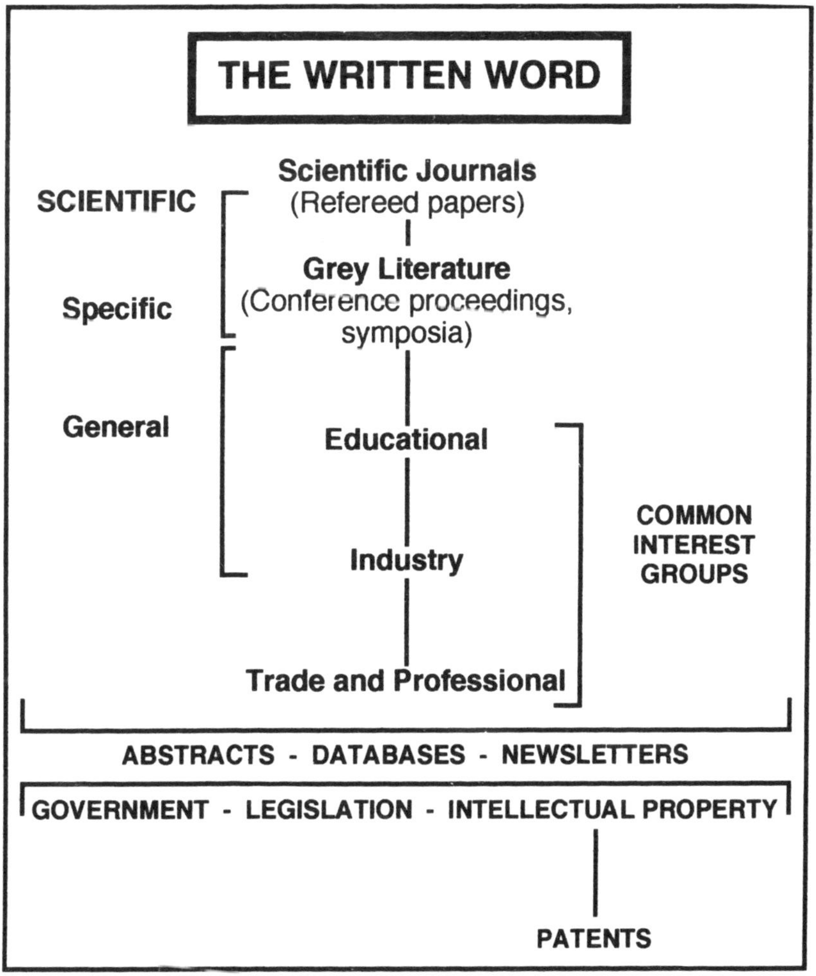

Bioscience: Information Sources & Services

The printed documentation covering the biosciences includes many different types of serials or periodicals. These range from scientific journals, through technical publications, society bulletins and official documents to rapid newsletters, often produced in-house using PC-based publishing systems. In addition the biosciences depend on catalogues of life forms, whether lists of herbarium acquisitions, seed catalogues, culture collection lists, breeding stock books, and so on. Initially these covered organisms where a listing of their name (Latin, binomial), general properties and descriptions (again often in Latin) were sufficient to establish identity. However, the development of biochemical and biophysical methods, linked to automated analysis and the subsequent ability to distinguish between strains, varieties or individuals on the basis of differences in performance, enzyme complement, chemical composition or yield and genetic makeup, including hybrids, and transgenic organisms, has reduced the value of the simple nomenclature and descriptive nature of taxonomy. This is especially true for information concerning nucleic acid sequences and protein structures as detailed in the previous section. However, the need to access such a broad range of information sources, as well as the costs of acquisition (covered in more detail below), generates an increasing need to handle all data in a way that can only realistically be searched, matched and worked upon using the power of modern computers.

As a result there is now a growing dichotomy between the printed word and the computer database as a source of bioscience information. However, this is not a case of competition between media, but rather the application of computer techniques as a means of handling the ever-increasing number of publications and other information now available, at a time when funds available for many institutions, the universities in particular, are decreasing. These aspects can be discussed first in terms of the written word and then with reference to the computer as a means of information storage and retrieval.

Reviewed (refereed) papers, published in recognized journals, run by a largely academic editorial board of acknowledged experts, appearing on a regular basis still form the mainstay of the scientific literature. Perhaps, regretfully, the output of such papers has also become the barometer whereby academic performance is monitored, in some quarters at least - hence the rise of publications such as *Science Citation Indexes*, which simply indicate the extent to which a given paper or author is mentioned by others. In part the growth in number of journals covering a given sector of science is also a barometer of the topicality of a particular subject. The changing emphasis of the biosciences, as well as the growth of the scientific literature over the last decades, is reflected in the number of new journals starting up each year (figure 3). This trend has been parallelled by an equal growth in trade magazines and commercial newsletters as well as by what has been referred to as the grey literature. That is scientific (academic) publications which have managed to by-pass the strict regime of peer review, the most common mechanism being through inclusion in the proceedings of a conference.

Obviously standards of such papers differ. However, it is not uncommon (and under the present economic climate becoming more common) for funding of attendance at a conference to be dependent on publication, whilst the financial viability of the conference depends on attendance, leading to a vicious circle capable of generating five-volume tomes covering hundreds of short (four to five page) contributions, presented to varying degrees of legibility, technical competence and linguistic perfection.

Figure 3. Growth in number and topics of journals in the life sciences.

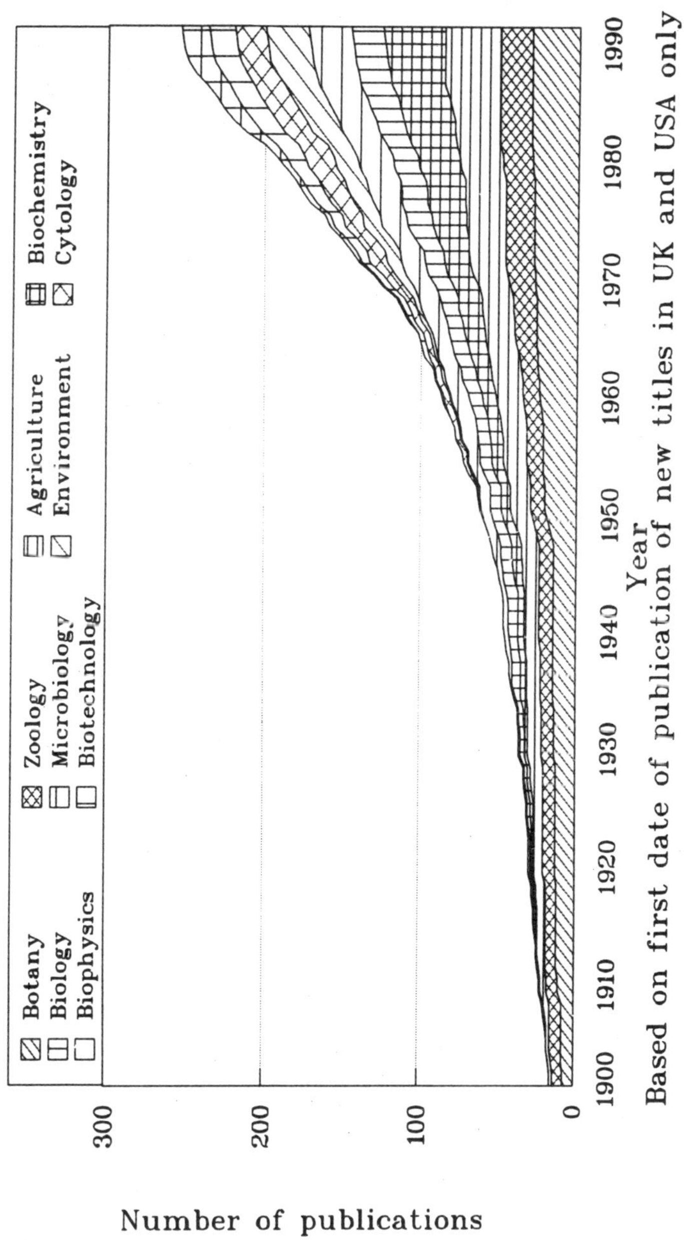

In part the dependence of the scientific literature on academia and research institutes reflects the need for industry to keep many of its results under restricted access - not least as a means of preventing prior disclosure which could invalidate subsequent patentability. In fact a succinct explanation of the differences in approach to publication of technical information can be summed up as:

Academic: publish or perish! Industry: patent or perish!

The basis of scientific information lies in the transactions of learned societies (eg *Science*, USA) and National Academies of Science, together with the more general scientific forum, including periodicals such as *Nature* (UK).

Earlier more specialist journals followed the classical division of science into physical, chemical and biological sectors, together with various branches of medicine. In particular the life sciences divided neatly into plants (botany), animals (zoology) and lower organisms (microbiology), reflecting the divisions of teaching at both elementary and in higher education. However, in the same way that much of modern biological sciences have shifted away from these classical areas, so has the emphasis of many journals.

In spite of changes in subject, the basic form of a scientific paper remains unchanged (Introduction, Methods, Results, Discussion, Conclusions and References). Often for the major journals these sections are further formatted according to detailed instructions following international guidelines, issued by organizations such as IUPAC - especially as far as chemical nomenclature is concerned. However, as yet the question of standardization of science, which would increase the value of computerized databases, remains unresolved. Although science has attempted to standardize units of time, measurement, electricity, force and so on the official SI units may become unwieldy for some uses, or not be compatible with commercial or manufacturing practice. Such problems increase as the biosciences increase in complexity, with nomenclature (particularly in terms of abbreviations adopted as shorthand for complex chemicals or procedures) a major problem. Through the offices of ICSU many international unions and committees have been set up to tackle such problems. However, many areas have yet to be resolved, especially in those areas of molecular biology showing the fastest growth in information output and hence where the need is greatest.

The biosciences have always been descriptive and dependent on terminology. Much of the earlier emphasis, covering taxonomy, morphology and anatomy, resulted in a technical vocabulary capable of distinguishing diverse forms on the basis of external appearance, coining words to cover shape, degree of hairiness and so on. The introduction techniques based on physics and chemistry in the biosciences led in turn to a desire to find the secrets of life itself and theability to describe many activities in terms of chemical equations. At the same time identification of the chemical structures of the major and minor classes of compounds, including proteins, carbohydrates, lipids, proteins, nucleic acids, vitamins, hormones and other components, increased the database of information further. The identification of proteins as biocatalysts and the advent of radiotracer techniques resulted in the ability to determine the nature and control of metabolic cycles, and the chemical basis of physiological responses.

Almost every facet of biology, as it developed, has spawned its own journal or journals, often reflecting the establishment of a common interest group, within a scientific society or association. A number of these have subsequently broken away from the parent to form their own more specialist society or association, often proliferating further journals and newsletters.

Various attempts have been made to reduce this rapidly expanding torrent of paper to more manageable levels through review papers, whilst, conversely, the opportunity to increase the speed with which new results can be made public has been enhanced by the development of both official rapid mechanisms and private circulation networks (often based on the papers presented at meetings, which were never officially published, but all too often appear as un-retrievable, un-traceable references).

The classic form of scientific review is typified by the various series produced by the Academic Press. These consist of detailed analysis of the literature, backed by detailed literature citation. However excellent the scholarship, and being aimed at the professional scientist in a given area, this type of review remains too detailed and discursive for many industrialists or newcomers to the area, such as students, who are catered for by more general, often well illustrated and journalistic articles in glossy publications such as *Bio/Technology, Biofuture* or *Agriculture Hi-Tech.*

An alternative mechanism of covering the proliferating literature is through abstracting journals (many of which are now available on-line, as discussed in more detail below). These vary in degree of specialization and the extent of information, which may be restricted to the title of the paper, provide a brief synopsis, give the author's abstract or be abstracted by professionals. However, once again they are primarily aimed at the active researcher and are of less value to the commercial or industrial sector. Hence, at the same time there is a growing market for publications which serve to bridge the gap between the detailed scholarship of academia and the needs of the decision maker in industry. These include technical newsletters, special studies and business bulletins, which may cover a technology or be aimed at a specific business sector. Such material varies greatly in depth, content, approach and presentation. Often the content extends beyond established science to include information derived from commercial press releases or snippets of gossip.

Trade magazines are a further considerable source of commercial information. In general these can be separated into two groups: those that deal with the interests of a particular industry based on the biosciences (for instance the pharmaceutical, dairy, or brewing industries) and those which are in effect catalogues of equipment and other services for sale. Another trend has been the development of rapid news services, the most impressive of which now offers to fax your copy to you on every business day.

This proliferation has caused a strain on library budgets, whilst presenting the user with an almost impossible task in keeping informed. Both problems have been overcome to some extent by the evolution of abstracting services and the provision of these on an on-line basis, or for personal or library use on CD-ROMs. This has been made increasingly possible due to the decrease in cost of personal computers, which may now, when fitted with a suitable modem, serve in place of the more conventional and historically costly terminals. At the same time familiarity with computer use, database structures and computer terminology has also fuelled the growth of this sector. However, once again it should be stressed that the content of such databases is still ultimately dependent on the written word, forming one link in the chain from author to user as illustrated in figure 4.

Many of these databases (some of which are described in more detail below) have their origin in systems designed by the publishers of large research indexes, such as *Chemical Abstracts* or *Biological Abstracts*. The primary objective was to reduce the

Figure 4. The network for production, construction and interrogation of on-line databases.

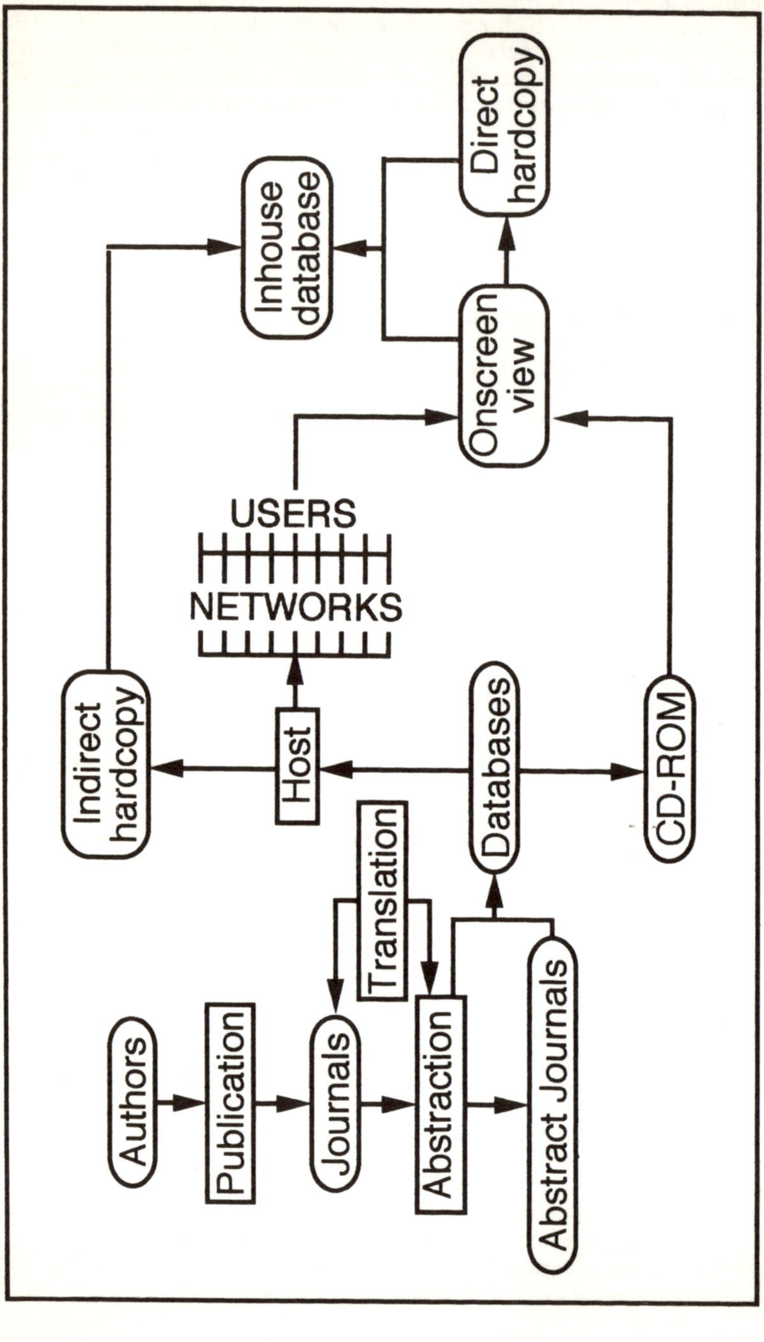

costs of manually producing hundreds of thousands of cross-referenced citations and such indexes have been available for the last 15 to 20 years. Many systems reflect the limitations of searches based on designated keywords coupled with an initial reliance on the skills of the abstracter in selecting the particular information to be included in the abstract.

The quality and precision with which this can be done differs quite markedly, reflecting the complexity of the technology and the level of knowledge of those employed, or where the abstract and keywords are supplied by the author the message emphasized. At present this can limit the value of such searches, resulting in some people who favour such systems and others who do not.

Improved data management and software will lead to a greater ability to store, retrieve and scan complete documents, and indeed this service is already available from some hosts. Such facilities would appear to be imperative for future growth in this area. However, in the meantime access to a wide range of databases is available to anyone equipped with a suitable computer, fitted with a modem linked through a suitable network to one of the many hosts.

The activity of producing a database can be divided into two areas: first compiling the data in a form in which it can be accessed, and second making available a multi-user system, often available 24 hours a day for the on-line user. This latter group, known as vendors of hosts, have formed a booming business selling access time through simplified network systems to more than 800 publicly available databases, many accessible through easy-to-use menu-driven systems a number of which may be free, other than the cost of the telephone call, to some users at least.

The vendor's computer is available for searching directly by telephone communication through a modem attached to a personal computer. Subscribing to a vendor service involves agreeing to a method of payment and receiving a password; most vendors offer user manuals or training classes. Each vendor may offer access to hundreds of databases. Vendors often seek out popular printed sources to sell on-line, including the complete text or abstracts of journals. Full text databases allow a searcher to have rapid access to a complete journal article printed out by a computer within minutes. In addition, some of the vendors offer a document-delivery service for articles that are referenced, but are not available in full text on-line. Searchers can use any microcomputer for this purpose, or they can use terminals which are mainly printer terminals with a modem connection. The choice of telecommunication software available for microcomputers allows searchers to 'download' information to a disk rather than printing it.

Over the last few years a growing number of more specialist databases have also become available. These include specific abstracting services, directed towards the interest of the bioscientist, the holdings of culture collections and strain banks and analytical data describing sequences, structures and characteristics of various types of molecule, including nucleic acids, proteins and complex carbohydrates. Many of these can now also be obtained on various types of media for direct access using a personal computer; smaller databases may be available on floppy disks, whilst large systems may be made available on compact discs (CD-ROMs).

Searching the databases is a simple matter of making a telephone call, typing a few numbers to get connected and then entering keywords or company names as required. Hosts include Dialog, Datastar, STN and others as detailed in Part III. Each host uses slightly different retrieval software, but the systems are generally easy to use.

Most hosts provide excellent training courses and full documentation including general manuals, specific details of databases and quick reference guides to commands. Many organizations run an on-line database service as part of their library or information facilities. However, any individual can obtain a network user identification (NUI) number from their national packet switching service, which is part of the telephone system. The companies providing such service include DIALOG, a major US information company, providing access to hundreds of publishers' databases. Its services are available internationally, 24 hours a day. The European office is based in Oxford UK. DATA-STAR is a Swiss company, also providing access to many databases. STN International is a German company, with offices also in Japan and the US.

The Directory for Biotechnology Information Resources (DBIR) is a central directory to international biotechnology information maintained by MEDLARS from the National Library of Medicine (NLM - USA). This directory lists information about databases (including the major bibliographic databases for literature references), computer networks, culture collections and specimen banks, biotechnology centres, publications and nomenclature committees in the biotechnology and molecular biology fields. ATCC is under contract to the NLM - US to develop and disseminate this product which is available on-line through NLM's MEDLINE and TOXNET systems.

DBIR contains four subfiles - Databases and Networks, Publications, Organizations, and Nomenclature Committees. The directory encompasses a broad spectrum of biosciences including applications in medicine, agriculture, pharmaceuticals, food technology, microbiology and molecular biology. The system includes resources such as computerized databases and their distributors, networks, electronic bulletin boards, culture collections and specimen banks, biotechnology centres and other organizations which stimulate biotechnology growth in academia and industry, publications focusing on general issues in biotechnology which include selected directories, serials, monographs, reviews, and compilations as well as nomenclature committees and standards-setting groups working on issues of nomenclature and terminology in biotechnology and molecular biology. The directory now contains information on some 1,000 information resources for the biosciences and continues to grow as new resources are identified. The search for new resources is international in scope. It is hoped that DBIR will be a comprehensive guide to worldwide sources of information covering all aspects of biotechnology. Potential resources are encouraged to submit data in the form of brochures, articles, information packets, and descriptive data sheets to ATCC for inclusion in the directory. All descriptive information is converted into standard format and standardized text word descriptors for entry into the databank following verification.

An alternative 'database of databases' is that provided by LiMB (Listing of Molecular Biology Databases) which is intended to aid in locating and accessing databases relevant to molecular biology. It was begun in October 1986; the first (Release 1.0) formal release, containing information on 53 databases, was distributed directly to over 300 people, with a greater number of researchers accessing it through six network servers that have mounted copies of LiMB for the communities they serve. Each entry in LiMB currently consists of 54 fields, including: database staff names and addresses; database maintenance hardware and software; scope of coverage and database goals; details about submission and access to the data sets, database size and types of data covered by the database. These data items are described in greater detail below. LiMB entries are based on questionnaires filled out by the database managers or, in the absence of a completed questionnaire, on secondary sources (eg a journal article).

LiMB serves both the potential users of these databases (by helping them find and contact the databases which contain the information they seek) and the designers and managers of the databases (by providing a centralized source of information that can aid coordination between the various databases). It was begun as a simple directory; in the future it will expand this role as a centralized information resource by providing a platform for designing front-ends for automatic access to distributed biological data sets.

Other media of value include:

BIOREP: Produced by the Royal Netherlands Academy of Arts and Sciences, on behalf of the Commission of the European Communities (CEC). This is a reference database for biotechnology research-in-progress containing descriptions of about 6000 research studies in biotechnology conducted in European Community (EC) member states.

ICGEB: The primary purpose of the ICGEB computer resource (ICGEBnet) is to disseminate the best of currently available computational technology in molecular biology to the ICGEB research community. This takes the form of four major objectives: to create an international computer network linking ICGEB New Delhi, the affiliated centres and member country laboratories to a central computer resource located at ICGEB Trieste; to provide computational assistance to ICGEB molecular biologistsin planning experiments, analysing protein and nucleic acid sequences and assembling sequencing projects; to provide on-line access to the major sequence data banks, includingGenBank, EMBL, PIR and SWISSPROT; and to serve as a focus for the development and sharing of new software developed by ICGEB scientists.

ICGEBnet: Provides access to the Intelligenetics Suite, an integrated package of molecular biology programs, supplying all the major functions for sequence analysis including: sequence entry and editing, DNA gel fragment assembly, sequencing project management, protein and nucleic acid sequence analyses, similarity searching, pattern recognition searching, restriction mapping and cloning simulation.

Biotechnology Abstracts: Contains about 75,000 citations, with abstracts and indexing, to the worldwide journal and patent literature on biotechnology, including microbiology, genetic manipulation, biochemical engineering, fermentation, down-stream processing, pharmaceuticals, industrial uses of microorganisms, plant breeding, cell hybridization, and industrial waste management.

Current Biotechnology Abstracts: A bibliographic reference database which contains about 27,000 citations, with abstracts, to the worldwide literature on biotechnology. Covers techniques (eg genetic manipulation, monoclonal antibodies, enzymology, fermentation technology) and relevant industry activities in such areas as pharmaceuticals, energy production, agriculture, chemicals and food. Also covers business, legal and safety issues and information on related conferences. Hosts: DIALOG, ESA-IRS, Telesystemes-Questel.

Pascal Biotechnologies: A bibliographic database which contains about 40,500 citations, with abstracts, to the worldwide literature on biotechnology. Covers biochemistry, cell biology, microbiology and applications of cultured microorganisms and cells in agriculture, energy, food processing, medicine, metallurgy and pollution control.

Bioscience: Information Sources & Services

Biobusiness: A bibliographic database covering biotechnology from the business and industrial viewpoint on an international basis extending into other aspects of the life-sciences. Contains about 164,000 citations, with abstracts to the worldwide periodical literature on business applications of biological and biomedical research. Covers agriculture and forestry, food technology, genetic engineering, pharmaceutical products and other industries affected by biotechnological developments. Also covers patents in such areas as immunological testing, food processes, and fishing. For each patent record, includes inventor's name and address, patent title and number, patent classes, date granted and assignee.

Abstracts in Biocommerce: A bibliographic database covering news and developments in commercial biotechnology worldwide. Contains all the information contained within the printed version (a twice monthly bulletin) as well as additional material in the form of directory references to specific companies worldwide. Contains over 200,000 citations covering industry news, reports of corporate financial developments, research and development projects, personnel changes, and relevant legislation derived from newsletters, journals, periodicals, newspapers, company results, press releases and analysts' reports.

TELEGEN: A Bibliographic database for biotechnology which contains over 23,000 citations, with abstracts, to the worldwide literature on all aspects of biotechnology and genetic engineering. Covers business and economics; social impacts and constraints; patents and legal issues; policy and regulatory issues; general and historical aspects; international discussions and general policy, industrial microbiology; applications to chemistry, energy and mining, the environment, food processing and production, crops, livestock, pharmacy and medicine; and general research and research by specific type of gene (ie human, animal, plants, yeast and fungus, bacterial or viral).

Life Sciences: A bibliographic reference database for life-sciences. Contains citations, with abstracts, to worldwide life sciences literature. Corresponds to the 17 abstracting journals published by CSA: Animal Behaviour Abstracts, Biochemistry Abstracts - Amino Acids, Peptides & Proteins; Biochemistry Abstracts -Biological Membranes; Biochemistry Abstracts - Nucleic Acids; Calcified Tissue Abstracts; Biotechnology Research Abstracts (from 1984 to date); Chemoreception Abstracts; Ecology Abstracts; Endocrinology Abstracts; Entomology Abstracts; Genetics Abstracts; Immunology Abstracts; Marine Biotechnology Abstracts; Microbiology Abstracts - Algology, Mycology & Protozoology; Microbiology Abstracts - Bacteriology; Microbiology Abstracts - Industrial & Applied Microbiology; Neurosciences Abstracts (from 1983 to date); Oncogenes and Growth Factors Abstracts; Toxicology Abstracts; Virology and AIDS Abstracts. Also includes Oncology Abstracts and Feeding, Weight, and Obesity Abstracts for the period they were published.

BIOSIS Previews: A bibliographic reference database covering the biosciences.Contains about 6 million citations, with abstracts, to the worldwide literature on research in the life sciences covering microbiology, plant and animal science, experimental medicine, agriculture, pharmacology, ecology, biochemistry, bioengineering and biophysics. Covers original research reports, reviews of original research, history and philosophy of biology and biomedicine, and documentation and retrieval of biological information. Also covers patents in such areas as immunological testing, food processes, and fishing.

Current Awareness in Biological Sciences: A bibliographic reference database for life-sciences. Contains over 1 million citations to the worldwide literature on biology. Subjects include biochemistry, cell and developmental biology, genetics and molecular biology, clinical chemistry, cancer research, immunology, endocrinology, microbiology, neurosciences, pharmacology and toxicology, physiology, ecology and environmental and plant sciences.

Biotechnology Newswatch: Contains full text of *Biotechnology Newswatch*, a newsletter on biotechnology and the international biotechnology industry. Technical coverage includes biomass conversion, enzyme and fermentation processes, genetic engineering, monoclonal antibodies and recombinant DNA technology. Business coverage includes company products and earnings, financing, new ventures and patent disclosures.

Genetic Technology News: Contains full text of *Genetic Technology News*, a monthly newsletter covering commercial applications in the chemical energy, food and pharmaceutical industries of products resulting from genetic engineering. Includes market forecasts, business news from leading biotechnology firms, announcements of investment opportunities, reports on research activities at universities and research institutions, and reports of new products. Also provides information on recently granted biotechnology patents.

ORIGIN, COST, VALUE AND USE OF INFORMATION

Why use information? In theory the answer could lie between an idealistic scientific search for truth and the commercial cost-effective search for profit. Both academic and commercial sectors would originally have been served by the same sources. Now there are growing departures, although both sectors make use of increasingly sophisticated data management systems. The primary objective of the scientific (academic approach) is validation, review, originality and discovery - often with the final product (a research paper) an end in itself.

In contrast commerce needs to get a product to the market place, without harming the consumer or contravening some other organization's intellectual property rights, and fulfilling the general requirement of the Trades Description Act and the general duty of care in that the product should be fit for the purpose intended, whilst complying with a myriad of health, safety, environmental, and fiscal regulations.

The easiest ways to ensure all these needs is to base each product on a sound technical and information base - whether this information covers economics, investment, patents markets or marketing. As a result a second tier of information has built up, ranging from that in the daily press, through trade journals to multi-client studies and other forms of consultancy.

The cost/benefit of such information can reflect the balance between cost and speed of publication as illustrated in figure 5. These suggested costs do not, with the exception of in-house activities and consultancy services include the cost of staff involvement. At the two extremes lie the cost of having all information services in house and the reliance on free handouts in the form of trade journals and catalogues- which for many small businesses may represent the major, if not only, source of technical information.

At the same time the cost of established scientific literature is increasing as publishers are forced to raise prices above the rate of inflation as a result of falling circulation as university libraries are forced to make cuts. The trend is illustrated in figure 6, which compares the growth of research spending in the United States (an indication of the rate at which new information is being generated) with the rate of growth in spending on library acquisitions. This is repeated throughout the world to varying degrees to such an extent that some journals are now ceasing publication, whilst the major publishers have merged following take-overs placing the future of scientific literature very much under the control of policy makers in a few large organizations. A similar trend can be seen in the development of databases, where the costs of maintaining these may become prohibitive for all but a few major organizations in the richer nations.

Figure 5. The cost and value of various information sources.

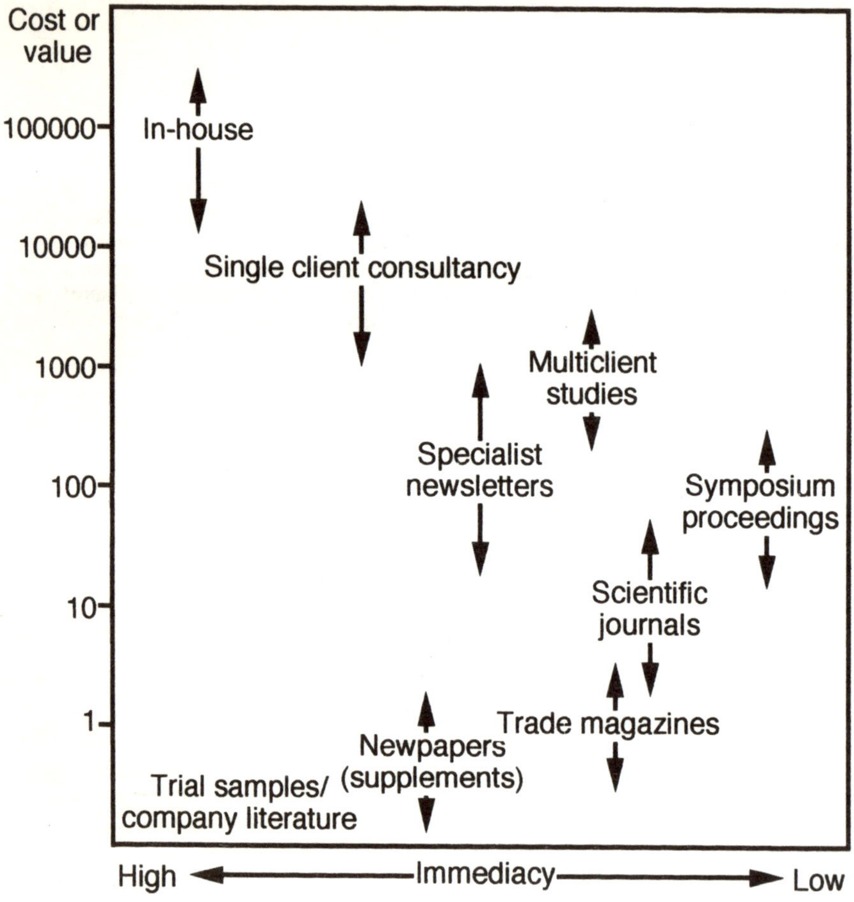

Figure 6. A comparison of research spending and library budgets in the USA.

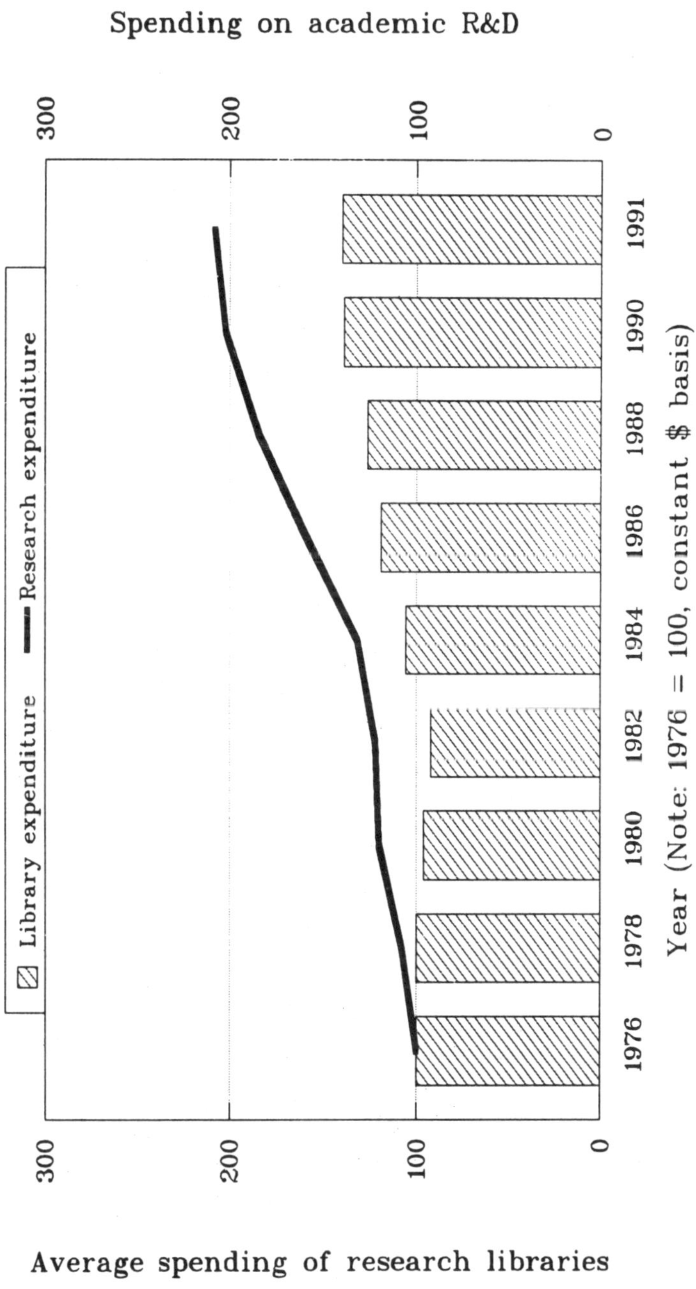

PART II

ORGANIZATIONS

INTERNATIONAL ORGANIZATIONS

Association of Biotechnology Companies - ABC

1666 Connecticut Ave NW, Suite 330,
Washington, DC-20009-1039, USA
Telephone: 202 234 3330
Fax: 202 234 3565

Commission on Biotechnology of the International Union of Pure and Applied Chemistry. Publishes proposals for standard procedures, standard units, nomenclature, symbols etc.

Association of the Cheese Processing Industries - ASSIFONTE

Avenue de Cortenbergh 172, Brussels,
B-1040, Belgium
Telephone: 2 735 81 70

Association of the Cider and Fruit Wine Industries of the EC

Avenue de Cortenbergh 172, Brussels,
B-1040, Belgium
Telephone: 2 735 81 70

Association of the Dairy Industries of the EC - ASSILEC

140 boulevard Haussmann, Paris,
F-75008, France
Telephone: 1 227 12 51

Association Europeenne pour la Biomasse - AEBIOM

APCA, 9 avenue George V, Paris,
F-75008, France
Telephone: 1 47 23 55 40

Established 1990 to further the utilization of biomass for energy in Europe by promoting and coordinating national initiatives and the activities of member biomass organizations and societies, organizing meetings, providing information etc and encouraging technology transfer.

Biomass Users Network - BUN

Washington DC Liaison Office,
PO Box 33308, Washington,
DC-20033, USA
Telephone: 202 778 9706/9665
Fax: 202 293 9211

A non-profit, non-governmental organization, with headquarters in Costa Rica, and regional office in Bangkok, Thailand and Harare, Zimbabwe. It was established to promote South-South cooperation in developing and strengthening national capabilities to increase production and use of biomass resources as a means of advancing economic, social and ecological well-being, primarily in the rural areas.

Biomass Users Network Headquarters - BUN

Central America/Caribbean Regional
Office, PO Box 1800-2100, Guadalupe,
San Jose, Costa Rica
Telephone: 506 40 89 97
Fax: 506 40 89 98

See Biomass Users Network, USA.

Biomass Users Network - BUN

Africa Regional Office, Private Bag
7768, Causeway, Harare, Zimbabwe
Telephone: 4 793 395 6
Fax: 4 793 313

See Biomass Users Network, USA.

Biomass Users Network - BUN

Asia/Pacific Regional Office, Wildlife Fund Thailand Building, 251/88-90 Paholyothin Road, Bangkhen, Bangkok, 10220, Thailand
Telephone: 662 5522111 Ext 133
Fax: 662 5526083

See Biomass Users Network, USA.

CAB International

Headquarters and Information Services, Wallingford, Oxfordshire, OX10 8DE, UK
Telephone: 491 32111
Fax: 491 33508

An international, inter-governmental organization owned by its member governments; these currently number 29, but membership is open to any others wishing to participate. The organization provides information, scientific and development services for agriculture and allied disciplines throughout the world. The main task of the Information Services is the creation of a bibliographic database (see CAB Abstracts, listed under Databases and Abstracts) of research on agriculture, forestry, the environment, applied social sciences and aspects of human medicine. Relevant scientific and technical literature is acquired worldwide for abstracting and indexing by almost 100 scientists/linguists in five divisions: Animal Health and Medical Parasitology; Animal Production and Human Nutrition; Crop Protection and Genetics; Crop Production; Natural Resources, Forestry and Economics. The Information Services also publish primary journals and books, and offer training for information professionals, especially from developing countries. Supports four institutes: CAB International Institute of Entomology; CAB International Mycological Institute; CAB International Institute of Parasitology; CAB International Institute of Biological Control. These institutes provide an authoritative identification service for agricultural pests and other organisms, a biocontrol service, and conduct research, implement projects and provide training in their fields of expertise. Formerly the Commonwealth Agricultural Bureau. Is a co-sponsor of the International Food Information Service.

CAB International - Asia

PO Box 11872, Kuala Lumpur, 50760, Malaysia
Telephone: 3 255 2922
Fax: 3 255 1888

CAB International - Caribbean and Latin America

Gordon Street, Curepe, Trinidad and Tobago
Telephone: 809 662 4173
Fax: 809 663 2859

CAB International Institute of Entomology

56 Queen's Gate, London, UK

Also located at the Natural History Museum, London, UK.

CAB International Institute of Biological Control

Imperial College, Silwood Park, Ascot, Berkshire, SL5 7PY, UK
Telephone: 890 28426

Provides information and consultancy services in techniques of biological control as well as carrying out research. Has stations in Trinidad and Tobago, Switzerland, Pakistan, Kenya and Malaysia.

CAB International Institute of Parasitology

Winches Farm, St Albans, UK

The field station of the London School of Hygiene and Tropical Medicine.

CAB International

Library Services Centre, Silwood Park, Buckhurst Road, Ascot, Berks, SL5 7TA, UK
Telephone: 344 872747
Fax: 344 872901

Provide copies of most articles cited in the CAB Abstracts journals and database.

CAB International Mycological Institute

Ferry Lane, Kew, Surrey, TW9 3AF, UK
Telephone: 81 940 4086
Fax: 81 581 1676

Offers a range of services to biotechnologists and mycologists. Has 300,000 reference specimens of microfungi plus a culture collection of 12,000 different isolates. Other facilities include a library, on-line database, electron microscopy facilities and biochemical facilities as well as expertise in chemotaxonomy. Most of the former activities of the Biodeterioration Centre of the University of Aston have been transferred to this institute. Also the official addressor secretariat for the International Mycological Association and the World Federation of Culture Collections.

CAB International - North America

845 North Park Avenue, Tucson, AZ-85719, USA
Telephone: 800 528 4841/602 621 7897
Fax: 602 621 3816

Centre for International Cooperation in Agricultural Research and Development

42 rue Scheffer, Paris, F-75116, France
Telephone: 1 47 04 32 15
Fax: 1 47 55 15 30

CIBA Foundation

41 Portland Place, London, W1N 4BN, UK

Promotes international cooperation in relation to medical and chemical research.

Commission on Biotechnology of the IUPAC

Seestrasse 13, D-1000, Germany
Telephone: 311 030 453011

Commission on Biotechnology of the International Union of Pure and Applied Chemistry. Publishes proposals for standard procedures, standard units, nomenclature, symbols etc.

Commission of the European Communities - CEC

200 rue de la Loi, Brussels, B-1049, Belgium
Telephone: 2 235 1111

Committee of Agricultural Organizations of the European Community

The European Community, 23/25 rue de la Science, PO Box 3, Brussels, B-1040, Belgium
Telephone: 2 230 39 45

Comite des Organisations Professionelles Agricoles de la Communaute Europenne.

Committee of the Bread Yeast Manufacturers of the EC - COFALEC

rue du Louvre 15, Paris, F-75001, France
Telephone: 1 508 54 82

Committee on Data for Science and Technology - CODATA

51 Boulevard de Montmorency, Paris, F-75016, France
Telephone: 1 45 25 04 96
Fax: 1 42 88 94 31

Interdisciplinary body of the International Council of Scientific Unions (ICSU). CODATA is concerned with all types of quantitative data resulting from experimental measurements or observations in the physical, biological, geological and astronomical sciences. Particular emphasis is given to data management problems common to different scientific disciplines and to data used outside the field in which they were generated. The general objectives are the improvement of the quality and accessibility of data, as well as the methods by which data are acquired, managed, and analyzed; the facilitation of international cooperation among those collecting, organizing, and using data; and the promotion of an increased awareness in the scientific and technical community of the importance of these activities.

Committee on Science and Technology in Developing Countries - COSTED

51 Boulevard de Montmorency, Paris, F-75016, France
Telephone: 1 45 25 03 29
Fax: 1 42 88 94 31

Interdisciplinary body of the International Council of Scientific Unions (ICSU). The objectives of COSTED are: to coordinate and encourage efforts by the International Scientific Unions to assist the developing countries; to work with the Special and Scientific Committees of ICSU in order to facilitate the greatest possible participation in their programmes by scientists in the developing countries; to foster affiliations with COSTED of national or regional committees which could identify scientific and technical problems related to developing countries and recommend programmes and other activities for COSTED; to provide liaison and advisory services, when requested, to international and regional scientific organizations; to consider the methodology of using science and technology to assist the developing countries; and to undertake other activities and programmes designed to assist developing countries through the use of science and technology.

Committee on Space Research

51 Boulevard de Montmorency, Paris, F-75016, France

Confederation of Industries of the European Community

6 rue de Loxum, PO Box 21, Brussels, B-1000, Belgium
Telephone: 2 513 45 61

Union des Industries de la Communaute Europeenne.

Consultative Group on International Agricultural Research - CGIAR

Agricultural Research, c/o World Bank, 1818 H St NW, Washington, DC-20433, USA
Telephone: 202 477 1234

Finances international research related to agriculture mainly in developing countries. Sponsored by the Food and Agriculture Organization (FAO) of the United Nations, the International Bank for Reconstructive and Development (World Bank) and the United Nations Development Programme (UNDP). The CGIAR consists of 50 donor countries, regional representatives, international and regional organizations and private foundations.

Economic and Social Council of the United Nations

United Nations Headquarters,
New York, NY-10017, USA

EUROBIOTA - COMETT

Gran Via de Colon 42 I-B, Granada,
E-18010, Spain
Telephone: 58 20 88 06
Fax: 58 29 59 53

Agencia de Transferencia de Investigacion. Set up by the Universities of Gent, Granada, Marseille and Sevilla together with the Spanish National Council for Scientific Research and a group of companies from the biotechnology sector of industry under the auspices of the European University Industrial Training Scheme COMETT with the aim of organizing a European pool of industries involved in health biotechnology and in food and nutrition to enter into exchange schemes between university researchers and industrial technologists, create mutually accessible databases of their specialities to enable joint scientific projects to be set up and create a fund of visual media for training technicians in biotechnology.

European Association for Research on Plant Breeding

Postbus 128, Wageningen,
The Netherlands

European Biotechnology Coordination Group - EBCG

Ave Louise 250, bte 71, Brussels,
B-1050, Belgium
Telephone: 2 640 20 95
Fax: 2 640 19 81

Formed in 1985 on the joint initiative of the Commission of the European Communities and CEFIC (the European chemical industry federation) to provide a vehicle for communication and advice on all matters concerning the application of biotechnology by community industries. Founding members of the industrial federation included the Association of Microbial Food Enzyme Producers, Confederation of the Food and Drink Industries of the EEC, European Chemical Industry Federation, European Federation of Pharmaceutical Industry Association and the International Group of National Associations of Manufacturers of Agrochemical Products.

European Committee for Standardization - Comite Europeen de Normalisation

2 rue Brederode, Brussels, B-1000,
Belgium
Telephone: 2 513 55 64

European Committee of Starch Products - CECPA

Corsa Venezia 47-49, Milan, I-20121,
Italy
Telephone: 70 55 43

European Council of Chemical Industry Federations - CEFIC

Federations, 250 avenue Louise,
PO Box 71, Brussels, B-1050, Belgium
Telephone: 2 640 20 95

European Federation of Biotechnology

DECHEMA, Postfach 970146,
Theodor Hause Alle 25,
Frankfurt A/M 97, D-6000, Germany
Telephone: 75 64 235

Europaische Federation Biotechnologie Federation Europeen du Genie Biologie.
Voluntary organization for the general advancement of biotechnology as a means for furthering commercial development. Federation of societies interested in biotechnology,

microbiology, engineering biology, public health, food science and fermentation. Around 40 member societies from 14 countries.

European Federation of Chemical Trade

49 square Marie-Louise, Brussels, B-1040, Belgium
Telephone: 2 230 40 90

European Federation of Manufacturers of Additives for Animal Nutrition

FEFANA, Additives for Animal Nutrition, Adenauerallee 170, Bonn, D-5300, Germany
Telephone: 734 39 70

European Federation of National Associations of Engineers - FEANI

4 rue de la Mission Marchand, Paris, F-75016, France

European Federation of Pharmaceutical Industries Associations - EFPIA

Ave Louise 250, Bte 91, Brussels, B-1050, Belgium
Telephone: 2 640 6815
Fax: 2 735 5222

European Molecular Biology Laboratory - EMBL

Postfach 102240, Heidelberg 1, D-6900, Germany

Has established a European nucleotide sequence data library.

European Molecular Biology Organization - EMBO

Postfach 102240, Heidelberg 1, D-6900, Germany

European Patent Office

Department 4.5.2, Erhardstrasse 27, Munchen 2, D-8000, Germany

European Petrochemical Association

Ave Louise 250, BP 74, Brussels, B-1050, Belgium
Telephone: 2 640 23 10

European Society for Animal Cell Technology

c/o PHLS CAMR Porton Salisbury, Wiltshire, SP4 0JG, UK
Telephone: 980 610391

Promotes the use of large-scale cell systems and provides a forum for organizations already involved in this area.

European Society for Opinion and Marketing Research

JJ Viottastraat 29, JP Amsterdam, NL-1070, The Netherlands

The European Solar Energy Association - EUROSOLAR

Plittersdorfer Strasse 103, PO Box 120618, Bonn, D-5300, Germany
Telephone: 228 362373
Fax: 228 361279

The European Solar Energy Association. An independent, non-governmental, non-commercial organization with individual and institutional members. Members are parliamentarians, scientists, architects and others, engaged in solar energy as well as legal entities like city administrations, scientific institutes, industrial corporations, associations and grass roots initiatives.

European Union for Alcohol Brandies and Spirits

avenue de Tervueren 182, PO Box 5,
Brussels, B-1150, Belgium
Telephone: 2 771 77 35

Federation of European Biochemical Societies

c/o Department of Biochemistry,
University College, London,
WC1E 6BT, UK

Secretary-General: Prof Vito Turk,
J Stefan Institute, Department of
Biochemistry, Jamova 39, 6100
Ljubljjana, Yugoslavia
Telephone: 38 61 159 ext 215, 365
(secretariat)

Unites biochemical societies in the European area with the aims of promoting the science, and bringing about closer contact between European biochemists. Promotes biochemistry in Europe through national and international meetings. Organizes advanced courses in specific areas, fellowships and prizes, youth travel funds, special lectures etc. Publishes proceedings of meetings and original research communications (FEBS Letters) and papers (European Journal of Biochemistry). Members include biochemical societies of Austria, Belgium, Denmark, Finland, France, Germany, Greece, Iceland, Ireland, Israel, Italy, The Netherlands, Norway, Portugal, Spain, Sweden, Switzerland, United Kingdom.

Food and Agricultural Organization - FAO

Information, David Lubin Memorial
Library, Via delle Terma di Caracalla,
Roma, I-00100, Italy
Telephone: 57971

Collects technical and socioeconomic literature on agriculture, fisheries, forestry and food, to assist FAO's programmes (available on request). Information includes biomass fuels, bioconversion, fuel wood and conservation.

International Agricultural Development Service - IADS

1133 Avenue of the Americas,
New York, NY-10036, USA
Telephone: 212 869 8500

Non-profit-making agricultural assistance organization. Aims to help developing nations expand production of crops and livestock and raise rural income. Offers assistance in agricultural research, training and implementation of production programmes. Interacts at government level.

International Association of Biological Standardization

CP 229, Geneva 4, CH-1211,
Switzerland

Concerned with the standardization of biological substances such as vaccines.

International Association for Hydrogen Energy

PO Box 248266, Coral Gables,
FL-33124, USA

Collects, compiles and disseminates information on all aspects of the use of hydrogen gas as a fuel for the future, as well as carrying out promotional work in this area.

International Association on Water Pollution Research and Control - IAWPRC

1 Queen Anne's Gate, London,
SW1H 9BT, UK
Telephone: 71 222 3848

International Association on Water Pollution Research and Control - IAWPRC

PO Box 248266, Coral Gables,
Florida, FL-33124, USA

Scientific associate of the International Council of Scientific Unions (ICSU). The purpose of IAWPRC is to exchange information and experience on all aspects, theoretical and practical, of water pollution control.

International Atomic Energy Agency

Vienna International Centre,
PO Box 100, Vienna, A-1400, Austria

International Bank for Reconstruction and Development

See World Bank

International Biosciences Networks - IBN

51 Boulevard de Montmorency, Paris,
F-75016, France
Telephone: 1 45 25 03 29
Fax: 1 42 88 94 31

Joint programme of the International Council of Scientific Unions (ICSU). Established by the Bio-Unions of ICSU and UNESCO to assist developing countries build up their capacities in the biosciences. IBN is overseen by an International Steering Committee and its regional activities are coordinated by regional networks.

International Biotechnology Suppliers Association

25 Science Park, New Haven,
CT-06511, USA
Telephone: 203 786 5600/800 243 2555
Fax: 203 624 3143

A non-profit association for biotechnology suppliers, affiliates and professionals formed to enhance the growth, image and economic status of commercial biotechnology. The association sponsors an annual International Biotechnology Exposition (IBEX) in San Francisco each October. This includes both a conference with seminars, posters and live demonstration workshops, and an exhibition.

International Cell Research Organization

UNESCO SC/SER, 7 Place de
Fontenoy, Paris, F-75700, France
Telephone: 1 4568 3886
Fax: 1 4567 2639

Scientific associate of the International Council of Scientific Unions (ICSU). ICRO's mission is to foster, in close cooperation with UNESCO, the development of basic research in cellular and molecular biology, mainly organizing international training courses in various countries.

International Centre for Genetic Engineering and Biotechnology

Padriciano 99, Trieste, I-34012, Italy
Telephone: 40 3757333
Fax: 40 226555

Consists of two component laboratories, one in Trieste, Italy, and the other in New Delhi, India, which constitute its permanent facilities as well as a network of Affiliated Centres in Member Countries. It maintains a permanent staff conducting research focusing mainly on agribiology and human health with special reference to the problems of the developing world.

International Centre of Insect Physiology and Ecology and Entomology

PO Box 30772, Nairobi, Kenya

Carrying out research in insect ecology, biostatistic physiology, biochemistry, pathology, taxonomy and morphology in relation to parasitic diseases and their transmission in man and other animals.

International Committee on Economic and Applied Microbiology

Massachusetts Institute of Biology,
Cambridge, MA-02139, USA
Telephone: 617 253 6430

International Council for Scientific and Technical Information - ICSTI

Technical Information - ICSTI,
51 Boulevard de Montmorency, Paris, F-75016, France
Telephone: 1 4525 6592

Scientific associate of the International Council of Scientific Unions. The purpose of ICSTI is to increase accessibility to, and awareness of, scientific and technical information. ICSTI aims at fostering communication and interaction among all participants in the information transfer chain, in order to develop appropriate tools to meet better the information requirements of the world community of scientists and technologists.

International Council of Scientific Unions - ICSU

51 Boulevard de Montmorency, Paris, F-75016, France
Telephone: 1 4525 0329
Fax: 1 4288 9431

An internation non-governmental scientific organization composed of 20 international Scientific Unions, 75 National Members, Associates and Observers and 26 Scientific Associates. The main objective of the council is to encourage international scientific activity for the benefit of mankind. This is achieved by: initiating, designing and coordinating international interdisciplinary research programmes; acting as a focus for the exchange of ideas, the communication of scientific information and the development of standards; holding scientific conferences, symposia, etc; producing a wide range of publications; assisting in the creation of international and regional networks of scientists with similar interests; initiating special studies.

International Council of Scientific Unions - ICSU

Advisory Committee on the Environment - ACE

Created by the Executive Board in 1989 for the purpose of providing the Board with advice on the status and development of all ICSU bodies' activities related to the environment and Global Change in order to promote coordination of efforts among members of the ICSU family, provide for a platform of discussion and exchange of information among relevant ICSU bodies and their external partners and set up and maintain an interface with external partners such as UN bodies, other international organizations including those in the social and engineering sciences and industry.

International Council of Scientific Unions - ICSU

Ad hoc Committee on Science in Central, and Eastern Europe - COMSCEE

Set up by the executive board in 1990, the terms of reference for the committee are to define the current problems affecting science in Eastern Europe and to explore ways in which ICSU might help to relieve these problems.

International Crop Research Institute for Semiarid Tropics - ICRISAT

Andhra Pradesh 802-324, India
Telephone: 224016

Undertakes research on farming systems for semi-arid tropics, including anaerobic digestion and fermentation for production of alcohol fuels from plant biomass.

International Development Association

1818 H Street NW, Washington, DC-20433, USA

International Development Research Centre

PO Box 8500, Ottawa, K1N 3H9, Canada
Telephone: 613 996 2321

Interests include development of bioconversion methods for production of fuel, fodder and fertilizers.

International Energy Agency

Chateau de la Muette, 2 rue Andre Pascal, Paris Cedex, F-75775, France
Telephone: 1 524 82 00

Aims to reduce oil dependency through development of alternative energy sources by coordination of the national activities of member countries and dissemination of information. Affiliated to the Organization for Economic Cooperation and Development (OECD).

International European Construction Federation

rue de Berri, Paris, F-75008, France
Telephone: 1 563 11 44

International Federation of Automatic Control - IFAC

Schlossplatz 12, Laxenburg, Austria, A-2361
Telephone: 2236 71 447
Fax: 2236 72 859

The purpose of IFAC is to promote the science and technology of control in the broadest sense, in all systems, in both theory and application. The primary object is to serve all those concerned with the theory and application of automatic control and systems engineering wherever situated.

International Federation for Cell Biology

c/o Imperial Cancer Research Fund, Lincoln's Inn Fields, London, WC2A 3PX, UK
Telephone: 71 242 0200

International Federation for Information and Documentation - FID

Contact: B G Goedegebuure, PO Box 90402, LK The Hague, NL-2509, The Netherlands
Telephone: 70 314 0671
Fax: 70 314 0667

Scientific associate of the International Council of Scientific Unions (ICSU). The aim of the FID is to promote, through international cooperation, research in and development of all aspects of the organization, storage, retrieval, dissemination and evaluation of information however recorded, in the fields of science, technology, social sciences, arts and humanities.

International Federation for Information Processing - IFIP

Processing - IFIP, c/o Bull SA,
101 Avenue de Malakof, Paris,
F-75016, France
Telephone: 1 40 65 03 83

Scientific associate of the International Council of Scientific Unions (ICSU). The basic aims of the IFIP are: to promote information science and technology; to advance international cooperation in the field of information processing; to stimulate research, development and application of information processing in science and human activity; to further the dissemination and exchange of information on information processing; to encourage education in information processing.

International Federation of Institutes for Advanced Study

168 rue de Grenelle, Paris, F-75007,
France

International Federation of Organic Agricultural Movements - IFOAM

Rodale Research Center, RDI,
Box 323, Kutztown, PA-19530, USA
Telephone: 215 683 6383

Interested in composting, nitrogen fixation and biological pesticides applicable to organic farming.

International Federation of Pharmaceutical Manufacturers' Associations

Postfach 328, Nordstrasse 15, Zurich,
CH-8035, Switzerland

International Federation of Societies for Electron Microscopy - IFSEM

Institute of Anatomy, Department of
Cell Biology, University of Aarhus,
Aarhus, DK-8000, Denmark
Telephone: 86 12 8808
Fax: 86 19 8664

Scientific associate of the International Council of Scientific Unions (ICSU). IFSEM is an independent federation, with the purpose of furthering international cooperation between microscopists.

International Food Information Service - IFIS

IFIS Publishing, Lane End House,
Shinfield, Reading, Berkshire,
RG2 9BB, UK
Telephone: 734 883895
Fax: 734 885065

IFIS compile and publish: Food Science and Technology Abstracts (FSTA), available in hardcopy and on CD-ROM; Packaging Science and Technology Abstracts (PSTA) compiled by the Fraunhofer-Institut fuer Lebsmitteltechnologie und Verpackung (Germany) in collaboration with IFIS, available in hardcopy and on-line.

International Food Information Service - IFIS GmbH

Melibocusstrasse 52, Frankfurt 71,
D-6000, Germany
Telephone: 69 669 007 0
Fax: 69 669 007 10

See International Food Information Service, UK.

International Food Information Service - IFIS - North American Desk

c/o National Food Lab Inc, 6363 Clark Avenue, Dublin, CA-94568, USA
Telephone: 800 336 3782
Fax: 510 833 8795

See International Food Information Service, UK.

International Food Policy Research Institute

1776 Massachusetts Avenue NW, Washington, DC-20036, USA

Collects information on and analyzes the world food situation. Promotes research on methods of increasing production and availability of food in respect of the less-developed countries in particular.

International Foundation for Science - IFS

Grev Turegatan 19, Stockholm, S-11438, Sweden

International Institute for Applied Systems Analysis - IIASA

Laxenburg, A-2361, Austria
Telephone: 2236 71 521/0
Fax: 2236 71 313

A non-governmental, interdisciplinary research institute whose member organizations comprise scientific and professional bodies in 16 countries representing much of Europe, both East and West, as well as Japan, Canada and the United States. IIASA's research programme focuses on issues for which solutions can be found through international research and cooperation. It includes studies on global environment issues, the growth and ageing of populations, the introduction of new technologies and the use of decision sciences.

International Institute for Environmental Technology and Management

The Stockholm Environment Institute, Stockholm, Sweden

International Life Sciences Institute

Contact: A Malaspina, 1126 Sixteenth Street NW, Suite 300, Washington, DC-20036, USA
Telephone: 202 659 0074
Fax: 202 659 3859

Covers all aspects of international cooperation in relation to global environmental matters. Activities include the INFOTERRA programme aimed at exchange of environmental information between member countries and GEM (Global Environment Monitoring) System which is an internationally evaluated effort to systematically collect, analyze and evaluate variables that determine changes in the environment.

International Organization for Biotechnology and Bioengineering - IOBB

c/o ICAITI, PO Box 1552, Guatemala City, Guatemala

International Organization for Standardization - ISO

1 rue de Varembe, Geneva 20, CH-1211, Switzerland
Telephone: 22 749 0111
Fax: 22 733 3430

International Patent Documentation Centre - INPADOC

Moellwaldplatz 4, Vienna, A-1040, Austria
Telephone: 1 50 155 53
Fax: 1 505 33 86

Supplies information on patents from more than 55 patent issuing authorities. Also specializes in all enquiries on patents in Japan. Available on-line etc.

International Patent Research Office

Moellwaldplatz 4, Vienna, A-1040, Austria
Telephone: 1 50 155 53
Fax: 1 505 33 86

International Research Centre on Environment and Development - CIRED

10 rue Monsieur le Prince, Paris, F-75006, France
Telephone: 1 633 78 59

Centre International de Recherche sur l'Environment et Developpement. Carrying out research on the scientific, technical and economic aspects of the development of renewable energy technologies, including the generation of biogas by anaerobic digestion.

International Society of Soil Science

Contact: J Van Baren, c/o ISRIC, 9 Duivendaal, PO Box 353, AJ Wageningen, NL-6700, Netherlands
Telephone: 8370 19 063
Fax: 8370 24 460

Collaborating Centre for Chemical Reference Standards.

International Union of Biochemistry and Molecular Biology - IUBMB

Techn Univ Berlin, Inst fur Biochem und Mol Bio, Franklinstrasse 29, Berlin 10, D-1000, Germany
Telephone: 30 3142 4205
Fax: 30 3142 4783

International Scientific Union member of the International Council of Scientific Unions (ICSU). IUBMB promotes international cooperation in biochemistry via research, discussion, publication and standardisation and is committed to promoting high standards of biochemistry throughout the world. Committees, charged with sponsoring symposia, educational activities and discussion of frontier research topics, sponsor meetings in all parts of the world, on average once each month.

Contact the General Secretary, H Kleinkauf, at the above address or care of the International Council of Scientific Unions (ICSU).

International Union of Biological Sciences - IUBS

51 Boulevard de Montmorency, Paris, F-75016, France
Telephone: 1 4525 0009
Fax: 1 4525 2029

International Scientific Union member of the International Council of Scientific Unions (ICSU). The objectives of IUBS are to promote the study of biological sciences, facilitate and coordinate research and other scientific activities that require international cooperation, ensure the discussion and dissemination of international conferences and to assist in the publication of their results.

International Union of Food Science and Technology - IUFoST

National Food Centre, Dunsinea, Castleknock, Dublin 15, Ireland
Telephone: 1 30 3222

Scientific associate of the International Council of Scientific Unions (ICSU). The primary purpose of IUFoST is the encouragement of international cooperation and exchange of scientific-technical information among scientists and specialists in development of research, encouraging progress in the fields of theoretical and applied food science, advancing technology in the processing, manufacturing, preservation, and distribution of food products, and in the education and training of food scientists and technologists.

International Union of Forestry Research Organizations - IUFRO

Contact: H Schmutzenhofer,
Seckendorf-Gudent-Weg 8, Vienna,
A-1131, Austria
Telephone: 222 82 0151
Fax: 1 82 9355

Scientific associate of the International Council of Scientific Unions (ICSU). The main objectives of IUFRO are to rationalize research techniques, to standardize systems of measurement, and to promote international cooperation in scientific studies, forestry and forest products, including: site and silviculture; forest plants and forest protection; forest operations and techniques; planning, economics, growth and yield, management, and policy; forest landscape, recreation and tourism; statistical methods, mathematics and computerization; information systems and terminology; remote sensing; management of forestry research; and history.

International Union of Immunological Societies - IUIS

Department of Immunology, Weizmann Institute of Science, Rehovot, 76100, Israel
Telephone: 8 34 3642
Fax: 8 46 9712

International Scientific Union member of the International Council of Scientific Unions (ICSU). The objectives of IUIS are to organize international cooperation in immunology and to promote communication between the various branches of immunology and allied subjects, to encourage within each scientifically independent territory cooperation between the societies that represent the interests of immunology and to contribute to the advancement of immunology in all its aspects.

Contact the Secretary General, R Arnon, at the above address or care of the International Council of Scientific Unions (ICSU).

International Union of Microbiological Societies - IUMS

IMBC/CNRS, Lab d'Immunochimie,
15 Rue Descartes, Strasbourg,
F-67000, France
Telephone: 8841 7022
Fax: 8861 0680

International Scientific Union member of the International Council of Scientific Unions (ICSU). The objectives of IUMS are to: promote the study of microbiological sciences internationally; initiate, facilitate and coordinate research and other scientific activities which involve international cooperation; ensure the discussion and dissemination of the results of international cooperative research; promote the organization of international conferences, symposia and meetings and assist in the publication of their reports; represent microbiological sciences in ICSU and maintain contact with other international organizations.

Contact the Secretary General, MHV van Regenmortel, at the above address or care of the International Council of Scientific Unions (ICSU).

International Union of Nutritional Sciences - IUNS

Dept of Human Nutrition,
Agricultural University, PO Box 8129,
EV Wageningen, NL-6700,
The Netherlands
Telephone: 8370 82 589
Fax: 8370 83 342

International Scientific Union Member of the International Council of Scientific Unions (ICSU). The objectives of IUNS are to promote international cooperation in the scientific study of nutrition and its application; to encourage research and the exchange of scientific information in the nutritional sciences by the holding of congresses and conferences, by publication and by other suitable means; to establish such commissions, committees and other bodies as may be required in the pursuit of the first two objectives; to provide a means of communication with other organizations and to encourage participation in the activities of ICSU; to develop activity regarded as helpful and appropriate in achieving the objectives of the Union.

Contact the Executive Secretary, R Janssen, at the above address or care of the International Council of Scientific Unions (ICSU).

International Union of Pharmacology - IUPHAR

Lab de Pharm, UCL-7350, Avenue E Mounier 73, Brussels, B-1200, Belgium
Telephone: 2 764 7350
Fax: 2 764 7308

International Scientific Union member of the International Council of Scientific Unions (ICSU). The objectives of IUPHAR are to: enhance collaboration between pharmacologists at an international level; to advance the science and appropriate applications in therapeutics; stimulate research in perceived growth areas of special significance; advance the training of teachers of pharmacology and the importance of pharmacology in the medical, dental and veterinary curricula of universities and specialist schools throughout the world.

Contact the Secretary General, T Godfraind, at the above address or care of the International Council of Scientific Unions (ICSU).

International Union of Physiological Sciences

Lab de Physiologie Nerveuese, Groupe de Labs du CNRS, Gif sur Yvette, F-91198, France
Telephone: 1 6907 6145
Fax: 1 6907 0538

International Scientific Union member of the International Council of Scientific Unions (ICSU). The objectives of IUPS are: to encourage the advancement of the physiological sciences; to facilitate the dissemination of knowledge in the field of physiological sciences; to foster and encourage research in the field of physiological sciences; to promote the International Congresses of Physiological Sciences; to promote such other measures as will contribute to the development of physiological sciences in developing countries.

Contact the Secretary, R Naquet, at the above address or care of the International Council of Scientific Unions (ICSU).

International Union for Pure and Applied Biophysics - IUPAB

Inst of Biophysics, Medical University, Szigeti ut 12, Pecs, 7643, Hungary
Telephone: 72 14 017
Fax: 72 26 244

International Scientific Union member of the International Council of Scientific Unions (ICSU). The objectives of IUPAB are to organize international cooperation in biophysics and promote communication between the various branches of biophysics and allied subjects, to encourage within each adhering body cooperation between the societies that are interested in the advancement of biophysics in all its aspects.

Contact the Secretary General, J Tigyi, at the above address or care of the International Council of Scientific Unions (ICSU).

International Union of Pure and Applied Chemistry - IUPAC

Bank Court Chambers, 2-3 Pound Way, Templars Square, Cowley, Oxford, OX4 3YF, UK
Telephone: 865 74 7744
Fax: 865 74 7510

International Scientific Union member of the International Council of Scientific Unions (ICSU). IUPAC is the international body which represents chemistry amongst the other disciplines of science. It is recognized as such by the Academies which are responsible for science at national levels. These National Academies of Science or their National Committees of Chemistry are members of IUPAC. Its objects are to promote continuing cooperation amongst the chemists of its member countries, to study topics of international importance to pure and applied chemistry which need regulation, standardization or codification, to cooperate with other international organizations which deal with topics of a chemical nature and to contribute to the advancement of pure and applied chemistry in all its aspects.

Contact the Executive secretary, M Williams, at the above address or care of the International Council of Scientific Unions (ICSU).

Liaison Committee of the Glucose Producers in the European Community

avenue de Cortenbergh 134, Brussels, B-1040, Belgium
Telephone: 2 735 68 17

Microbiological Resource Centre - MIRCEN

Department of Bacteriology, Karolinska Institute, Stockholm, S-10401, Sweden
Telephone: 8 340560

MIRCEN Biotechnology and Agriculture Unit

University of Waterloo, Ontario, N2L 3G1, Canada

MIRCEN Biotechnology Unit - Commonwealth Mycological Institute

University of Waterloo, Ontario, N2L 3G1, Canada

MIRCEN Fermentation Technology Unit

ICME University of Osaka, Suitashi 565 Osaka, Japan

MIRCEN Rhizobium Unit

Niftal Project, University of Hawaii, PO Box 'O', Paia, HI-96779, USA

Organization for Economic Cooperation and Development - OECD

Development, Chateau de la Muette, 2 rue Andre Pascal, Paris Cedex 16, F-75775, France
Telephone: 1 524 81 07

The Packaging Group Inc - TPG

TPG Publications, PO Box 345, Milltown, NJ-08850, USA
Telephone: 201 636 0885
Fax: 201 390 1402

Publish conference proceedings, multi-client studies and short course texts in packaging related fields.

Permanent International Committee of the Vinegar Industry of the EC

Reuterstrasse 151, Bonn, D-5300, Germany
Telephone: 21 20 17

Pharmaceutical Group of the European Community

rue Archimede 11, Brussels, B-1040, Belgium
Telephone: 2 733 98 20

Regional Programme of Biotechnology for Latin America and the Caribbean

Universidad Nacional de La Plata, Calles 47 y 115, 1900 La Plata, Argentina
Telephone: 21 21 5556

Aims to develop biotechnology in Latin America with the active participation of Argentina, Colombia, Costa Rica, Cuba, Chile, Guatemala, Mexico, Peru, Uruguay and Venezuela. Supported by UNESCO and UNIDOl.

Scandinavian Society of Plant Physiology

c/o Department of Biology NLHT, Trondheim University, Dragvoll, N-7055, Norway

Promotes all aspects of plant physiology; achieved by organization of national and international meetings and dissemination of information, including publication of the journal Physiologia Plantarum. Membership, although largely Scandinavian, is open to all those interested in the subject.

Scientific Committee on Biotechnology - COBIOTECH

Contact: Dr K G Skyryabin, Centre of Bioengineering, USSR Academy of Sciences, Vavilov Street 32, Moscow, Russia, 117984, CIS
Telephone: 95 135 7319
Fax: 95 135 7319

Interdisciplinary body of the International Council of Scientific Unions (ICSU). The objectives of COBIOTECH are: to facilitate cooperation between the members of the ICSU family for research and education in the various fields of biotechnology; to cooperate with other appropriate organizations in order to advance international aspects of research and education in biotechnology; to stimulate collaboration between the scientific and the industrial communities involved in biotechnology; to gather and transmit information about opportunities for education and training in biotechnology; to provide a forum for information exchange and the coordination of international and, where appropriate, national congresses, meetings, symposia, workshops, etc, on aspects of biotechnology; to stimulate and assist culture collections, gene banks, molecular probe banks, data banks, etc, and to facilitate accessibility to such collections and banks by the international community; and to review, evaluate, and provide information on regulations, safety requirements, risk assessment, standards and nomenclature for biotechnology.

Scientific Committee on Genetic Experimentation - COGENE

Contact: Dr G Bernardi, Laboratoire de Genetique Moleculaire, Institut Jaques Monod, 2 Place Jussieu, Paris, F-75005, France
Telephone: 1 4329 5824
Fax: 1 4427 7977

Interdisciplinary body of the International Council of Scientific Unions (ICSU). Among COGENE's purposes are: to review, evaluate and make available information on the practical and scientific benefits, safeguards, containment facilities and other technical matters; to consider environmental, health-related and other consequences of any disposal of biological agents constructed by recombinant DNA techniques; to foster opportunities for training and international exchange; and to provide a forum through which interested national, regional and other international bodies may communicate.

Scientific Committee on Problems of the Environment - SCOPE

Contact: Dr V Plocq-Fichelet, 51 Boulevard de Montmorency, Paris, F-75016, France
Telephone: 1 4525 0498
Fax: 1 4288 1466

Interdisciplinary body of the International Council of Scientific Unions (ICSU). The committee was created to assemble, review and assess the information available on man-made environmental changes and the effects of these changes on man; to assess and evaluate the methodologies of measurement of environmental parameters; to provide an intelligence service on current research; and by the recruitment of the best available scientific information and constructive thinking to establish itself as a corpus of informed advice for the benefit of centres of fundamental research and of organizations and agencies operationally engaged in studies of the environment.

Senior Advisory Group on Biotechnology - SAGB

Avenue Louise 250, bte 71, Brussels, B-1050, Belgium
Telephone: 2 640 2095
Fax: 2 640 1981

Operating within the European Chemical Industry Federation (CEFIC), which consists of board members of Feruzzi, Hoechst, ICI, Monsanto, Rhone-Poulenc, Sandoz and Unilever, provides an industrial forum for debating policy issues affecting biotechnology in the European Community.

UNEP/UNESCO/ICRO Panel on Microbiology

Swedish University of Agricultural Science, Uppsala, S-75007, Sweden

Union of International Association

rue aux Laines, Brussels, B-1000, Belgium

Promotes international trade to accelerate economic development. Studies the transfer and development of technology.

United Nations Conference on Trade and Development - UNCTAD

Palais des Nartions, Geneve 10, CH-2111, Switzerland
Telephone: 34 60 11

Promotes international trade to accelerate economic development. Studies the transfer and development of technology.

United Nations Environment Programme - UNEP

PO Box 30552, Nairobi, Kenya
Telephone: 333930

Covers all aspects of international cooperation in relation to global environmental matters. Activities include the INFOTERRA programme aimed at exchange of environmental information between member countries and GEM (Global Environment Monitoring) System which is an internationally evaluated effort to systematically collect, analyze and evaluate variables that determine changes in the environment.

United Nations Industrial Development Organization - UNIDO

Postfach 707, Vienna, A-1011, Austria

Coordinates international activities and disseminates information on advances in all aspects of industrial technology.

United Nations Industrial Development Organization - UNIDO

Vienna International Centre,
PO Box 300, Vienna, A-1400, Austria
Telephone: 21131 ext 3728
Fax: 2307355

Biotechnology and Genetic Engineering Unit Industrial Technology Development Division.

United Nations Institute for Training and Research - UNITAR

Via Panama 12, Roma, I-00198, Italy
Telephone: 845 95 48

Centre on Small Energy Resources. Promotes the development of small energy resources particularly for rural areas of developing countries, by facilitating the exchange of information on those resources and on related technology.

World Association of Industrial and Technical Research Organizations

3650 Westbrook Crescent, Vancouver, V6S 2L2, Canada

World Bank - International Bank for Reconstruction and Development

1919 H St NW, Washington, DC-20433, USA
Telephone: 202 477 1234

Among its other activities, publishes occasional documents on new and renewable energy, agricultural developments etc as well as funding some research in these areas which includes production of alcohol fermentation.

World Federation for Culture Collections

International Mycological Institute, Ferry Lane, Kew, Surrey, TW9 3AF, UK

An interdisciplinary commission of the International Union of Biological Sciences.

World Health Organization - WHO

Centre for Chemical Reference Standards, Box 3045, Solna, S-17103, Sweden
Telephone: 468783 9500

Collects and disseminates information. Promotes research and collaboration needed between governments for the understanding of the control of pests, diseases, etc, related to human health on a global basis. Part of the United Nations. Collaborating Centre for Chemical Reference Standards.

World Intellectual Property Organization

34 chemin des Columbettes,
Geneve 20, CH-2111, Switzerland
Telephone: 99 91 44

Specialist agency of the United Nations Organization dealing with international cooperation in the area of intellectual property and the administration of the International Patent Classification. Covers all fields of technology accessible for patent protection. All services are exclusively for the benefit of institutions in developing countries.

GOVERNMENT ORGANIZATIONS

AUSTRALASIA

Australian Government Analytical Laboratories

PO Box 385, Pymble, Sydney,
NSW 2073, Australia
Telephone: 02 449 0111

International depository for microorganisms under the Budapest Treaty.

CSIRO - Division of Horticulture

Hartley Grove, Urrbrae, Adelaide,
SA 5064, Australia
Telephone: 8 274 9244
Fax: 8 274 9247

Plant breeding and genetic manipulation of horticultural crops including development of DNA probes.

CSIRO - Division of Animal Health

Private Bag 1, Parkville, Melbourne,
Victoria 3052, Australia
Telephone: 3 342 9700
Fax: 3 347 4042

Development of diagnostics and vaccines for animal health.

CSIRO - Division of Animal Production

Great Western Highway, Prospect,
Sydney, NSW 2148, Australia
Telephone: 2 631 8022
Fax: 2 688 2340

Production of hormones and growth factors as well as genetic engineering of domestic animals.

CSIRO - Division of Biotechnology

Bayview Avenue, Clayton, Melbourne,
Victoria 3168, Australia
Telephone: 3 542 2244
Fax: 3 543 6613

Industrial biotechnology, microbiology and fermentation technology.

CSIRO - Division of Entomology

Clunies Ross Street, Black Mountain,
Canberra, ACT 2601, Australia
Telephone: 62 464 911
Fax: 62 470 217

Development of insect viruses and entomopathogenic bacteria to increase pathogenicity as well as production of pheromones and pheromone analogues.

CSIRO - Division of Food Processing

39-51 Delhi Road, North Ryde,
Sydney, NSW 2113, Australia
Telephone: 2 887 8333
Fax: 2 887 3107

Pilot plant food processing activities.

CSIRO - Division of Forestry

Banks Street, Yarralumla, Canberra,
ACT 2601, Australia
Telephone: 62 818 211
Fax: 62 818 312

Cell culture and micropropagation of forest trees.

CSIRO - Division of Plant Industry

GPO Box 1600, Canberra, ACT 2601, Australia
Telephone: 62 46 4911
Fax: 62 47 3748

CSIRO - Division of Soils

Private Bag No. 2, Glen Osmond, Adelaide, SA 5064, Australia
Telephone: 8 274 9311
Fax: 8 338 1636

Department of Science and Technology

PO Box 65, Belconnen, ACT 2616, Australia
Telephone: 3 641155

DSIR - Biotechnology Division

Fitzherbert West, Palmerston North, New Zealand
Telephone: 63 68019
Fax: 63 61130

Covers both animal and plant biotechnology as well as monoclonal antibody production, fermentation and controlled release of GMOs.

DSIR - Crop Research

Ellesmere Junction Road, Lincoln, Caterbury, New Zealand
Telephone: 3 252 511
Fax: 3 252 074

Covers agricultural biotechnology and plant genetics including plant tissue culture and rapid *in vitro* plant propagation.

DSIR - Entomology

120 Mount Albert Road, Mount Albert, Aukland, New Zealand
Telephone: 9 893 660
Fax: 9 863 330

Control of insect pests in agriculture and horticulture.

DSIR - Horticulture

120 Mount Albert Road, Mount Albert, Aukland, New Zealand
Telephone: 9 893 660
Fax: 9 863 330

Plant breeding and food technology.

DSIR - Plant Diseases

Mount Albert Research centre, Auckland, New Zealand
Telephone: 9 893 660
Fax: 9 863 330

Diagnosis of plant disease and plant disease resistance.

DSIR - Plant Physiology

Fitzherbert West, Palmerston North, New Zealand
Telephone: 63 680 19
Fax: 63 61 130

Genetic conservation, plant health and micropropagation.

DSIR - Water Quality Centre

100 Aurora Terrace, Hamilton, New Zealand
Telephone: 71 670 26
Fax: 71 601 51

Water treatment and quatic environmental processes.

National Biological Standards Laboratory

PO Box 462, Canberra, ACT 2601, Australia
Telephone: 69 891 555
Fax: 69 897 724

Regulatory body for submissions for marketing of therapeutics and associated analytical facilities.

New Zealand Department of Health

National Health Institute, PO Box 50 348, Porirua, Wellington, New Zealand
Telephone: 4 370 149
Fax: 4 378 983

Technology Development Corporation

formerly Technology Park Adelaide Corporation
Innovation House, First Avenue, Technology Park, SA 5095, Australia
Telephone: 8 260 8111
Fax: 8 260 8100

Established as a specialized development agency to encourage the growth of new technology-based industries and the application of new technology in existing industry.

Technology and Industry Development Authority

PO Box D160, Perth, WA 6001, Australia
Telephone: 9 327 5555
Fax: 9 327 5542

Supplies information on funding and liaises between government and industry on regulatory matters.

AUSTRIA

Austrian Scientific Research Fund

Fonds zur Foederung der Wissenschaftlichen Forschung, Garnisongasse 7, Vienna, A-1090,
Telephone: 1 421236 0

Federal Cabinet for Health

Sektion VII, Radezkystrasse 2, Vienna A-1031
Telephone: 2 22 11 58-0

Responsible for genetic manipulation aspects of biotechnology.

Federal Ministry for Agriculture and Forestry

Bundesministerium fuer Land- und Forstwirtschaft, Stubenring 1, Vienna, A-1011

Federal Ministry for Health and Environment

Radetskystrasse 2, Vienna, A-1030
Telephone: 1 71158
Fax: 1 71158 4221

(Bundesministerium fur Umwelt, Jugend und Familie)

Federal Ministry for Public Science and Communication

Bundesministerium fuer oeffentliche Wirtshaft und Verkehr, Renngasse 5, Vienna A-1010
Telephone: 1 534 64-0
Fax: 1 534 64 20 13

Provides information concerning innovation in biotechnology and genetic engineering.

Federal Ministry for Science and Research

Bundesministerium fuer Wissenschaft und Forschung, Freyung 1, Vienna, A-1014
Telephone: 222 531 20
Fax: 1 53120 4499

Federal Ministry for Trade and Industry

Bundesministerium fuer Wirtschaftliche Angelegenheiten, Stubenring 1, Vienna, A-1010
Telephone: 1 71100
Fax: 1 71393 11

Patent Office

Oesterreichisches Patentamt, Kohlmarkt 8-10, Vienna, A-1014
Telephone: 1 53424 0
Fax: 1 53424 520

Plant Breeding

Landesanstalt fuer Pflanzenzucht und Samenpruefung, Versuchsfeld 4, Rinn

BELGIUM

Customs and Excise

59 rue Ducale, Brussels, B-1000
Telephone: 233 71 11

Department of Energy

23 square de Meuse, Brussels, B-1040
Telephone: 511 19 30

Department of Industry

23 square de Meuse, Brussels, B-1040
Telephone: 511 19 30

European Communities

Press and Information Office, 73 rue Archimede, B-1049 Brussels
Telephone: 2 235 1111

Information and Documentation Institute

3 rue Montoyer, Brussels, B-1040

The government publications and sales office for Belgium.

Institute pour l'Encouragement de la Recherche Scientifique dans l'Industrie et Agriculture - IRSIA

6 rue de Crayer, Brussels
Telephone: 648 81 45

Ministere des Technologies Nouvelles et des Relations Exterieures

Avenue des Arts 13-14, Brussels, B-1040
Telephone: 2 211 55 11

Ministry of Agriculture

Plant Pathology Institute, 8 avenue Marechal Jiun, Gembloux, B-5800
Telephone: 8 16 20 99

Studying plant fungal and viral diseases.

Ministry of Development

rue des Quatre-Bras, Brussels, B-1000

Ministry of Economic Affairs

Central Laboratory, 17A rue de la Senne, Brussels, B-1000
Telephone: 13 23 26

Ministry of Economic Planning and Scientific Policy

30 rue Montoyer, Brussels, B-1040
Telephone: 230 94 90

Ministry of Education (Flemish)

Kunstlann 58, Bur 5, Brussels, B-1040
Telephone: 219 93 85

Ministry of Education (French)

30 rue Joseph II, Brussels, B-1040
Telephone: 219 02 37

Ministry of Foreign Affairs

2 rue des Quartre-Bras, Brussels, B-1000
Telephone: 513 62 40

Ministry of Health

Ministere de la Sante, Cite Adminstrative de l'Etat, quartier Esplanade, Brussels, B-1000

Ministry of Public Health and Environment

Rijkdam Centrum, Esplanade 7, Brussels, B-1010

Patent Office

Service de la Proprietarie Industrielle et Commerciale, 24-26 rue J A de Mot, Brussels, B-1040

BRAZIL

CNPq

Conselho Nacional de Desenvolvimento Cientifico e Tecnologico, Avenida W3/Norte Q 507-B, CP 111142, Brasilia, 70740
Telephone: 61 272 2450

Responsible for direction and funding of biotechnology research through the Program Nacionale e Biotechnologia (PRONAB).

EMBRAPA

Empresa Brasileira de Pesquisa Agropecuaria, Coronel Pacheco, Minas Gerais
Telephone: 32 212 8550

Carrying out research in cell and tissue culture to produce new varieties of crop plants including genetically improved seed potatoes.

FINEP

Financiadora de Estudos e Projetos, Avenue Rio Branco 124, Rio de Janeiro, RJ 20042
Telephone: 21 291 3993

The national agency for funding of applied studies.

FIPEC

Fundo Incentivo Pesquisa Technico Cientifico, Banco do Brasil Sede III/11, Sector Bancario Sul, Brasilia, DF 70070

Funding developments in biotechnology.

IBICT

Instituto Brasilerio de Informacao em Ciencia e Technologia, Rua General Argolo 90, Rio de Janeiro, RJ 20921
Telephone: 21 228 3790

Information service.

INCQS

Instituto Nacional de Controle de Qualidade em Saude, Avenida Brasil 4365, Manguinhos, Rio de Janeiro, RJ 21040
Telephone: 21 270 1073

Federal institute responsible for nationwide quality control of food, drugs, cosmetics, clinical diagnostics and related products.

Ministry of Agriculture

Esplenada dos Ministerios, 8 Andar, Brasilia, 7000
Telephone: 21 225 5203

Ministry of Industry & Commerce

Instituto Nacional d Propredode Industrial, Praca Maua, 7-11 Andar, Rio de Janeiro, RJ 20083

Patent office.

Ministry of Industry & Commerce

Instituto do Acucar e Alcoool - IAA, Praqa XV de Novembro, 42-7 Andar, Rio de Janeiro, RJ 20010

Ministry of Industry & Commerce

Comissao Executiva Nacional do Alcool - CENAL, Esplanada dos Ministerios, Bloco B-Sala 920, Brasilia, DF 70053

Ministry of Industry & Commerce Brasilia

Coordenacao de Biomassa, STI SAS Q 02, Lotes 1-3, Brasilia, DF 70070
Telephone: 61 224 8740

Coordinates the energy from biomass programme.

PROMOCET

Promocao de Pesquisa Ciencia e Technologia do Estado de Sao Paulo, Avenida Angelica 2632, 4 Andar, Sao Paulo, SP 01228

Government agency supporting science and technology in the state of Sao Paulo.

CANADA

Agriculture Canada

Saanichton Plant Quarantine Station, Sir John Carling Building, 930 Carling Avenue, Ottawa, Ontario, K1A 0C5
Telephone: 613 995 7084

National plant quarantine and introduction facility for fruit trees, grapevines and woody ornamentals. Carries out virus detection using serological and biochemical laboratory diagnostic techniques and inoculation on indicators. Carries out virus elimination from plants via thermotherapy and chemotherapy. Carries out research in plant virus purification and identification, plant tissue culture techniques and the development of biotechnological virus detection techniques. Has a collection of plant viruses infecting fruit trees and grapevines. Maintains a large collection of virus-tested fruit tree and grapevine cultivars for international distribution. Also maintains a large collection of Kiwi and *Ilex* sp cultivars.

Agriculture Canada

Priorities and Strategies Research Branch, 3851 Fallowfield Rd, PO Box 11300, Station II, Nepean, Ontario, K2H 8P9
Telephone: 613 998 9320
Fax: 613 952 8072

Carries out plant research (crop improvement, nitrogen fixation), animal research (production, health), food research (food processing development, technology innovation) and regulatory support (monitoring, mycotoxins, bio-inoculants). Biotechnology research is carried out by 19 research stations and centres across Canada and by the Animal Diseases Research Institute and Chemistry and Biology Research Institute.

Alberta Environmental Centre

Bag 4000, Vegreville, Alberta,
T0B 4LO
Telephone: 403 632 6761

Government-funded agency performing analytical and diagnostic services as well as applied research and technical developments related to pollution and the environment. Activities include studies on the microbial metabolism of toxic materials.

Alberta Research Council

Biotechnology Department, Box 8330, Station F, Edmonton, Alberta,
T6H 5X2
Telephone: 403 450 5303
Fax: 403 450 2328

Research Facilities: 250 Karl Clark Rd, Edmonton, Alberta

Carry out contract R&D and commercial production including: industrial microbiology and strain improvement; fermentation and scale up from bench to 15,000 litres for biomass production; large-scale protein production and purification; *in vitro* plant tissue culture and somatic embryogenesis; and plant cell fermentation.

Alberta Research Council

Industrial and Research Division, 11315, 87th Avenue, Edmonton, Alberta, T6G 2C2
Telephone: 403 438 0666

Alberta Research Council

Applied Sciences Division, PO Box 8330, Station F, Edmonton, Alberta,
T6H 5X2
Telephone: 403 450 5218

Provincial government agency promoting development of biotechnology applications in local industry. Current activities include fermentation and scale up, cell culture and crop improvement, production and synthesis of proteins, genes, growth hormones etc, and genetic engineering.

Alberta Research Council

7th Floor, Terrace Plaza, 4445 Calgary Trail, South Edmonton, Alberta,
T6H 5R7

Canadian Centre for Occupational Health and Safety

250 Main St East, Hamilton, Ontario,
L8N 1H6
Telephone: 416 572 2981
Fax: 416 572 2206

Promotes a healthy and safe working environment by disseminating computerized information available on-line (CCINFOline) and on CD-ROM (CCINFOdisc), an inquiries service and publications.

Canadian Commercial Corp

11th Floor, 50 O'Connor Street, Ottawa, Ontario, K1A 0S6
Telephone: 613 996 0034
Fax: 613 995 2121

Federal Crown Corporation which serves as prime contractor in Canadian exports to foreign governments and international agencies.

Centre Quebecois de Valorisation de la Biomasse - CQVB

3180 chemin Sainte-Foy, Sainte-Foy, Quebec, G1X 1R4
Telephone: 418 657 3853
Fax: 418 657 7934

Promote the use of biomass such as wood, peat and wastes, for energy and other uses such as chemicals etc; using biological and physico-chemical processes. Coordinates and stimulates research and development and exchange of information within the biomass sector.

Consumer and Corporate Affairs

Intellectual Property Directorate Patents and Trademarks, Documentation Division, 10th Floor, Zone 7, Place du Portage, Phase I, 50 Victoria Street, Hull, Quebec, K1A 0C9
Telephone: 613 994 4922

Administers the act that grants a 17-year exclusivity right to an inventor.

Department of Agriculture

PO Box 1600, Charlottetown, Prince Edward Island, C1A 7N3
Telephone: 902 892 5465

Government department providing information and technical services involved in technology development and transfer with specific interests in tissue testing and nutrition of potatoes, embryo techniques for use in swine and microbiology services.

Department of Development and Tourism

Small Business and Technology, PO Box 4750, St John's, Newfoundland, A1C 5T7
Telephone: 709 576 3980

Provincial government agency promoting development of biotechnology applications in local industry. Current activities include fermentation of fish meal and potatoes to produce high-protein animal feed, aquaculture and forest genetic and productivity research.

Department of Regional Industrial Expansion - DRIE

Food and Consumer Products Branch, Grocery Products Division, C D Howe Building, 10th Floor East, 235 Queen Street, Ottawa, Ontario
Telephone: 613 954 3089

Gives financial support to industry for commercial developments and special expenditures for biotechnology facilities in Halifax, Montreal and Vancouver.

Economic Diversification and Trade

Science and Technology Division, 206-15 Innovation Boulevard, Saskatoon, Saskatchewan, S7N 2X8
Telephone: 306 933 7209
Fax: 306 933 8244

Provincial government agency promoting development of biotechnology applications in local industry. The agency delivers a number of programmes to assist biotechnology industries complete all phases of commercial development from R&D to marketing.

Energy Mines and Resources Canada

Canada Centre for Mineral and Energy Technology - CANMET, Mineral Sciences Laboratories, 555 Booth Street, Ottawa, Ontario, K1A 0G1
Telephone: 613 996 5619
Fax: 613 992 9389

Research activities include: mechanistic studies on pipeline biocorrosion; characterization and evaluation of biosorbants for metal recovery; mechanisms for bacterial leaching of sulphide ores; development of biometallurgical processes to extract and recover metals from deposits which are low-grade or refractory;

characterization and application of biological phenomena of benefit to mineral processing operations; and development of biological processes to mitigate the environmental effects of industrial waste streams.

Environment Canada

Conservation and Protection Service, Biotechnology Centre, 15th Floor, Place Vincent Massey, 351 St Joseph Boulevard, Hull, Quebec, K1A 1C8
Telephone: 613 953 1652

Research is carried out in support of regulation (biomonitoring of toxicants, biodegradation, sewage treatment, bacterial leaching), waste treatment (Burlington Wastewater Technology Centre) and animal health care (Canadian Wildlife Service).

Environment Canada

St Lawrence Center, 105 McGill St, Montreal, Quebec, H2Y 2E7
Telephone: 514 283 9997
Fax: 514 283 1719

Promotes biotechnology for the effective treatment of water, wastes and toxic spills.

External Affairs Canada

Science Technology and Communications Division, Tower C, 5th Floor Lester B Pearson Building, 125 Sussex Drive Ottawa, Ontario, K1A 0G2
Telephone: 613 996 0675
Fax: 613 995 6319

Supports and coordinates international activities in biotechnology.

Fisheries and Oceans Canada

Aquaculture and Resource Development Branch, Fisheries Research Directorate, Centennial Towers, 200 Kent Street Ottawa, Ontario, K1A 0E6
Telephone: 613 990 0275

Currently research is being carried out on the development of transgenic fish and on growth hormones.

Forestry Canada

21st Floor Place Vincent Massey, 351 St Joseph Boulevard, Hull, Quebec, K1A 1G5
Telephone: 613 997 1107
Fax: 613 990 5437

Government of the North West Territories

Department of Justice and Public Services, Yellowknife, NWT X1A 2L9
Telephone: 403 873 7619

Provincial government agency promoting development of biotechnology applications in local industry. Current activities include the recently established Science Institute as a first step toward the introduction of biotechnology to the region.

Health and Welfare Canada

Sir Frederick G Banting Research Centre, Bureau of Drug Research, Ross Avenue Tunney's Pasture, Ottawa, Ontario, K1A 0L2
Telephone: 613 957 1059

Industry, Science and Technology Canada

1 rue Front Ouest, 4th Floor, Toronto, Ontario, M5J 1A4
Telephone: 416 973 5056
Fax: 416 973 8714

Industry, Science and Technology Canada

Strategic Technologies Branch, Biotechnology Unit, 5th Floor West C D Howe Building, 235 Queen St, Ottawa, Ontario, K1A 0H5
Telephone: 613 941 0613
Fax: 613 996 7887

800 Victoria Square, Suite 3800, PO Box 247, Montreal, Quebec, H4Z 1E8
Telephone: 514 283 4168
Fax: 514 283 3302

Promotes the better management and coordination of government programmes that assist industry to develop and exploit new technologies, and improved linkages between government, industry and academic institutions. Responsible for allocation of funds to support the National Biotechnology Strategy and provides secretariats for the National Biotechnology Advisory Committee and interdepartmental committees. The information centre on Canadian biotechnology developments, safety and regulations and national networks.

Industry, Science and Technology Canada - ISTC

800 Place Victoria, Suite 3800, PO Box 247, Montreal, Quebec, H4Z 1E8
Telephone: 514 283 8185
Fax: 514 283 3302

International Development Research Centre - IDRC

Research Programmes, PO Box 8500, Ottawa, Ontario, K1G 3H9
Telephone: 613 236 6163 ext 2569
Fax: 613 238 7230

Supports research designed to adapt science and technology to the needs of developing countries such as agriculture, medicine, waste treatment, biomass and basic research.

Labour Canada

Technical Services Division, Occupational Safety and Health Branch, Ottawa, Ontario K1A 0J2
Telephone: 819 997 3100
Fax: 819 997 1664

Concerned with occupational safety and health of labour force working in biotechnology.

Manitoba Energy and Mines

Renewable and Alternative Energy, 550-330 Graham Avenue, Winnipeg, Manitoba, R3C 4E3
Telephone: 204 945 3417

Supporting the development of industrial processes for the manufacture of fuel alcohol from cellulosic raw materials.

Manitoba Industry Trade and Technology

Industrial Technology Division, 214-155 Carlton Street, Winnipeg, Manitoba, R3C 3H8
Telephone: 204 945 0127

Provincial government agency promoting development of biotechnology applications in local industry. Current activities include membrane separation and food waste recovery techniques in the food and beverage industry, animal health care and breeding, plant strain development, fermentation and downstream processing and fuel alcohol from biomass.

Medical Research Council of Canada

Tower B, 5th Floor, Holland Cross, 1600 Scott St, Ottawa, Ontario, K1A 0W9
Telephone: 613 954 1959
Fax: 613 954 1800

Support for university research and training in biotechnology is found within the regular programmes of operating grants, including funds for equipment and salaries.

Ministere des Affaires Internationales

Program de Recherches en Biotechnologies, Cooperation Franco-Quebecois, 1225 Place Georges V 1er etage, Quebec, G1R 4Z7
Telephone: 418 644 9719

Promotes cooperation in biotechnical research between France and French-speaking regions of Canada, enabling exchange of visits to take place between universities, medical institutes or industrial laboratories.

Ministere du Commerce Exterieur et du Developpement Technologique

Edifice H 3e etage 875 est Grande-Allee, Quebec, G1R 4Y8
Telephone: 418 644 1088

Provincial government agency promoting development of biotechnology applications in local industry. Current activities include biomass conversion and fermentation and interests in the agri-food industry and health care.

Ministere de l'Enseignement Superieur et de la Science

1000 route de l'Eglise, 5e etage Sainte-Foy, Quebec, G1V 3V9
Telephone: 418 644 4731

Ministry of Economic Development

Science & Technology Division, Policy Research and Information Systems, Legislative Bldg, 818 Fort Street, Victoria, British Columbia, V8V 1X4
Telephone: 604 387 2034

Provincial government agency promoting development of biotechnology applications in local industry. Current activities include projects related to natural resource extraction and processing, food fermentation, biological waste treatment (pulp mill effluents and domestic sewage) and tissue culture and micropropagation of conifers.

National Research Council - NRC

Industrial Research Assistance Program - IRAP, Laboratory Network Industrial Development Office, Bldg M-55 Montreal Road, Ottawa, Ontario, K1A 0R6
Telephone: 613 993 1790

Assists Canadian companies to undertake new biotechnology R&D projects through its IRAP Biotechnology Programme. In addition, encourages collaboration with Canadian universities and public sector laboratories.

National Research Council - NRC

Division of Biological Services, 100 Sussex Drive, Ottawa, Ontario, K1A 0R6
Telephone: 613 990 0884

One of four NRC centres of biotechnology research. Fundamental research is carried out in molecular genetics, immunochemistry, microbiology, physiology, molecular biophysics and biochemistry and biomathematics.

National Research Council - NRC

6100 Royalmount Avenue, Montreal, Quebec, H4P 2R2
Telephone: 514 496 6102

One of four NRC centres of biotechnology research. 'State of the art' facility for generic research (cell fusion, enzyme technology, genetic engineering, fermentation and biochemical and process engineering). In partnership with industry, brings basic research through the developmental stages between laboratory and commercial application. Research is focused on biological systems and development of new processes and products in areas of food, health care, waste treatment, pulp and paper and forestry.

National Research Council of Canada

Division of Energy Research and Development, Building M-50 Montreal Rd, Ottawa, Ontario, K1A 0R6
Telephone: 613 993 0357/0331
Fax: 613 993 4270

Coordinates the energy research programmes which include biological application to renewable energy production from forest products in particular.

Natural Sciences and Engineering Research Council - NSERC

5th Floor Centennial Towers, 200 Kent Street, Ottawa, Ontario, K1A 1H5
Telephone: 613 995 5849/6424
Fax: 613 992 5337

Its role is to secure a healthy base in universities and as a corollary to secure a sound balance between this diversified research base and the more targeted research programmes; to secure an adequate supply of highly qualified personnel in basic science as well as trained 'state of the art' facilities; and to facilitate collaboration between R&D performing sectors in Canada.

New Brunswick Department of Economic Development & Tourism

Science and Technology Secretariat, PO Box 6000, Fredericton, New Brunswick, E3B 5H1
Telephone: 506 453 2489
Fax: 506 453 7904

Provincial government agency promoting development of biotechnology applications in local industry. Current activities include plant tissue culture and micropropagation for tree improvement, aquaculture and a bioenergy programme.

Nova Scotia Research Foundation Corporation - NSRFC

101 Research Drive, Woodside Industrial Park, PO Box 790, Dartmouth, Nova Scotia, B2Y 3Z7
Telephone: 902 424 8670
Fax: 902 424 4679

Provincial government agency promoting development of biotechnology applications in local industry. Current activities include straw fermentation for ruminant food stock, mycorrhizal inoculum for agricultural crops and tree seedlings, increasing efficiency of nutrient and water intake and resistance against soil-borne diseases.

Ontario Ministry of Energy

Department of Renewable Energy, 56 Wellesley Street West, Toronto, Ontario, M7A 2B7
Telephone: 416 965 4365

Ontario Ministry of Industry Trade and Technology

Premier's Council Technology Fund, 900 Bay Street 8th Floor Hearst Block, Toronto, Ontario, M7A 2E1
Telephone: 416 963 3721

Provincial government agency promoting development of biotechnology applications in local industry. Current activities include waste stream treatment (fermenter/purifier and rotating biological contactors), mineral leaching and processing, fluidized bed anaerobic digestion, health care (eg cardiovascular drug: carionatrin) and regeneration of plantlets from explants of forest trees.

Prince Edward Island Department of Industry

Shaw Building, PO Box 2000,
Charlottetown, Prince Edward Island,
C1A 7N8
Telephone: 902 368 4240
Fax: 902 368 4224

Provincial government department promoting development of biotechnology applications in local industry. Current activities include funding for demonstration projects, technology transfer and research and development.

Quebec Ministry of Industry, Trade and Technology - MICTQ

Place Mercantile, 10th Floor,
770 Sherbrooke St W, Montreal,
Quebec, H3A 1G1
Telephone: 514 982 3012
Fax: 514 864 3755

Saskatchewan Research Council

15 Innovation Blvd, Saskatoon,
Saskatchewan, S7N 2X8
Telephone: 306 933 5400
Fax: 306 933 7446

Standards Council of Canada

1200-45 O'Connor St, Ottawa, Ontario,
K1P 6N7
Telephone: 613 238 3222
Fax: 613 995 4564

Promotes voluntary standardization in fields of construction, manufacture, production, performance and safety. Furthers international cooperation in the field of standards.

Supply and Services Canada

Ottawa, Ontario K1A 0S6
Telephone: 819 997 6368

Encourages private sector research and development, through grants, in areas related to existing general programmes. Publishes monthly research and development bulletin which highlights its activities.

Yukon Executive Council Office

Intergovernmental Relations,
Box 2703, Whitehorse, Yukon
Territory, Y1A 2C6
Telephone: 403 667 5331
Fax: 403 668 4936

Provincial government agency promoting development of biotechnology applications in local industry. Current activities include the recently established Science Institute of the North West Territories as a first step toward the introduction of biotechnology to the region.

DENMARK

Customs Directorate

Direktorate for Tolduaesenet,
Amaligade 44, Copenhagen
Telephone: 33 15 73 00

Danish Council for Scientific Policy and Planning

Holmens Kanal 7, Copenhagen K,
DK-1060

Danish Energy Agency

Landemaerket 11, Copenhagen K,
DK-1119
Telephone: 33 92 67 00
Fax: 33 11 47 43

Danish Standards Association

Baunegardsvej 73, Hellerup, DK-2900
Telephone: 39 77 01 01
Fax: 39 77 02 02

Department for Development - DANIDA

Asiatisk Plads 2, Copenhagen K, DK-1448
Telephone: 33 92 00 00

European Communities

Press and Information Office, Hojbrohus Ostergade 61, PO Box 144, Copenhagen K, DK-1004
Telephone: 33 14 41 40

European Communities

Publications, J H Schulz Information Services A/S, Ottiliavej 18 A, Valby, DK-2500
Telephone: 36 44 23 00
Fax: 36 44 01 41

Ministry of Agriculture

Landbrugsministeriet, Slotholmsgade 10, Copenhagen K, DK-1216
Telephone: 33 12 37 17

Ministry of Education and Research

Department of Research, H C Andersens Boulevard 40, Copenhagen V, DK-1553
Telephone: 33 11 43 00
Fax: 33 32 35 01

Ministry of Energy

1 Slotsholmsgade, Copenhagen K, DK-1216
Telephone: 33 92 75 00
Fax: 33 12 87 07

Ministry of Energy

Energiministeriet, Strandgade 29, Copenhagen K, DK-1401
Telephone: 31 54 36 11

Ministry of Environment

Miljostyrelsen, Slotholmsgade 12, Copenhagen K, DK-1216
Telephone: 1 12 76 88

Responsible for water resources, waste water and sewage, recycling of wastes and valuable materials.

Ministry of Foreign Affairs

Udenrigsministeriet, Asiatisk Plads 2, Copenhagen K, DK-1448
Telephone: 33 92 00 00

Ministry of Health

Herluf Trollesgade 11, Copenhagen K, DK-1052
Telephone: 33 92 33 60
Fax: 33 93 15 63

Ministry of Industry

Industriministeriet, Slotholmsgade 12, Copenhagen K, DK-1216
Telephone: 33 12 11 97

Patent Office

Patentdirektorate, Helgeshoej Alle 81, Taastrup, DK-2630
Telephone: 43 71 71 71
Fax: 43 71 71 70

FINLAND

Agricultural Research Centre Library

Maatalouden tutkimuskeskus Kirjasto, SF-31600 Jokioinen
Telephone: 16 1881
Fax: 16 188339

Finnish Council for Scientific and Technical Information - TINFO
Otaniemi, SF-02150

Ministry of Agriculture and Forestry
Hallituskatu 3, Helsinki, SF-00170

Ministry of Education
Meritullinkatu 10, Helsinki, SF-00170

Ministry of Environment
Ratakatu 3, Helsinki, SF-00120
Telephone: 0 19911
Fax: 0 1991 499

Ministry of Foreign Affairs
Merikasarmi, Helsinki, SF-00160

Ministry of the Interior
Department of Environmental Protection, PL 257, Helsinki 17, SF-00171

Interest includes recycling of wastes and bacterial and organic pollution.

Ministry of Social Affairs and Health
Snellmaninkatu 4-6, Helsinki, SF-00170

Ministry of Trade and Industry
Ministry of Foreign Trade, Aleksanterinkatu 10, Helsinki, SF-00170

National Board of Health
Siltasaarenkatu 18A, Helsinki 53, SF-00530

National Board of Patents and Registration
Patent-och registerstyrelsen Patentti- ja rekisterihallitus, Albertinkatu 25, Helsinki 18, SF-00180
Telephone: 0 695 31
Fax: 0 695 3204

National Board of Water
PL 250, Helsinki 10, SF-00101

VTT Information Service
Technical Research Centre of Finland, Vuorimiehentie 5, PO Box 42, Espoo, SF-02151
Telephone: 0 4561
Fax: 0 456 4374

Research fields include malting and brewing, production and use of microbial enzymes, plant biotechnology, molecular biology, protein engineering, bioreactors, mathematical modelling and downstream processing.

FRANCE

Administration for Energy and Raw Materials
Direction General de l'Energie et des Materieres Premiers - DGEMP, 101 rue de Grenelle, Paris, F-75700

Agency for Energy Conservation
Agence pour les Economies d'Energie, 26 rue de Cambronne, Paris, F-75015
Telephone: 1 306 44 90

Agency for Environment and Energy Control - ADEME

Agence de l'Environnement et de la Maitrise de l'Energie, 27 rue Louis-Vicat, Paris, F-75015
Telephone: 1 47 65 20 00
Fax: 1 46 45 52 36

Activities include implementation of research and development in renewable energies including biomass energy (anaerobic digestion and fermentation for liquid fuel production).

ANRED

Agence Nationale pour la Recuperation et Elimination des Dechets, Boite Postale 406, 2 Square Lafayette, Angers, F-49004
Telephone: 41 87 29 24

National agency concerned with disposal of wastes. Carrying out joint project with Gaz de France on production of methane from landfills. This is part of its overall role to assist development of facilities for treatment and use of waste, and to encourage new technologies. Also aims to promote educational campaigns for the general public and industries. Functions under the auspices of the Ministries of Environment and Industry.

ANRED

Agence Nationale pour la Recuperation et l'Elimination des Dechets, 2 square Lafayette BP 406, Angers, F-49004
Telephone: 41 87 29 25

National agency for waste recovery and pollution control.

ANVAR

Agence Nationale de Valorisation de la Recherche, 43 rue Caumartin, Paris, F-75436
Telephone: 42 66 93 10

Promotes scientific technologies, supports research and the development of innovation over a wide area of disciplines including aspects of biotechnology.

Department of Environment

14 boulevard Leclerc, Neuilly sur Seine Cedex, F-92521
Telephone: 1 758 12 12

Interested in aquatic biology, providing information and education.

Department of Pollution Control

Direction de la Prevention des Pollutions, 14 boulevard Leclerc, Neuilly sur Seine Cedex, F-92521
Telephone: 1 758 12 12

Concerned with industrial environment and pollution from industry, effluents and water, especially energy aspects.

Directorate for Energy and Raw Materials

Direction Generale de l'Energie et des Matieres Premieres, 101 rue de Grenelle, Paris, F-75700
Telephone: 1 555 93 00

Directorate General for Customs and Indirect Taxes

Direction Generale des Douanes et Droits Indirects, 8 rue de la Tour-des-Dames, Paris Cedex 09, F-75436
Telephone: 1 280 67 22

Directorate for Scientific and Technical Research

Delegation Generale a la Recherche Scientifique et Technique - DGRST, 35 rue St Dominique, Paris, F-75007

European Communities

Press and Information Office, 61 rue des Belles Feuilles, Paris Cedex 16, F-75782
Telephone: 1 501 58 85

European Communities Publications

Service de Vente en France des Publications des Communautes Europeennes, 26 rue Desaix, Paris Cedex 15, F-75732
Telephone: 1 578 61 39

Forestry Institute

Institut du Developpement Forestier, 23 avenue Bosquet, Paris, F-75008
Telephone: 1 555 23 49

French Association for Standardization

Association Francaise de Normalisation - AFNOR, 2 tour Europe, Courbevoie Cedex 7, F-92400
Telephone: 788 11 11

Government Publications

La Documentation Francaise, 29 quai Voltaire, Paris, F-75007
Telephone: 1 261 50 10

Interministerial Committee of Scientific and Technical Research

35 rue St-Dominique, Paris, F-75700
Telephone: 1 550 32 50

Ministry of Agriculture

78 rue de Vasenne, Paris, F-75700
Telephone: 1 555 95 50

Ministry of Agriculture

Energy Mission, 3 rue Barbet de Jouy, Paris, F-75700
Telephone: 1 555 95 50

Ministry of Agriculture

Direction des Industries Agro-Alimentaires - DIAA, 35 rue Saint-Dominique, Paris, F-75007
Telephone: 1 45 55 95 50

Ministry of Cooperation (Development)

20 rue Monsieur, Paris, F-75700
Telephone: 1 567 55 90

Ministry of Economic Affairs

rue de Rivoli 93, F-75056
Telephone: 260 33 00

Ministry of Education

110 rue de la Grenelle, Paris, F-75357
Telephone: 1 550 10 10

Ministry for the Environment

Ministere de l'Environment et Cadre de Vie, 246 boulevard St Germain, Paris Cedex 16, F-75775
Telephone: 1 544 39 93

Ministry of Foreign Affairs

37 Quai d'Orsay, Paris, F-75700
Telephone: 1 555 95 40

Ministry of Health

8 ave de Segur, Paris, F-75700
Telephone: 1 567 55 44

Ministry of Industry

101 rue de Grenelle, Paris, F-75700

National Agricultural Research Institute

Institute National de la Recherche Agronomique - INRA, 149 rue de Grenelle, Paris Cedex 07, F-75341
Telephone: 1 550 32 00

National Centre of Agricultural Mechanisation for Forestry & Water

Management, Centre National du Machinsme Agricole du Genie Rural des Eaux, et des Forets - CEMAGREF, Parc de Tourvoie, Antony, F-92160
Telephone: 1 46 66 21 07

Carrying out research on energy production in the areas of agricultural and forest biomass, combustion and gasification and biofuels.

National Centre of Scientific Research

Centre National de la Recherche Scientifique (CNRS), 15 quai Anatole France, Paris, F-75007

Office of Research and Statistical Studies

Boite Postale 16, Pontoise, F-95301

Patent Office

Institut Nationale de la Propriete Industrielle, 24 bis rue de Leningrad, Paris, F-75008

Scientific and Technical Documentation Centre

CNRS Centre de Documentation Scientifique et Technique, 26 rue Boyer, Paris Cedex 20, F-75971
Telephone: 1 358 35 59

Information centre providing service in abstracting and indexing, database supply and documentation dissemination for researchers and engineers in all areas of science.

Technical Institute for Cattle Breeding

Institut Technique de l'Elevage Bovin, Maison Nationale des Eleveurs, 149 rue de Bercy, Paris Cedex 12, F-75595
Telephone: 1 346 12 20

Technical Institute for Cereals and Fodder

Institut Technique de Cereales et des Fourrages (ITCF), 8 avenue du President Wilson, Paris, F-75016
Telephone: 1 723 55 13

GERMANY

Central Commission for Biology

Zentrale Kommission fuer die Biologische Sicherheit, Columbiadamm 3, Berlin 61, D-1000
Telephone: 30 6 9 01-1 77
Fax: 30 6 9 01-1 73

Concerned with the regulation of genetic engineering.

European Communities

Press and Information Office, Zitelmannstrasse 22, D-5300 Bonn
Telephone: 30 892 40 28

European Communities

Publications, Bundesanzeiger, Postfach 108 006, Breite Strasse 78-80, D-5000 Koln 1
Telephone: 221 2029-0
Fax: 221 2029 278

Federal Bureau of Statistics

Statistisches Bundesamt, Postfach 5528,
Gustav Stresemann Ring 11,
Wiesbaden 1, D-6200
Telephone: 611 75 1
Fax: 611 75 3425

Federal Environmental Agency

Umweltbundesamt, Bismarckplatz 1,
Berlin 33, D-1000
Telephone: 30 8903 0
Fax: 30 8903 2285

Federal Institute for Agriculture and Forestry

Biologische Bundesanstalt fuer Land- und Forswirrtshaft,
Dokumentationsstelle fuer Phytomedizin, Koenigin Luise Strasse,
Berlin 33 D-1000
Telephone: 30 83 04-2 15

Manitains a databank on phytommedicine available on-line under STN and DIMDI.

Federal Ministry for Food, Agriculture and Forestry

Bundesministerium fuer Ernaehrung,
Landwirtschaft und Forsten,
Rochstrasse 1, Bonne D-5300
Telephone: 2 28 59-1
Fax: 2 28 59 36 01

Federal Office of Economics

Frankfurter Strasse 29-31,
Eschborn/Taunus 1, D-6236
Telephone: 6196 1404 0
Fax: 6196 1404 212

German Patent Office

Deutsches Patentamt,
Zweibrueckenstrasse 12, Muenchen 2,
D-8000
Telephone: 89 2195 0
Fax: 89 2195 2221

Institute for Agricultural Information

Institut fuer Landwirtschaftliche
information und Dokumentation,
Elecktronische Fachinformation,
Krausenstrasse 38-39, Berlin D-1086
Telephone: 20 35 92 16
Fax: 20 35 92 19

Institute of Genetics and Plant Science

Institut fuer Genetik und
Kulturpflanzenforshung, Abteilung
Information und Dokumentation,
Correnstrasse 3, Gatersleben D-4325

Ministry for Economic Cooperation

Bundesministerium fuer wirtschaftliche
Zusammenarbeit, Referat 220,
Karl-Marx-Strasse 4-6, Bonn 1, D-5300
Telephone: 228 5350

Ministry of Education and Science

Bundesministerium fuer Bildung und
Wissenschaft, Heinemann Strasse 2,
Bonn 2, D-5300
Telephone: 228 570
Fax: 228 57 2096

Ministry of Finance

Faculty of Medicine, Graurheindorfer
Strasse 108, Bonn 1, D-5300
Telephone: 228 6820
Fax: 228 682 4420

Ministry of Finance

Customs Department, Graurheindorfer
Strasse 108, Bonn 1, D-5300
Telephone: 228 6820
Fax: 228 6824420

GREECE

European Communities

Press and Information Office,
2 Vassilissis Sofias, Athens, GR-10674
Telephone: 1 724 1515
Fax: 1 724 4620

European Communities

Publications, G C Eleftheroudakis Ltd,
4 Nikis Street, Athens 126
Telephone: 1 322 22 55

General Secretariat for Press and Information

Odos Zalokosta Kriezotou 8, Athens
Telephone: 1 363 0911

Hellenic Organization for Standardization

Didotout 15, Athens
Telephone: 1 360 9717

Ministry of Agriculture

2-6 Acharnon St, Athens, GR-10176
Telephone: 1 529 1111

Ministry of Coordination

National Council for Planning and Environment, Platia Syntagmatos, Athens

Ministry of Culture and Sciences

14 Aristidou St, Athens, GR-10186
Telephone: 1 324 3015

Ministry of Education and Religion

15 Mitropoleos St, Athens, GR-10185
Telephone: 1 323 0861

Ministry of Energy and Natural Resources

Ministry of Industry and Technology,
80 Michalakopoulou St, Athens,
GR-10193
Telephone: 1 770 8615

Ministry of Finance

10 Karageorgi Servias St, Athens,
GR-10184
Telephone: 1 322 4071

Ministry of Finance

Customs Department, Odos Karageorgi Servias 10, Athens

Ministry of Foreign Affairs

2 Zalokosta St, Athens, GR-10671
Telephone: 1 361 0581

Ministry of Health and Welfare

17 Aristotelous St, Athens, GR-10187
Telephone: 1 523 2821

Ministry of Industry and Energy

Odos Mihalakopoulo 80, Athens

Ministry of Planning and Environment

Armalidos Pouliou 17, Ambelokipi, Athens
Telephone: 1 642 8611

Ministry of Research & Technology

2 Hermou St, Athens, GR-10193
Telephone: 1 325 1312

National Energy Council

Academic Street 42, Athens

IRELAND

Council for Development in Agriculture - ACOT

39 Northumberland Road, Dublin 4
Telephone: 1 682 455

Customs and Excise

Collectors Office, Custan House, Dublin 1
Telephone: 1 742961

Department of Agriculture

Kildare Street, Dublin 2
Telephone: 1 789011

Department of Education

An Roinn Oideachais, Marlborough Street, Dublin 1
Telephone: 1 717101

Department of Energy

Clare Street, Dublin 2
Telephone: 1 715233

Department of Finance

Government Buildings, Upper Merion Street, Dublin 2
Telephone: 1 767571

Department of Fisheries and Forestry

22 Upper Merion Street, Dublin 2
Telephone: 1 789211

Department of Foreign Affairs

80 St Stephen's Green, Dublin 2
Telephone: 1 780822

Department of Industry Commerce and Energy

Oisin House, Pearse Street, Dublin 2
Telephone: 1 688888

Department of Industry Trade Commerce and Tourism

Kildare Street, Dublin 2
Telephone: 1 789411

European Communities

Publications, Stationery Office, St Martin's House, Waterloo Road, Dublin 4
Telephone: 1 789644

European Communities

Press and Information Office, 39 Molesworth Street, IRL-Dublin 2
Telephone: 1 712244

Government Publications Stationery Office

Beggar's Bush, Dublin 4
Telephone: 1 688433

Industrial Development Authority

Lansdowne House, Dublin 4

Irish Peat Development Authority

Bord Na Mona, Baggot Street, Dublin 2
Telephone: 1 688555

Droichead Nua, Co Kildare
Telephone: 045 31 201

National Board for Science and Technology

Sherbourne House, Sherbourne Road, Dublin 4
Telephone: 1 683311

Promotes science and technology including new and renewable energy technologies and conservation with formulation of research and development and policy of recommendations. Active in the International Energy Agency information programmes in biomass conversion technology.

ISRAEL

Ministry of Agriculture

8 Arania St, Hakirya, Tel Aviv 61070
Telephone: 3 211211
Fax: 3 268899

Ministry of Education and Culture

34 Shivtei Israel Street, PO Box 292, Jerusalem 91911
Telephone: 2 278211
Fax: 2 278223

Ministry of Energy and Infrastructure

234 Yaffo St, PO Box 13106, Jerusalem 91130
Telephone: 2 551211
Fax: 2 381444

Ministry of Environment

2 Kaplan St, Kiryat Ben-Gurion, PO Box 6158, Jerusalem 91061
Telephone: 2 660151/523629
Fax: 2 385038

Ministry of Health

2 Ben-Tabai St, PO Box 1176, Jerusalem 91010
Telephone: 2 692212

Ministry of Industry and Trade

30 Agron St, Jerusalem 94190
Telephone: 2 210111
Fax: 2 245110

Ministry of Information and Infrastructure

National Centre of Scientific and Technological Information, 84 Hachashmoniam Street, PO Box 21125, Tel Aviv 61201
Telephone: 3 29 77 81

Ministry of Science and Technology

Kiryat Hamemshala, 3rd Bldg, PO Box 18195, Jerusalem 91181
Telephone: 2 277096/277400

National Council for Research & Development

Ministry for Science and Technology, New Government Building Complex, PO Box 18195, Jerusalem 91181
Telephone: 2 277062

Office of the Chief Scientist

4 Mevo H'Matmid, Jerusalem 91020
Telephone: 2 226368
Fax: 2 248159

Research and Development Authority

PO Box 1025, Ben Gurion University of Negev, Beer Sheva 84361
Telephone: 57 78382

ITALY

Biotechnology Centre

Centro per il Transferimento delle Biotecnologie, Via Sardegna 38, Roma, I-00100
Telephone: 6 485 451
Fax: 6 465 6114

Directorate of Research and Development

Via G B Martini 3, Roma, I-00100

European Communities

Press and Information Office, Via Poli 29, Roma, I-00187
Telephone: 6 678 97 22

Corso Magenta 61, I-20123 Milano
Telephone: 02 80 15 05

European Communities

Publications, Istituto Poligrafico e Zecca dello Stato, Libreria dello Stato Piazza G Verdi 10, Roma, I-00100
Telephone: 6 85 08 22 76

Corso Magenta 61, I-20123 Milano
Telephone: 02 80 15 05

Government Publications

Servizio dell Informazione e Ufficio della Proprieta Letteraria, Via Boncompagni 15, Roma, I-00100
Telephone: 6 48 95 06

Institute of Health

Department of Chemistry, Viale Regina Elena 299, Roma, I-00161
Telephone: 6 49 90

Working on separation and analysis of proteins and nucleic acids.

Institute of Health

Department of Physics, Via Regina Elena 299, Roma
Telephone: 6 49 90

Studying nucleic acids of ribosomes, viruses and viral proteins, cell membranes and energetics.

Institute of Plant Biochemistry

Istituto di Biosintesi Vegetale, Via Bassini 15, Milano
I-20100
Telephone: 2 29 21 70

Studying various aspects of plant biotechnology, including research on the regulatory genes of storage protein sysnthesis in maize.

Ministry of Agriculture and Forestry

Ministero dell'Agricoltura e della Foreste, Via XX Settembre, Roma, I-00100
Telephone: 46 65

Ministry for Culture and Environmental Affairs

Ministero per i Beni Culturali e Ambientali, Via del Collegio Romano 27, Roma, I-00100
Telephone: 6 67 23

Ministry of Education

Ministero della Pubblica Istruzione, Viale Trastavere, Roma, I-00100
Telephone: 6 58 59

Ministry of Finance

Ministero delle Finanze, Viale America, Roma, I-00100
Telephone: 6 36 91 1

Ministry of Finance

Directorate General of Customs and Excise, Viale America, Roma, I-00100

Ministry of Foreign Affairs

Ministero degli Affari Esteri, Farnesina, Roma, I-00100
Telephone: 6 36 91 1

Ministry of Health

Ministero Inistero Dellanita, Via dell'Industrie 20, Roma, I-00100
Telephone: 6 59 94

Ministry of Industry

Via Vittorio Veneto 33, Roma, I-00100
Telephone: 6 47 05

Ministry of Industry and Commerce Patent Office

e Via Molise 19, Roma, I-00187

National Biotechnology Committee

Comitato Nazionale per le Biotechnologie, c/o Ministero per l'Univ e la Ricerca Sci e Tecnologica, Lungotevere Thaon di Revel 76, Roma, I-00196
Telephone: 6 3969941

National Consultative Committee on Biotechnology

Comitato Nazionale Consulenza del CNR per le Biotecnologi, Consiglio Nazionale delle Ricerche, Piazzale Aldo Moro 7, Roma, I-00196
Telephone: 6 49 933 713

National Research Council

Cibsiglio Nazionale delle Richerche, Piazzale Aldo Moro 7, Roma, I-00100
Telephone: 6 49 93

National Standards Bureau

Ente Nazionale Italiano di Unification, Piazza Diaz 2, Milano, I-20123
Telephone: 2 87 69 14

JAPAN

Administration Bureau

Science and International Affairs Division, 36 Nishigonaka, Myodaiji, Okasaki 444

European Communities

Publications, Kinokuniya Company Ltd, 17-7 Shinjuku-3-chome Shiniuku-ku, Tokyo 160-91
Telephone: 3 354 0131

For sales and subscriptions to publications from the EC.

Japan Information Centre of Science and Technology

CPO Box 1478, 5-2 Nagatacho-2-chome, Chiyoda-ku Tokyo

Ministry of Agriculture, Forestry and Fisheries

2-1 Kasumigaseki-1-chome, Chiyoda-ku, Tokyo 100
Telephone: 3 502 8111

Ministry of Education Science and Culture

2-2 Kasumigaseki-3-chome, Chiyoda-ku, Tokyo 100

Ministry of Health and Welfare

Pharmaceutical Affairs Bureau, Office of Advanced Research and Technology, 2-2 Kasumigaseki 1-chome, Chiyoda-ku, Tokyo 100-45
Telephone: 3 503 1711

Also: Minister's Secretariat, The Health Science Division.

Ministry of International Trade and Industry

Basic Industries Bureau, Biochemical Industry Division, 3-1 Kasumigaseki-1-Chome, Chiyoda-ku, Tokyo 100
Telephone: 3 501 1511

Patent Information

Japan Institute of Invention and Innovation, 9-14 Toranomon-2-chome, Minato-ku, Tokyo 105

Supplies copies of Japanese patents. Offers translation and abstracting services.

Patent Office

Kokkyocho, 4-3 Kasumigaseki-3-chome, Chiyoda-ku, Tokyo 100

Science and Technology Agency

Research and Development Bureau, The Life Science Division, 2-1 Kasumigaseki-2-chome, Chiyoda-ku, Tokyo 100
Telephone: 3 503 1711

THE NETHERLANDS

Directorate General for International Cooperation

Postbus 20061, EB Den Haag, NL-2500
Telephone: 70 76 77 67

Responsible for overseas aid and development.

Directorate General for Taxation

Customs and Excise, Casuariestraat 32, VB Den Haag, NL-2511

European Communities

Publications, Staatsdrukkerij en Uitgeverijbedrijf, Christophel Plantijnstraat, Den Haag
Telephone: 70 78 99 11

European Communities

Press and Information Office, Lange Voorhout 29, Den Haag
Telephone: 70 46 93 26

Federation of Netherlands' Industry

Postbus 93093, Prinses Beatrixlaan 5, Den Haag

Government Printing Offices

Postbus 2001, EA Den Haag, NL-2500
Telephone: 70 79 89 11

Ministerie Van Volksuisvesting En Ruimtelijke Ordening

van Alkemadelaan 85, AC Den Haag, NL-2597
Telephone: 70 79 39 11

Ministry of Agriculture and Fisheries

Postbus 204001, EK Den Haag, NL-2500
Telephone: 70 79 39 11

Ministry of Economic Affairs

Ministerie van Economische Zaken, Bezuiddenhoutsweg 70, Den Haag
Telephone: 70 79 89 11

Ministry of Education and Science

Postbus 20551, EN Den Haag, NL-2500
Telephone: 70 74 27 42

Ministry of Finance

Casuariestraat 32, VB Den Haag, NL-2511
Telephone: 70 76 77 67

Ministry of Foreign Affairs

Postbus 20081, EB Den Haag, NL-2500
Telephone: 70 76 77 67

Ministry of Public Health and Environment Hygiene

Postbus 439, AK Leidschendam, NL-2260
Telephone: 70 20 92 60

The Netherlands Foreign Investment Agency

91 Bezuidenhoutseweg, PO Box 20101, EC Den Haag, NL-2500
Telephone: 70 797645

Central agency responsible for promoting foreign investment, offering assistance at all stages of the investment planning process from preliminary research through implementation, including site selection, fact-finding visits and advice on joint-venture and takeover opportunities. Has offices in Western Europe, the US, Canada, Japan, South Korea and Taiwan.

Patent Office

Postbus 5818, Octrooivaad
Patentlaan 2, HV Rijswijk (ZH), NL-2280

Programme Bureau Biotechnology

Scheveningseweg 10, KT Den Haag, NL-2517
Telephone: 3 17 06 31938

Stichting Voorlilchting Energisbesparing Nederland

Postbus 503, AM Apeldoorn, NL-7300
Telephone: 55 33 07 50

NORWAY

Agricultural Research Council of Norway

Norges Landbruksvitenskapelige Forskningrad - NLVF, Postboks 8154 Dep, Oslo 1
Telephone: 2 20 57 35

Royal Ministry of Agriculture

Akersgt 42, PO Box 8007 Dep, Oslo 1, N-0030
Telephone: 2 34 90 90

Royal Ministry of Education and Research

Akersgt 42, PO Box 8119 Dep, Oslo 1, N-0032
Telephone: 2 34 90 90
Fax: 2 34 95 40

Royal Ministry of Environment

Myntgt 2, PO Box 8013 Dep, Oslo 1, N-0030
Telephone: 2 34 90 90

Royal Ministry of Fisheries

Ovre Slottsgate 2, PO Box 8118 Dep, Oslo 1, N-0032
Telephone: 2 34 90 90
Fax: 2 34 95 85

Royal Ministry of Industry

Ploensgt 8, PO Box 8014 Dep, Oslo 1, N-0030
Telephone: 2 34 90 90

Royal Ministry of Petroleum and Energy

Ploensgt 8, PO Box 8148 Dep, Oslo 1, N-0033
Telephone: 2 34 90 90

Royal Ministry of Social Affairs

Grubbegt 10, PO Box 8011 Dep, Oslo 1, N-0030
Telephone: 2 34 90 90

Royal Norwegian Council for Scientific and Industrial Research - NTNF

Postboks 70, Tasen, Sognsveien 70, Oslo, N-0801
Telephone: 2 23 76 85
Fax: 2 181139/184137

PORTUGAL

Centro de Energia e Tecnologia Lda - CENERTEC

125 Esc 33, Porto, P-4300
Telephone: 561159
Fax: 563968

European Communities

Publications, Livraria Bertrand Sarl, Rua Joao de Deus, Venda Nova Amadora
Telephone: 97 45 71

For sales and subscriptions to publications from the EC.

Ministry of Industry Energy and Exports

Ministerio da Industria Energia e Exportacoes, Rua de Beneficiencia 2415, Lisbon, P-1600

Secretariat of Commerce and Industry

Secretaria Regional do Comercio e Industria, Rua Dr Caetano de Audrade 11, Ponta Delgada-Acores, P-9500

SPAIN

Centro para el Desarrollo Tecnologico Industrial - CDTI

Edificio Cuzco IV, P Castellana 141, Madrid, E-28046
Telephone: 581 55 00

Implements the technological innovation policy of the Ministry of Industry and Energy of Spain.

European Communities

Publications, Mundi Prensa Libros SA, Castello 37, Madrid 1
Telephone: 1 275 4655

For sales and subscriptions to publications from the EC.

Ministry of Agriculture, Fisheries and Food

Ministerio de Agricultura, Pescas y Alimentacion, Paseo Infanta Isabel 1, Madrid, E-28014
Telephone: 1 228 50 22
Fax: 1 347 56 23

Ministry of Education and Science

Ministerio de Educacion y Ciencia, Alcala 34, Madrid, E-28014
Telephone: 1 522 11 00

Ministry of Energy and Industry

Ministerio de Industria y Energia, Paseo de la Castellana 160, Madrid, E-28046
Telephone: 1 458 80 10

Ministry of Foreign Affairs

Ministerio de Asuntos Exteriores, Plaza de la Provincia 1, Madrid, E-28012
Telephone: 1 265 86 05

Ministry of Health

Ministerio de Sanidad y Consumo, Paseo del Prado 18-20, Madrid, E-28071
Telephone: 1 420 20 34
Fax: 1 468 45 54

National Council for Research & Development

Edificio Torre Rioja, Calle Rosario Pino 14-16, Madrid 20
Telephone: 341 450 1663

Patent Office

Registro de la Propiedad Industrial, Avenida del Generalisimo 59, Madrid 16

Plan National de I+D

Rosario Pino 14-16, Madrid, E-28020
Telephone: 572 00 98
Fax: 571 57 81

SWEDEN

Commission for Environmental Protection

Miljovardsberedningen, S-103 33 Stockholm
Telephone: 8 763 10 00
Fax: 8 24 16 29

Department of Industry

Industridepartmentet, S-103 33 Stockholm
Telephone: 8 763 10 00
Fax: 8 11 36 16

National Environment Protection Board

Statens Naturvardsverk, Smidesvagen 5, Solna, S-171 85
Telephone: 8 799 10 00
Fax: 8 29 23 82

National Swedish Board for Energy Source Development

Energiforskningsnamnden, Box 43020, Stockholm, S-100 72
Telephone: 8 744 97 25
Fax: 8 19 48 22

Undertakes assessment studies. Provides an information centre for energy including aspects of renewable energy and production and use of forestry products in particular. Covers anaerobic digestion and ethanol fermentation as well as possible environmental impacts of such technologies.

National Swedish Board of Occupational Health and Safety

Arbetarskyddsstyrelsen,
Ekelundsvagen 16, Solna, S-171 84
Telephone: 8 730 90 00
Fax: 8 730 19 67

National Swedish Board for Technical Development

Styrelsen for Teknisk Utveckling,
Box 47300, Stockholm, S-100 74
Telephone: 8 775 40 00
Fax: 8 19 68 26

National Swedish Industrial Board

Statens Industreverk, Liljeholmsvagen 30, Stockholm, S-117 43
Telephone: 8 744 90 00
Fax: 8 18 22 05

Central government board responsible for policy and legal aspects of industry and energy supply, as well as providing advice, training and information.

Parliamentary Committee for Environmental Protection

Koncessionamnden for Miljoskydd,
Box 2121, Stockholm, S-103 13
Telephone: 8 791 51 00

Patent Office

Kungliga Patent-och Registreringsverket, Box 5055, Valhallavagen 136, Stockholm, S-102 42
Telephone: 8 782 25 00

Swedish Natural Science Research Council

Naturvetenskapliga Forskningsradet,
Box 6711, Stockholm, S-113 85
Telephone: 8 610 07 00
Fax: 8 610 07 40

Responsible for fundamental research programmes.

SWITZERLAND

European Communities

Publications, FOMA, 5 avenue de Longemalle CP 367, Renens-Lausanne, CH-1020
Telephone: 21 35 13 61

For sales and subscriptions to publications from the EC.

Federal Office for Education and Science

Bundesamt fur Bildung und Wissenschaft, Postfach 5675, Wildhainweg 9, Bern, CH-3001
Telephone: 31 61 96 91
Fax: 31 61 78 54

Federal Office for Energy

Bundesamt fur Energiwirtschaft,
Kapellenstrasse 14, Bern, CH-3003
Telephone: 31 61 56 11

Federal Office of Environment, Forest and Landscape

Bundesamt fur Umwelt, CH-3003 Bern
Telephone: 31 61 93 11
Fax: 31 61 99 81

Federal Office for Health

Bundesamt fur Gesundheitswesen,
Bollwerk 27, Bern, CH-3001
Telephone: 31 61 95 11

Federal Office for Industry and Labour

Bundesamt fuer Industrie, Gewerbe und Arbeit, Bundesgasse 8, Bern, CH-3003
Telephone: 31 61 29 44

Swiss Intellectual Property Office

Einsteinstrasse 2, Bern, CH-3003
Telephone: 31 61 49 67
Fax: 31 61 48 95

UK

Agricultural and Food Research Council - AFRC

Polaris House, North Star Avenue, Swindon, Wilts, SN2 1VH
Telephone: 793 413200
Fax: 793 413201

Manages research activites related to agriculture and food in England and Wales through the Institutes for Animal Health, Animal Physiology and Genetics, Arable Crops, Engineering, Food, Grassland and Environmental Research and Plant Science as well as supporting research in universities through grants.

AFRC Institute for Animal Health - IAH

Compton, Nr Newbury, Berks, RG16 0NN
Telephone: 635 578 411

AFRC Institute of Animal Physiology and Genetics Research - IAPGR

Babraham Hall, Babraham, Cambridge, CB2 4AT
Telephone: 223 832312

Developing detection systems for specific RNA and DNA sequences and automated oligonucleotide synthesis and labelling. Studying use of recombinant DNA in bacteria to increase the digestibility of forages in ruminants. Developing veterinary diagnostics using monoclonal antibodies, three-dimensional computer-based analysis of molecular structure and molecular modelling. Carries out transfection studies for modification of cellular activity and molecular expression. Maintains close links with Animal Biotechnology Cambridge Ltd for the purpose of technology transfer. Maintains informal links with Agricultural Genetics Co. Acts as host to start up biotechnology companies.

AFRC Institute of Arable Crops Research - IACR

Rothamsted Experimental Station,, Department of Biochemistry and Physiology, Harpenden, Herts, AL5 2JQ
Telephone: 582 763133
Fax: 582 760981

Investigating the cellular and molecular processes that affect crop productivity, particularly nutrient transport, nitrogen metabolism and photosynthesis. Other research is studying the biosynthesis of metabolites that confer immunity to pests and diseases.

AFRC Institute of Engineering Research - IER

Wrest Park, Silsoe, Bedford, MK45 4HS
Telephone: 525 60000

AFRC Institute of Food Research - IFR

Reading Laboratory, Earley Gate, Whiteknights Road, Reading, RG6 2BZ
Telephone: 734 357000
Fax: 734 884763

Norwich Laboratory, Norwich Research Park, Colney, Norwich, NR4 7UA
Telephone: 603 56122
Fax: 603 507723

Laboratories at Norwich and Reading carry out fundamental research into food safety and biotechnology, nutrition and food acceptability and food structure. Biotechnology-related activities include protein and enzyme engineering, genetic manipulation, rapid analytical methods, novel polysaccharides, natural preservation and plant biotechnology. The institute maintains National Collections of Yeast Cultures and Food Bacteria.

AFRC Institute for Grassland and Animal Production - IGAP

Hurley, Maidenhead, Berks, SL6 4LR
Telephone: 62 882 3631

AFRC Institute of Horticultural Research - IHR

Bradbourne House, East Malling, Maidstone, Kent, ME19 6BJ
Telephone: 732 844 565

AFRC Institute of Plant Science Research

John Innes Centre, Norwich Research Park, Colney, Norwich, NR4 7UH
Telephone: 603 52571
Fax: 603 56844

British Standards Institution - BSI

2 Park Street, London, W1A 2BS
Telephone: 71 629 9000

Recognized body for preparation of standards for all fields of industry. Secures agreement among those concerned in the setting of standards, negotiates industrial agreements, provides information services and certifies and approves products as complying with the British Standards.

Central Veterinary Laboratory

Woodham Lane, New Haw, Weybridge, Surrey
Telephone: 932 34 1111
Fax: 932 34 7046

Supports the activities of MAFF in maintaining animal health, as an executive agency, operating as an independent business. Caries out diagnosis of statutory and other disease of public health importance as well as the development of novel immunoassays, vaccines and diagnostics for the monitoring and control of animal disease.

Department of Agriculture and Fisheries for Scotland

Agricultural Scientific Services, Chesser House, 500 Georgie Road, Edinburgh, EH11 3AW
Telephone: 31 443 4020

Department of Agriculture for Northern Ireland

Dundonald House, Upper Newtownards Road, Belfast, BT4 3SB
Telephone: 232 650111

Department of Education and Science

Elizabeth House, York Road, London, SE1 7PH
Telephone: 71 928 9222

Department of the Environment - DoE

2 Marsham Street, London, SW1P 3EB
Telephone: 71 276 3000

Department of Health and Social Security

Alexander Fleming House, Elephant & Castle, London, SE1 6BY
Telephone: 71 928 9222

Energy Technology Support Unit - ETSU

Renewable Energy Enquiries Bureau, Harwell, Oxon, OX11 0RA
Telephone: 235 432450

Undertakes assessment studies for the Department of Trade and Industry (DTI) Renewable Energy Research and Development Programmes, in areas including biological conversion and biomass production.

European Communities

Publications, H M Stationery Office, PO Box 569, London, SE1 9NH
Telephone: 71 928 6977

European Communities

Press and Information Office, 8 Storey's Gate, London, SW1 P3
Telephone: 71 222 8122

Foreign and Commonwealth Office

Downing Street, London, SW1A 2AL
Telephone: 71 928 9222

Forestry Commission

231 Corstorphine Road, Edinburgh, EH12 7AT
Telephone: 31 334 0303

Statutory forest authority. Manager of the national forestry enterprises in Great Britain. Directly responsible to the Ministry of Agriculture, Fisheries and Food in England and to the Secretaries of State in Scotland and Wales. Research areas include genetics, pathology and entomology, as well as experimental plantations and field trials.

Health and Safety Commission

Regina House, 259-269 Old Marylebone Road, London, NW1 5RR
Telephone: 71 723 1262

Responsible to the Secretary of State for Employment for the administration of the Health and Safety at Work Act. Members are drawn from industry and from local authorities. Formed by theamalgamation of six inspectorates, including those for factories and agriculture.

Health and Safety Executive

Baynards House, 1 Chepstow Place, London, W2 4TF
Telephone: 71 229 3456

Operational part of the Health and Safety Commission. Divided into 21 regional areas, each of which has inspectors specializing in particular technical subjects.

H M Customs and Excise

King's Beam House, 39-41 Mark Lane, London, EC3R 7HE
Telephone: 71 626 1515

H M Stationery Office

51 Nine Elms Lane, London SW8 5DR
Telephone: 71 873 0022/0011
Fax: 71 873 8643

Home Grown Cereals Authority - HGCA

Hamlyn House, Highgate Hill, London, N19 5PR
Telephone: 71 263 3391

Statutory body that provides market information for the UK cereal industry and sponsors research related to important marketing of cereals. Acts as the Intervention Board's agent in relation to cereals. Produces publications concerned with cereal marketing.

MAFF - Ministry of Agriculture, Fisheries and Food

Nobel House, 17 Smithh Square,
London, SW1P 3HX
Telephone: 71 238 3000
Fax: 71 238 6215

MAFF - Agricultural Science Services

Plant Pathology Laboratory, Hatching Green, Harpenden, Herts, AL5 2BD
Telephone: 58 27 5241

Holds national collection of plant pathogens and carries out tests on new agrochemicals.

MAFF - Environmental Prortection Division

Room 139, Nobel House, 17 Smith Square, London SW1P 3JR
Telephone: 71 238 3000
Fax: 71 238 6215

MAFF - Food Science

Room G23, Nobel House, 17 Smith Square, London SW1P 3JR
Telephone: 71 238 5996
Fax: 71 238 6591

Provides advice on the composition of foods, food microbiology, food additives and contaminants, methods of analysis and agricultural chemicals in relation to food.

MAFF - Pesticides and Pest Infestation Control

Plant Pathology Laboratory, Hatching Green, Harpenden, Herts, AL5 2RD
Telephone: 5827 5241

Responsible for the Pesticide Safety Precautions Scheme and the Agricultural Chemicals Approval Scheme.

MAFF - Plant Variety Rights

White House Lane, Huntingdon Road, Cambridge CB3 0LF
Telephone: 223 277 151
Fax: 223 342 386

Controls seed certification scheme regulations for England and Wales.

MAFF - Publications

Government Building, Tolcane Drive, Pinner, Middlesex, HA5 2DT
Telephone: 81 868 7161

MAFF - Veterinary Services

Garrison Lane, Chessington, Surrey
Telephone: 81 397 9121

Activities include licensing of veterinary medicines and the use of antibiotics in animal feedstuffs.

Medical Research Council - MRC

20 Park Crescent, London, W1N 4AL
Telephone: 71 636 5422

Main government agency for the promotion of medical research in the UK.

The Patent Office

Marketing and Information Services, State House, 66-71 High Holborn, London, WC1R 4TP
Telephone: 71 829 6512/6306
Fax: 71 831 9094

The Patent Office

Cardiff Rd, Newport, Gwent, NP9 1RH
Telephone: 633 814000

An Executive Agency of the Department of Trade and Industry.

The Patent Office

Sales Branch, Unit 6, Nine Mile Point, Onmfelin Fach, Cross Keys, Newport, Gwent, NP1 7HZ
Telephone: 495 201100
Fax: 495 200970

Supplies copies of UK patents.

The Patent Office Search and Advisory Service

Industrial Property and Copyright Department, Hazlitt House, 45 Southampton Buildings, Chancery Lane, London, WC2A 1AR
Telephone: 71 438 4747
Fax: 71 438 4750

Uses the expertise of its patent examining staff to provide consultative access to a number of commercial on-line patent databases as well as its own extensive hard copy collection.

USA

California Energy Commission

Synthetic Fuels Office, 1111 Howe Avenue, Sacramento, CA-95825
Telephone: 916 920 6031

Activities include partial financial support of a number of projects aimed at production of ethanol for use as fuel using raw materials such as poor-grade citrus fruit, grapes, peaches, almond hulls etc as well as purpose-grown corn or sweet sorghum. Such projects may also generate stillage suitable for use as animal feed supplement.

California State Water Resources Control Board

901 P Street, Sacramento, CA-95814
Telephone: 916 322 0200

Commissioner of Patents and Trade Marks

Patents Information, Washington, DC-20231

Committee on Science & Technology

388 House Annexe 11, Washington, DC-20515
Telephone: 202 226 3641

Congress of the United States

Office of Technology Assessment, 119 D Street NE, Washington, DC-20510

Carries out studies and reviews on the state of the art and impact from social, technical and scientific aspects of new areas of technology and publishes the results of such studies which include: regulations concerning biotechnology; genetic engineering and the potential of biomass (plant material) as a source of energy in the US; reports on commercial biotechnology.

Department of Agriculture

Agricultural Research Service, Plant Genetics and Germplasm Institute, Beltsville, MA-20705

Maintains a seed bank as well as carrying out research on plant breeding.

Department of Commerce

National Technical Information Service, 5285 Port Royal Road, Springfield, VA-22161

Department of Energy

17th and Pennslyvania Avenue NW, Washington, DC-20006

Department of Energy

Forrestal Building 1000, Independence Avenue SW, Washington, DC-20585
Telephone: 202 252 5810

Includes Office of the Environment and Office of Public Affairs.

Department of Energy

Office of Energy Research, 20 Massachusetts Avenue NW, Washington, DC-20545

Department of Food and Agriculture

California Pesticide Registration, 1220 N Street Room A-447, Sacramento, CA-95814
Telephone: 916 322 5031

Department of Health and Human Services

Public Health Service Food and Drug Administration, Bureau of Foods, 200 C Street SW, Washington, DC-20204

Special Assistant to the Commissioner for Science: 5600 Fisher's Lane, Room 14-90, Rockville, MD-20857
Telephone: 301 443 4650

Department of Health and Human Services

Center for Disease Control, 1600 Clifton Road, North-East Atlanta, GA-30333

Department of Health and Human Services

National Institutes of Health, 9000 Rockville Pike, Bethesda, MD-20205

The following National Institutes are located at Rockville Pike: National Cancer Institute, National Eye Institute, National Heart Lung and Blood Institute, Allergy and Infectious Diseases, Arthritis, Metabolism and Digestive Diseases, Child Health and Human Safety, General Medical Sciences, Neurological Communicative Disease and Stroke and Division of Research Resources.

Department of Health Services

California Food and Drug Branch, 714 P Street Room 400, Sacramento, CA-95814
Telephone: 916 445 2263

Department of the Interior

Office of Water Research and Technology, 18th and C Streets, Washington, DC-20240

Department of Treasury

Bureau of Alcohol, Tobacco and Firearms, 114 Commerce Street Room 701, Dallas, TX-75242

Administration related to fermentation alcohol, distillation and use as a fuel.

Environmental Protection Agency - EPA

401 M Street SW, Washington, DC-20460

European Community Information Service

2100 M Street NW, Suite 707, Washington, DC-20037
Telephone: 202 862 9500

For sales and subscriptions to publications from the EC.

National Institutes of Health

Office of Recombinant DNA Activities, Building 31, Room 3B10, Bethesda, MD-20892
Telephone: 301 496 6051

National Institutes of Health

National Cancer Institute, Bethesda, MD-20892
Telephone: 301 496 4164

National Institutes of Health

National Institute of Allergy and Infectious Diseases, Rocky Mountain Laboratories, Hamilton, MT-59840
Telephone: 406 363 3211

Developing vaccines by conventional methods as well as through use of recombinant DNA technology. Activities include attempts to produce a safer vaccine against whooping cough.

National Library of Medicine

8600 Rockville Pike, Bethesda, MD-20209

National Science Foundation

1800 G Street NW, Washington, DC-20550
Telephone: 202 357 9854

Includes: Directorate for Biological Behavioural and Social Sciences: Division of Physiology, Cellular and Molecular Biology; Division of Social and Economic Sciences; Division of Environmental Biology. Directorate for Engineering; Division of Applied Research; Division of Electrical, Computer and Systems Engineering; Division of Problem-Focused Research. Directorate for Science Education. Directorate for Scientific Technology and Internal Affairs.

Washington State Department of Trade and Economic Development

312 First Ave, North, Seattle, WA-98109
Telephone: 206 464 7143
Fax: 206 464 7222

NATIONAL ORGANIZATIONS

AUSTRALASIA

Agricultural and Veterinary Chemicals Association of Australia

12th Floor, Denison Bldg, 65 Berry St, P O Box 938, North Sydney, NSW 2060, Australia
Telephone: 2 963 7690
Fax: 2 929 0213

Australian Academy of Sciences

GPO Box 783, Canberra, ACT 2601, Australia
Telephone: 6 247 5777
Fax: 6 257 4620

An independent organization of Australia's pre-eminent research scientists, applied scientists and research engineers elected to Fellowship on the basis of their contribution to research as assessed by leading scientists in Australia and overseas. The academy's principle objectives include the recognition of scientific excellence, promotion of science education and public awareness and the provision to Parliament, government and the community of advice on science policy.

Australian Academy of Technological Science

191 Royal Parade, Parkville, Victoria, 3052, Australia
Telephone: 3 347 3166

Australian Biotechnology Association

PO Box 303, Clayton, Victoria, 3168, Australia
Telephone: 3 524 2393
Fax: 3 543 6613

Australian Pharmaceutical Manufacturers Association

Level 2, 77 Berry St, North Sydney, NSW 2060, Australia
Telephone: 2 922 1157
Fax: 2 959 4860

Australian Scientific Industries Association

9 Queens Road, Melbourne, Victoria, 3004, Australia
Telephone: 3 268 7111

Australian Society for Microbiology

Room 508A, Clunies Ross House, 191 Royal Parade, Parkville, Victoria, 3051, Australia
Telephone: 3 348 1441
Fax: 3 347 0605

Australian Water and Wastewater Association

PO Box A232, Sydney South, NSW 2000, Australia

Council of Australian Food Technology Associations

11th Floor, 65 Berry St, North Sydney, NSW 2060, Australia
Telephone: 2 963 7672
Fax: 2 954 4327

Institute of Patent Attorneys of Australia

207 Riversdale Rd, Hawthorn,
Victoria, 3122, Australia
Telephone: 3 819 2515

Pharmaceutical Society of Australia

44 Thesiger Court, Deakin,
(PO Box 21, Curtin), ACT 2600
(ACT 2605), Australia
Telephone: 62 81 1366

Royal Australian Chemical Institute

Secretariat, Clunies Ross House, 191
Royal Parade, Parkville, Victoria 3052,
Australia
Telephone: 3 347 1577

Royal Society of New Zealand

PO Box 598, Wellington, New Zealand
Telephone: 4 72 7421
Fax: 4 73 1841

The society has objects and functions consistent with its role as New Zealand's national academy of science and is charged with the statutory responsibility to promote scientific endeavour and achievement. The organization comprises Fellows, Branches and affiliated Scientific Societies.

AUSTRIA

Academy of Sciences

Die Oesterreichische Akademie de
Wissenschaft, Dr-Ignaz-Seipel-Platz 2,
Vienna, A-1010
Telephone: 222 51581

Association of Chemical Trades

Bundesgesinnung der Chemischen
Gewerbe, Wiedner Haupstrasse 63,
Postfach 69, Vienna, A-1045
Telephone: 1 50 10 50

Association of Pharmaceutical Industries

Fachverband der Chimischen Industrie,
Gruppe Pharmazeutika, Bauermarkt 13,
Schliessfach 69, Vienna, A-1011
Telephone: 222 52 84 94

Association of Water Interests

Osterreichisches Wasserwirtschafts
Verband, Marc Aurel Strasse 5,
Vienna, A-1010
Telephone: 1 5355 5720 0

Austrian Biochemical Society

Osterreichische Biochemische
Gesellschaft, Inst Biochem Technologie
und Mikrobio, Getreidemarkt 9,
Vienna, A-1060

Austrian Biophysical Society

Osterreichische Biophysokalische
Gesellschaft, Wahringer Strasse 13,
Vienna, A-1090

Austrian Bioprocessing Association

Gesellschaft fuer Bioprozesstechnik,
Schloegelgasse 9, Graz, A-8016
Telephone: 3 16 8 73 84 05
Fax: 3 16 8 11 05 0

Austrian Biotechnology Association

Oesterreichische Gesellschaft fuer
Biotech, Kleine Sperlgasse 1/37,
Vienna, A-1020
Telephone: 222 26 22 42

Austrian Brewing Society

Bund Osterreichischer Braumeister, und Brauereitechniker, Michaelerstrasse 25, Vienna, A-1180

Austrian Chemical Association

Eschenbachgasse 9, Vienna, A-1010
Telephone: 2 22 57 42 49

(Gesellschaft Oestereichisher Chemikar)

Austrian Pharmacists Association

Osrterreichischer Apothekerverband, Spitalgasser 31, Vienna, A-1094
Telephone: 1 42 56 76

Austrian Society for Chemical Industry

Fachverband der Chemischen Industrie, Oesterreichs, Wiedner Haupstrasse 63, Vienna, A-1045
Telephone: 2 22 5 01 05-0
Fax: 2 22 5 02 06 2 80

Austrian Society for Immunology

Osterreichische Gesellschaft fur Allergologie, und Immunologie, 2nd Dept Medizin, Garnisongasse 13, Vienna, A-1090

Plant Science Association

Gesellschaft fuer Phytotechnologie, Sieveringer Strasse 150, Vienna, A-1190

BELGIUM

Agricultural Engineering Association

Assoc des Ingenieurs des Sciences Agronomique, c/o Faculte des Sciences Agronomique, Passage des Deportes 2, Gembloux, B-5800
Telephone: 81 61 22 40

Promotes the development of agricultural science.

Association of Belgium Industries

Association des Industriels de Belgique, 29 ave Andre Drourt, Brussels, B-1160
Telephone: 2 673 80 90

Association of the Drug Industries

Assoc Generale d'Industrie du Medicament, 49 Square Marie-Louise, Brussels, B-1040
Telephone: 2 230 40 90

Belgium Bioindustries Association

c/o Institut Superior de Commerce, Saint-Louise, rue du Marais 113, Brussels, B-1000
Telephone: 2 219 67 00

Belgium Pharmaceutical Association

Association Pharmaceutical Belge, 11 rue Archimede, Brussels, B-1040

Biochemical Society

Genetique Moleculaire, ICP-UCL 74.49, Avenue Hippocrate 75, Brussels, B-1200
Telephone: 2 764 7451
Fax: 2 764 53 22

Federation of Belgium Chemical Industries

49 Square Marie-Louise, Brussels, B-1040

Federation of Belgium Industries

49 rue Ravenstein, Brussels, B-1000
Telephone: 2 511 58 80

Royal Academy of Belgium

Academie Royale de Belgique, Palais des Academies, 1 Rue Ducale, Brussels, B-1000
Telephone: 2 514 4064/4256
Fax: 2 502 0424

Standards Institute

Institut Belge de Normalisation, 29 avenue de Brabanconne, Brussels, B-1040
Telephone: 2 734 92 05

Wastes Trade Association

Confederation de la Recuperation, BP 5, 6 Place du Samedie, Brussels, B-1000
Telephone: 2 217 99 93

BRAZIL

ABRABI - Association Brasileria das Empresas de Biotechnologia

Av Almirantem Barroso 91, Sala 520, Rio de Jeniero, CEP 20031,
Telephone: 21 240 4430

CANADA

Canadian Chemical Producers Association

350 Sparks Street, Suite 805, Ottawa, Ontario, K1R 7S8
Telephone: 613 237 6215

DENMARK

Danish Organization for Renewable Energy - OVE

Willemoesgade 14, Copenhagen O, DK-2100
Telephone: 3142 9091
Fax: 3142 9095

Involved in Danish development of renewable energy.

Agricultural Science

Danmarks Jordbrugsvidenskalbelige, Kandidatforbund, Strandvejen 863, Klampenborg, DK-2830
Telephone: 1 631166
Fax: 1 638811

Association for graduates in agricultural science.

Danish Biochemical Society (Biokimisk Forening)

Institute of Microbiology, Oster Farimagsgade 2A, Copenhagen K, DK-1353
Telephone: 33 125907
Fax: 33 935220

Danish Biology Society

Biologisk Selskab, c/o Biokenisk Institut, Blegdamsvej 3, Copenhagen, DK-2200
Telephone: 357900

Danish Botanical Society

Dansk Botanisk Forening, Solvgade 83, Copenhagen K, DK-1307
Telephone: 1 14 17 03

Danish Microbiology Society

Danmarks Mikrobiologiske Selskab, O Farimagsgade 2D, Copenhagen K, DK-1353
Telephone: 1 14 51 81

Danish Pharmaceutical Association

Danmarks Apoterkerforening, Bredgade 54, PO Box 2181, Copenhagen K, DK-1017
Telephone: 33 11 02 76
Fax: 33 32 91 17

Danish Pharmaceutical Society

Danmarks Farmaceutiske Selskab, Universitats Parken 2, Copenhagen O

Danish Standards Institute

Dansk Standardiseringsraad, Postbox 77, Aurehojvej 12, Copenhagen
Telephone: 31 62 93 15

Danish Water Supply Association

Danks Vandteknish Forening, Vilh Becks Vej 60, Viby J, DK-8260
Telephone: 86 11 23 33

Royal Agricultural Society of Denmark

Rolighedsvej 26, Copenhagen V, DK-1958
Telephone: 1 35 02 27

Royal Danish Academy of Sciences and Letters

HC Andersens Boulevard 35, Copenhagen V, DK-1553
Telephone: 33 11 3240
Fax: 33 91 0736

A non-governmental organization whose objective is the promotion of science and the humanities.

FINLAND

Agricultural Society of Finland

Risto Tahvonen/MTTK, Jokionen, SF-31600

Association of Finnish Chemical Societies

Suomen Kemian Seura, P Hesperiankatu 3, Helsinki 26, SF-00260
Telephone: 0 40 80 22

Association of Finnish Pharmacies

Suomen Apteekkariliitto, Pieni Roobertinkatum 14C, Helsinki, SF-00120
Telephone: 0 647 212
Fax: 0 647 167

Association of Finnish Water Supply

Vesihououltoliitto RY, Arkadiakatu 15C, Helsinki 10, SF-00100
Telephone: 0 49 54 62

Association of Pharmaceutical Distributors

Aptekkitavaratukkukauppiait RY,
PL 150 Mannerheimintie 76A,
Helsinki, SF-00251
Telephone: 0 441 651
Fax: 0 496 142

Association of Pharmaceutical Industries

Laaketeollisuuyhdistys RY, PL 316,
Iso Roobertinkatu 4-6, Helsinki 12,
SF-00121
Telephone: 0 64 20 01
Fax: 0 60 32 86

Central Union of Agricultural Producers

PL 510, Simonkatu 6, Helsinki 10,
SF-00101
Telephone: 0 69 55 1

Delegation of Finnish Academies of Science and Letters

5 Mariankatu, Helsinki, SF-00170
Telephone: 0 63 30 05

A joint organ of the Finnish Academy of Science and Letters, the Finnish Society of Sciences and Letters and the Finnish Academy of Technology. The Delegation's main functions are to cooperate and maintain contact with international, preferably non-governmental, organizations and with scientific academies in other countries, and take part in the planning and execution of international research programmes. It promotes debate on scientific findings so as to foster interdisciplinary research, and distributes information on foreign academies of science. It popularizes research results, proposes improvements to research conditions and has organized several symposia for experts and members of the public.

Finnish Agricultural Association

c/o Tyotehoseura, Melkonkatu 16A,
Helsinki 21, SF-00210
Telephone: 0 692 24 45

Societas Biochemica, Biophysica et Microbiologica Fenniae

Marja Paloheimo, Alko Ltd, POB 350,
Helsinki, SF-00101
Telephone: 0 13311
Fax: 0 1332781

Finish Biochemical Society.

FRANCE

Academie des Sciences

23 quai Conti, Paris, F-75006
Telephone: 1 4326 6621
Fax: 1 4354 6399

One of the five Academies forming the Institut de France. It is supported by various sources such as returns from different bequests and donations. Its main activity is geared to the advancement of science through hearing weekly presentations and producing scientific and technical reports on major issues on a basis of competence, objectivity and independence.

Agricultural Society of France

Societe des Agriculteurs de France,
8 rue d'Athenes, Paris, F-75009
Telephone: 1 42 85 72 27

Provides advisory services to both farmers and foresters.

Assemble Permanante Chambres d'Agriculture - APCA

9 ave George - V, Paris, F-75008
Telephone: 1 47 23 55 40

Association of Agricultural Engineers

Association Francaise de Genie Rural, 30 rue Las Cases, Paris Cedex 07, F-75340
Telephone: 1 45 55 95 50

Association pour le Developpement de la Bioindustrie - ADEBIO

1 rue Favart, Paris, F-75002
Telephone: 1 42 61 80 88

French Federation of Biotechnology.

Association of Engineers in Food and Agricultural Industries

Association des Ingeniers des Industries, et Alimentaires, 9-11 ave Franklin D Roosevelt, Paris, F-75008
Telephone: 1 42 25 92 48

Association for the Promotion of Agroindustry

Assoc pour la Promotion Industrie Agriculture, 35 rue du General Foy, Paris, F-75008
Telephone: 1 42 93 19 24

Biology Society

Societe de Biologie, College de France, 11 place M Berthelot, Paris Cedex 05, F-75231
Telephone: 1 43 29 12 11

Biotechnica - Assoc pour la developpement des biotechnologies en Aquitaine

Comite d'expansion 2, Place de la Bourse, Bordeaux Cedex, F-33076
Telephone: 56 52 65 47
Fax: 56 44 32 69

French Biomass Association

Association Francaise pour la Biomasse, APCA 9, avenue Georges V, Paris, F-75008
Telephone: 1 45 32 27 19

French Botanical Society

Societe Botanique de France, 4 ave de l'Observatoire, Paris, F-75006
Telephone: 1 46 83 55 20

French Microbiology Society

Societe Francaise de Microbiologie, c/o Institut Pasteur, 28 rue du Dr Roux, Paris, F-75015
Telephone: 1 45 68 80 00
Fax: 1 45 67 46 98

French Society for Biochemistry and Molecular Biology

Societe Francaise de Biochimie et de Biologie, Centre Uni Pharmaceutique, Lab de Bioc, 5 rue JB Clément, Tour D4-115, Chatenay-Malabry, F-92296
Telephone: 1 46 83 17 31

Microbiology Research Association

Association de Recherche en Microbiologie, 7 rue Jean-Baptiste Clement, Chatenay Malabry, F-92290
Telephone: 1 46 61 33 25

National Association for the Food Industry

Assoc Nat des Industries Agro-Alimentaires, 52 rue du Fg Saint Honore, Paris, F-75008
Telephone: 1 42 66 40 14
Fax: 1 47 42 22 45

National Association for Plant Protection

Assoc Nationale de Protection des Plantes, 6 bd de la Bastille, Paris, F-75012
Telephone: 1 43 44 89 64
Fax: 1 43 44 29 19

Organisation Nationale Interprofessionnelle des Bioindustries - ORGANIBIO

28 rue Saint Dominique, Paris, F-75007
Telephone: 1 47 53 09 12
Fax: 1 45 55 98 62

French organisation of bioindustries.

Plant Physiology Society

Societe Francaise de Physiologie Vegetale, 4 Place Jussieu, Paris Cedex 05, F-75230

PROBIOTECH

5 quai Voltaire, Paris, F-75007
Telephone: 1 42 60 25 00

Association Francaise pour la Promotion et le Developpement des Biotechnologies et de la Biochimie. French association for the promotion and development of biotechnology and biochemistry.

Producers of Cereals

Association Generale des Producteurs de Ble, et autres Cereals, 8 avenue du President Wilson, Paris, F-75116
Telephone: 1 44 31 10 00
Fax: 1 47 20 44 03

Producers of Maize

Association Generale des Producteurs de Mais, 8 Avenue du President Wilson, Paris, F-75116
Telephone: 1 47 23 48 32
Fax: 1 40 70 93 44

PROMOTECH

Assoc pour la Promotion de la Technologie, 6 alle Pelletier Doisy, Villers les Nancy, F-54600
Telephone: 83 44 16 16
Fax: 83 44 04 82

Society for the Chemical Industry

Societe de Chimie Industrielle, 28 rue Saint Dominique, Paris, F-75007
Telephone: 1 45 55 69 46

Society for Food Hygiene

Societe Scientifique d'Hygiene Alimentaire, ISHA, rue du Chemie Blanc, BP 138, Champlan, F-91163
Telephone: 64 48 52 10
Fax: 64 48 82 49

GERMANY

German Biomass Association

c/o Technical University Munchen, Freising, D-8050
Telephone: 8161 713

Association of Agricultural Engineers

Bundesverbund de Agraringenieure, Adenauralle 238, Bonn 1, D-5300

Represent the interests of agricultural engineers and promotes training for engineers and managers.

Asssociation of German Engineers

Postfach 1139, Graft Recke Strasse 84, Duesseldorf, D-4000
Telephone: 211 62141

Association of the Water Treatment Industry

Bismarkstrasse 154, Duisburg 1, D-4100
Telephone: 203 350 655

Brewing Association

Deutscher Braumeister und Malzmeister Bund, Arndtstrasse 47, Dortmund, D-4600
Telephone: 231 57 11 21

Deutsche Forschungsgemeinschaft - DFG

Kennedyallee 40, Postfach 20 50 04, Bonn 2, D-5300
Telephone: 228 885 2346
Fax: 228 885 2221

An autonomous body of the German scientific community that formulates its own statutes and itself elects the scientific and academic members of its agencies. Members include universities, other research institutions, academies of science and scientific associations. The DFG finances research projects, promotes cooperation among researchers and coordinates basic research. It also advises parliaments and public authorities on scientific matters, cultivates relations between science and economy and promotes cooperation between German and foreign scientists.

German Agricultural Society

Deutsche Landwirtschafts-Gesellschaft, Zimmerweg 16, Frankfurt a/Main, D-6000
Telephone: 611 71 680

German Biochemical Society

Gessellschaft fur Biologische Chemie, Bahnhofstr 9-15, Tutzing, D-8132
Telephone: 815 8 22 4249
Fax: 815 8 22 4119

German Botanical Society

Deutsche Botanische Gesellschaft, Untere Karspule 2, Gottingen, D-3400
Telephone: 551 39 78 78

German Microbiology Society

Deutsche Gesellschaft fur Hygiene, und Mikrobiologie, Ratzeburger Allee 160, Lubeck, D-2400

German Pharmaceutical Society

Frankfurterstrasse 250, Darmstadt

GREECE

Academy of Athens

28 Rue Panepistimiou, Athens, GR-10679
Fax: 1 362 6721

Center for Renewable Energy Sources

6 Frati Street, Koropi, GR-19400
Telephone: 1 662 6460
Fax: 1 662 6462

Hellenic Biochemical and Biophysical Society

National Hellenic Research Foundation, 48 Vas Constantinou Ave, Athens, GR-11635
Telephone: 1 7239 965
Fax: 1 7212 729

Hellenic Biochemical and Biophysical Society

University of Athens, Dept of Biochemistry, Cell & Molecular Biology, & Genetics, Panepistemiopolis, Kouponia, Athens, GR-15701

ICELAND

Icelandic Biochemical Society

University of Iceland, Department of Biochemistry, Faculty of Medicine, Vatnsmyrarvegur, Reykjavik, IS-101
Telephone: 1 694846
Fax: 1 694884

IRELAND

Biochemical Society, Irish Area Section

Arthur Guinness and Co, St James's Gate, Dublin 8

Royal Irish Academy

19 Dawson Street, Dublin 2
Telephone: 1 76 4222/2570
Fax: 1 76 2346

A society for the advancement of science and learning. Its membership includes most of the leading authorities in the various branches of science and humanities in Ireland.

ISRAEL

Israel Academy of Sciences and Humanities

Albert Einstein Square, Talbieh, PO Box 4040, Jerusalem, 91040
Telephone: 2 63 6211/6214
Fax: 2 66 6059

An elected body of the country's leading scholars whose principal aims and functions are to foster and promote work on international bodies and represent Israeli sciences on international bodies and at international conferences and to publish basic documents and works contributing to the advancement of sciences. The academy administers a research fund on behalf of the universities, as well as a number of other funds established by private donations. It maintains close relations with Ministerial Offices in matters relating to national science policy and higher education in Israel. The academy represents Israel on international scientific bodies and has agreements with several academies and other academic bodies abroad for joint scientific endeavours.

Israel Biochemical Society

Institute of Science, Dept of Membrane Research, 76100 Rehovot
Telephone: 8 342716
Fax: 8 466966

ITALY

Associazione Nazionale per lo Svuluppo delle Biotechnologie - ASSOBIOTEC

via Accadamia 33, Milano, I-20131
Telephone: 2 636 2306
Fax: 2 465 6114

Trade association for the biotechnology industry.

Biochemical Society

Department of Biochemisty,
Via Bassi 21, Pavia, I-27100
Telephone: 382 392237
Fax: 382 932240

Biology Society

Consiglio Nazionale dei Biologi,
Via Arenula 71, Rome, I-00186
Telephone: 6 68 75 145

Italian Association of Agricultural Engineers

Associazione Italiana di Genie Rurale,
Via Gradenigo 6, Padova, I-35131
Telephone: 49 807 1065

Italian Association of Clinical Microbiology

Associazione Microbiologi Clinici Italiani, c/o Laboratorio Microbiologia, Ospedale S Carlo Borromeo, Via Pia, Secondo 3, Milano, I-20153
Telephone: 40 84 016

Italian Biomass Association - ITABIA

161 Via Archimede, Rome, I-00197
Telephone: 6 8036 17

Italian Botanical Society

Societa Botanica Italiana, Via Giorgia La Pira 4, Firenze, I-50121
Telephone: 55 21 07 55

Italian Brewing Association

Associazione Italiana dei Tecnici della Birra, Via Savoia 29, Rome, I-00198
Telephone: 6 865161

Italian Chemical Society

Societa Italiana Chimici,
Viale Liegi 48, Rome, I-00198
Telephone: 6 854 969 18
Fax: 6 854 8734

Italian Society for Microbiology

Societa Italiana di Microbiologia,
Inst di Microbiologia-Spedoli Civili,
Bresca, I-25100

National Federation of Farmers and Foresters

Federazione Nazionale Dottori in Scienze, Agrarie e Forestali,
Via Livenza 6, Rome, I-00198
Telephone: 6 85 60 36

Society of General Microbiology and Biotechnology

Societa Italiana di Microbiologia Generale, e di Biotecnologia, Dipt di Genetica, et di Biologia, via Celoria, Milano, I-20133
Telephone: 2 230 823

JAPAN

Association of Chemical Engineers

Nippon Building, Room 427,
6-2 Ohte-machi-2-chome, Chiyoda-ku,
Tokyo, 100
Telephone: 3 279 3375

Association of the Chemical Industry

Tokyo Club Building, 1-21 Nishi, Shimbashi-1-chome, Minato-ku,
Tokyo, 106
Telephone: 3 585 3375

Association for the Promotion of International Trade

Nippon Building 6-2,
Ohte-machi-2-chome, Chiyoda-ku,
Tokyo, 100
Telephone: 3 270 3631

Japan Antibiotics Research Association

20-8 Kamiosaki-2-chome, Bunkyo-ku,
Tokyo

Japan Association for Tissue Culture

c/o Japan Academic Societies,
4-16 Yayoi-2-chome, Bunkyo-ku,
Tokyo, 113

Japan Bioindustries Association

10-5 kaminiosaki-2-chome,
Shinagawa-Ku, Tokyo,

Science Council of Japan

22-34 Roppongi, 7-Chome, Minato-ku,
Tokyo 106,
Telephone: 3 3403 6291
Fax: 3 3403 6224

The council comprises 210 members with three-year terms of office representing all fields of science. It acts as the representative of Japanese scientists, both nationally and internationally, for the promotion and permeation of science. The government may seek opinions of the council, and the council has the right to take the initiative in making recommendations to the government. It also conducts such activities as participation in international scientific organizations, sending delegates to important international meetings overseas, dispatching delegations to foreign countries for the purpose of bilateral scientific exchanges, and organizing and supporting important international conferences in Japan.

THE NETHERLANDS

Dutch Microbiology Society

Nederlands Vereniging voor
Microbologie, Rijksinstituut voor de
Volksgezondheid, Postbus 1,
BA Bilthoven, NL-3720
Telephone: 30 742 971

Netherlands Biochemical Society

Nederlands Vereninging voor
Biochimie, Biochemische
Laboratorium, Vrije Unive,
de Boelelaan 1083, HV Amsterdam,
NL-1081
Telephone: 20 5485606
Fax: 20 461479

Netherlands Biochemical Society.

Netherlands Biophysics Society

Vereniging voor Biofysica,
Laboratorium voor Algemene
Natuurkunde, Rijksuniversiteit,
CM Groningen, NL-9718
Telephone: 50 634 785

Netherlands Biotechnology Society

PO Box 108, AL Zeist, NL-3700
Telephone: 3404 55 4444

Netherlands Pharmaceutical Association

Alexandestraat 11, Den Haag

Netherlands Society of Plant Pathology

PO Box 31, AA Wageningen, NL-6700
Telephone: 8370 83594

Royal Botanical Society of the Netherlands

Korninklijke Nederlands Botanische, Vereniging, Botanisch Lab, Nieuwstraat 106, PN Utrecht, NL-3512

Royal Netherlands Academy of Arts and Sciences

Kloveniersburgwal 29, PO Box 19121, GC Amsterdam, NL-1000
Telephone: 20 551 0700
Fax: 20 620 4941

The functions of the academy are: advisory body of the government in all fields of science; science promotion, communication, exchange and collaboration of Dutch and foreign scientists; management of institutes for fundamental research.

Royal Netherlands Society of Agricultural Science

Koninklijk Genootschap voor Landbouwwetensch, Foulkesweg 1A, BG Wageningen, NL-6703
Telephone: 8 370 835 37

Society of Agricultural Engineering

Nederlands Vereniging Techniek in de Landbouw, Mansholtlaan 10-12, PA Wageningen, NL-6708
Telephone: 8 370 94911

NORWAY

National Commission on Agricultural Engineering

Den Norske Avdling av CIGR, Box 15, As-NLH, Oslo, N-1432
Telephone: 2 94 90 60

Norwegian Academy of Sciences and Letters

Drammensveien 78, Oslo 2, N-0271
Telephone: 2 86 7010
Fax: 2 56 2656

Norwegian Biochemistry Society

Norsk Biokjemisk Selskap, Institute for Nutrition Research, University of Oslo, POB 1046 Blindern, Oslo 3, N-0316
Telephone: 2 856638
Fax: 2 854194

Norwegian Botanical Association

Norsk Botanisk Forening, Botanisk Museum, Oslo 5,
Telephone: 2 68 69 60

Norwegian Farmers Union

Norges Bondelag og Landbrukssamvirket, Postboks 3723, Gamlebyer, Oslo 1, N-0135
Telephone: 2 68 18 00

Norwegian Society for Microbiology

Norsk Forening for Mikrobiologi, Norges Veterinaerhorgskole, Inst for Indremedisin, Boks 8146, Oslo 1, N-0033
Telephone: 2 69 36 90

PORTUGAL

Academy of Sciences of Lisbon

Rua da Academia das Ciencias 19,
Lisbon, P-1200
Telephone: 1 346 3866

The academy promotes the advancement of science and learning for the progress and prosperity of the country. It was one of the first National Members of the International Council of Scientific Unions (ICSU), is one of the three Portuguese Members of the European Science Foundation and is the Portuguese partner of the European Science Exchange Programme of the Royal Society, London.

Agricultural Association

Associacao Central de Agricultura Portuguesa, Rua D Dinis, Lisbon 2,
P-1300
Telephone: 1 63 67 19

Biomass Center for Energy - CBE

Centro da Biomassa para a Energia,
3220 Miranda do Corvo
Telephone: 39 52436/52488
Fax: 39 52452

Develop the use of plants and forestry, agroindustrial and other organic wastes, for energy purposes.

Botanical Society

Sociedade Boteriana, c/o Instituto Botanico, Universidade de Coimbra,
Coimbra
Telephone: 39 2 28 97

Portuguese Biochemistry Society

Sociedade Portuguesa de Bioquimica,
Department of Zoology, University of Coimbra, Coimbra codex, P-3049
Telephone: 39 35812
Fax: 39 35812

Portuguese Biomass Association

APOBIOM, Rua da Horta Seca 15,
Lisboa
Telephone: 1 347 22 50
Fax: 1 346 98 16

Portuguese Biotechnology Association

Sociedade Portuguesa de Biotechnologia, Instituto Superior Tecnico, Av Rovisco Pais,
Lisboa Cedex, P-1069

Portuguese Microbiology Society

Socidade Portuguesa de Microbiologia,
Dept of Chemical Indistries, LNETI,
Queluz de Baixo, Queluz, P-2745

SPAIN

Agricultural Engineers

Asociacion Nacional de ingenieros Agronomes, Calle General Arrando 38,
Madrid,
Telephone: 1 419 74 21

Spanish Brewing Association

Asociacion Espanola de Tecnicos de Cerveza, Juan de la Cierva 3, Madrid,
Telephone: 1 262 29 00

Spanish Biochemical Society

Sociedad Espanola de Bioquimica, Hortaleza 104-2°izda, Madrid, E-28004
Telephone: 1 308 49 34

Spanish Biomass Association - ADABE

Asociacion para la Difusion del, Aprovechamiento de la Biomasa en, Espana, c/o ETSIA, Ciudad Universita, Madrid, E-41080
Telephone: 5 4 23 55 11
Fax: 5 4 23 62 68

Spanish Microbiology Society

Socidad Espanola de Microbiologia, Vitravia 8, Madrid, E-28006
Telephone: 1 261 9800

SWEDEN

Association of Agricultural Engineers

Jordbrukstekniska Foreningen, Box 7033, Uppsala 7, S-75007

Federation of Swedish Farmers

Klara Ostra Kyrkagata 12, Stockholm, S-10533
Telephone: 8 787 50 00

Royal Swedish Academy of Agriculture and Forestry

Box 6806, Stockholm, S-11386
Telephone: 8 736 09 00

Royal Swedish Academy of Sciences

Box 50005, Stockholm, S-10405
Telephone: 8 15 0430
Fax: 8 15 5670

The academy primarily promotes mathematics and natural sciences. Its programme is based on its scientific classes/sections and committees, with main programme elements such as international cooperation and scientific exchange programmes, science publishing and dissemination of scientific information to the general public and provision of scientific advice to government and parliament. It runs seven research institutes, grants scholarships and awards prizes.

Swedish Biochemical Society

Sveriges Biokemiska Forening, Institutionen för Medicinsk, Kemi och Biofysik, Umeå Universitet, Umeå, S-90187
Telephone: 90 166743
Fax: 90 136310

Swedish Botanical Society

Svenska Botaniska Foreningen, c/o Naturhistorika Riksmuseet, Botaniska Sektionera, PO Box 50007, Stockholm, S-10405
Telephone: 8 666 4000

Swedish Brewmasters Association

Sveriges Bryggmastare Forening, AB Pripps Bryggerier, Bromma, S-16186
Telephone: 8 757 74 80
Fax: 8 733 96 16

Swedish Society for Microbiology

Svenska Foreningen for Mikrobiologi, Microbiology Department, University of Agricultural Sciences, Uppsala, S-75007

SWITZERLAND

Association of Agricultural Engineers

Ave des Jordils 3, Lausanne 6,
CH-1000
Telephone: 21 27 36 52

Association of Swiss Speciality Pharmaceutical Manufacturers

98 rue de St Jean, Geneve 11,
CH-1211

Swiss Academy of Sciences

Hirschengraben 11, CP 8160, Bern,
CH-3001
Telephone: 31 22 3375
Fax: 31 21 3291

An association of private law, the academy is one of the offical national institutions designated to encourage research. Most of its funds are government grants. It represents the Swiss scientific community in the fields of exact and natural sciences. Its tasks cover communication, exchange, coordination and funding of scientific fields on a national and international level. It has established more than 60 scientific commissions and national committees in order to execute special duties and to guarantee the connection to other organizations in the international scientific community.

Swiss Association of Agricultural Engineers - SVIAL

Zolikofen, CH-3052
Telephone: 31 57 06 68

Swiss Pharmaceutical Society

Schweizerischer Apothekervrein,
Stationstrasse 12, Postfach 226, Bern,
CH-3097
Telephone: 31 53 58 58
Fax: 31 59 15 69

Swiss Society for Biochemistry

ISREC, Chemin des Boveresses 155,
Epalinges, CH-1066

Swiss Society for Microbiology

Schweizerische Gesellschaft fur
Mikrobiologie, Institue d'Hygiene,
CP 166, Geneva 4, CH-1211

Swiss Union of the Societies for Experimental Biology

Dept of Microbiology, CMU,
9 avenue de Champel, Geneva 4,
CH-1211
Telephone: 22 22 98 53
Fax: 22 46 72 37

Swiss Water Pollution Control Association

Postfach 607, Gruetliststrasse 44,
Zurich, CH-8027
Telephone: 1 202 69 92

UK

Association of Applied Biologists

Horticulture Research International,
Wellesbourne, Warks, CV35 9EF
Telephone: 789 470382
Fax: 789 470234

Association for the advancement of practice, growth and development of applied biology including production and preservation of food and improvement of the environment. Interests include entomology, nematology, pesticide applications, plant breeding, plant microbial interactions, plant physiology, post harvest biology, virology and weeds.

Association of British Health Care Industries

Consort House, 26/28 Queensway, London, W2 3RX
Telephone: 71 221 4612
Fax: 71 229 4708

Association of the British Pharmaceutical Industry

12 Whitehall, London, SW1A 2DY
Telephone: 71 839 3961

Association of Cheese Processors

19 Cornwall Terrace, London, NW1 4QP
Telephone: 71 486 7244
Fax: 71 487 4734

Association of Clinical Biochemists

Burlington House, Piccadilly, London, W1V 0BN
Telephone: 71 437 8656

Association for the advancement of clinical biochemistry in the United Kingdom.

Association of Consulting Engineers

Alliance House, 12 Caxton Street, London, SW1 0LQ
Telephone: 71 222 6557
Fax: 71 222 0750

Association of Consulting Scientists

Owles Hall, Buntingford, Herts,
Telephone: 763 72665

The Biochemical Society

59 Portland Place, London, W1N 3AJ
Telephone: 71 580 5530
Fax: 71 323 1136

BioIndustry Association

1 Queen Anne's Gate, London, SW1H 9BT
Telephone: 71 222 2809/235 510629
Fax: 71 222 8876

Aims to promote the commercial interests of companies actively involved in biotechnology in the United Kingdom. Technologies include biotechnology, genetic engineering, monoclonal antibodies, fermentation, pharmaceuticals, recombinant DNA.

Biological Engineering Society

Royal College of Surgeons, Lincoln's Inn Fields, London, WC2A 3PN
Telephone: 71 242 7750

Promoting a professional partnership between physical and life sciences. Collaborates with industry.

Biomass And Biofuels Association - BABA

PO Box 7, Southend, Reading, RG7 6AZ
Telephone: 635 862131

British Agrochemicals Association

Alembic House, 93 Albert Embankment, London, SE1 7TU
Telephone: 71 735 8471

British Association for the Advancement of Science

Fortress House, 23 Saville Row,
London, W1X 1AB
Telephone: 71 734 6010

British Association of Pharmaceutical Physicians

1 Wimpole Street, London, W1M 8AE
Telephone: 71 491 8610
Fax: 71 499 2405

British Biophysical Association

Department of Physiology/
Pharmacology, Southampton
University, Hampshire, SO9 3TU

Association for the advancement of the science of biophysics. Holds meetings for the discussion of physical biochemistry and inorganic biochemistry.

British Chemical Distributors and Trade Association Ltd

12b Westminster Place Gardens,
Artillery Row, London, SW1P 1RL
Telephone: 71 222 5292

British Chemical Engineering Contractors Association

1 Regent Street, London, SW1Y 4NR
Telephone: 71 839 6514
Fax: 71 930 3466

British Coordinating Committee for Biotechnology

c/o Vaccine Research & Production
Lab, CAMR (PHLS), Porton Down,
Salisbury, Wilts, SP4 0JG
Telephone: 980 610 391

British Crop Protection Council

2A Kidderminster Road, Croydon,
CR0 2UE
Telephone: 81 683 0211

British Effluent and Water Association

5 Castle Street, High Wycombe,
Bucks, HP13 6RZ
Telephone: 494 444 544

British Medical Association

BMA House, Tavistock Square,
London, WC1H 9JP
Telephone: 71 387 4498

British Pest Control Association

Alembic House, 93 Albert
Embankment, London, SE1 7TU
Telephone: 71 582 8268

British Society for Immunology

2 Triangle House, 2 Broomhill Road,
London, SW18 4HX

British Society for Plant Growth Regulation

Long Ashton Research Station,
University of Bristol, BS18 89AF
Telephone: 275 392 181
Fax: 275 394 007

British Society of Toxicological Pathologists - BSTP

ICI Pharmaceuticals, Mereside,
Alderley Park, Macclesfield, Cheshire,
SK10 4TG

Society for the advancement of pathology in all aspects pertaining to the effects of biologically active substances and environmental agents to which man and other species are exposed by design or inadvertently.

British Veterinary Association

7 Mansfield Street, London, W1M 0AT
Telephone: 71 636 6541
Fax: 71 436 2970

Institute of Biology

20 Queensbury Place, London, SW7 2DZ
Telephone: 071 581 8333
Fax: 071 823 9409

Institute of Clinical Molecular Biology

c/o John Radcliffe Hospital, Headington, Oxford, OX3 9DU
Telephone: 865 222359
Fax: 865 222501

Interested in application of techniques of molecular and cell biology to medical research; the training of young clinical and non-clinical scientists in this field; development of links with the biotechnology industry. Units include: molecular genetics; molecular and cellular immunology; neuroscience.

Institute of Food Science and Technology

5 Cambridge Court, 210 Shepherds' Bush Road, London, W6 7NL
Telephone: 071 603 6316

Institute for Life Sciences

35 Cartmel Drive, Carlisle, CA2 7SF

Interested in the study of aspects of dermatology, drug therapy and alternative medicine. Carries out research into effects of changing populations on crop production.

Royal Society

6 Carlton House Terrace, London, SW1Y 5AG
Telephone: 71 839 5561
Fax: 71 930 2170

The Royal Society is an independent learned society, self-governing under a Royal charter, for the promotion of the natural sciences, including mathematics and the scientific aspects of engineering, agriculture and medicine. Each year 40 new Fellows and 6 new Foreign Members are elected from among the most distinguished scientists in the British Commonwealth and other countries. The society encourages scientific endeavour by providing research appointments in British universities and grants for national and international scientific activities, hosts meetings and lectures, promotes the public understanding of science and provides independent scientific advice. It has exchange agreements with scientific bodies in 42 countries.

Society of Chemical Industry

14-15 Belgrave Square, London, SW1X 8PS
Telephone: 71 235 3681
Fax: 71 823 1698

Society to advance the application of chemistry and related sciences for public benefit, including agriculture, biotechnology, environmental technology, food engineering and water.

Society for Experimental Biology

Burlington House, Piccadilly, London, W1V 0LQ
Telephone: 71 439 8732
Fax: 71 287 4786

Society for General Microbiology

Marlborough House, Basingstoke Road, Reading, RG7 1AE
Telephone: 734 88 55 77
Fax: 734 88 56 56

USA

American Association for the Advancement of Science

1176 Massachusetts Avenue NW, Washington, DC-20036

American Association for Clinical Chemistry

1725 K Street, Suite 1010, Washington, DC-20005
Telephone: 202 857 0717

American Association of Immunologists

428E Preston Street, Baltimore, MD-21202

American Chemical Society

1155 16th Street NW, Washington, DC-20036
Telephone: 202 872 4600

American Medical Association

535 North Dearborn Street, Chicago, IL-60610

American Pharmaceutical Association

2215 Constitution Avenue NW, Washington, DC-20037

American Phytopathological Society

3340 Pilot Knob Road, St Paul, MN-55121
Telephone: 612 454 7250

American Society of Agricultural Engineers

PO Box 410, 2950 Niles Rd, St Joseph, MI-49085

American Society of Biological Chemistry

9650 Rockville Pike, Bethesda, MD-20814

American Society for Biotechnology

PO Box 2820, Sausalito, CA-94966

American Society for Cell Biology

9650 Rockville Pike, Bethesda, MD-20814

American Society of Clinical Pathologists

2100 W Harrison St, Chicago, IL-60612
Telephone: 312 738 4890

American Society for Horticultural Science

701 North St Asaph Street, Alexandria, VA-22314

American Society for Microbiology

1325 Massachusetts Avenue NW, Washington, DC-20005
Telephone: 202 737 3600

American Society of Plant Physiologists

PO Box 1688, Rockville, MD-20850
Telephone: 301 251 0560

Association of Biotechnology Companies - ABC

1220 L Street NW, Suite 615, Washington, DC-20005

Founded in 1983. International association composed of corporate members from 24 countries serving the biotechnology industry. Members fall into 4 categories: Core - a firm with lab facilities; Support - serving the biotech industry; Affiliate - non-profit and government organizations; and Council: state, university or regional biotech centre.

Crop Science Society of America

677 South Segoe Road, Madison, WI-53711

Federation of American Societies for Experimental Biology

FASEB, 9650 Rockville Pike, Bethesda, MD-20814

Industrial Biotechnology Association

2115 E Jefferson Street, Rockville, MD-20852
Telephone: 301 984 9598

National Academy of Sciences - NAS

CIOP, OIA, FO-2020, NRC, 2101 Constitution Avenue NW, Washington, DC-20418
Telephone: 202 334 2807
Fax: 202 334 2231

An honorific scientific society and an adviser of the government with a membership of about 1500 scientists and engineers, NAS publishes a scholarly journal, organizes symposia and meetings, bestows prizes and scientific awards and maintains contact with scientific academies in other countries. It operates through the National Research Council (NRC) in conducting policy and technical studies.

National Pharmaceutical Association

College of Pharmacy, Howard University, Washington, DC-2001

Pharmaceutical Manufacturers Association

1100 15th Street NW, Washington, DC-20005
Telephone: 202 463 2000

Society for Experimental Biology and Medicine

630 West 168th Street, New York, NY-10032

Society for Industrial Microbiology

1401 Wilson Boulevard, Arlington, VA-22209

PART III

PUBLICATIONS

JOURNALS

Acta Biotechnologica

Akademie Verlag, Leipzigerstrasse 3-4, Berlin, D-1086, Germany

ISSN: 0138-4988
(German, summaries in English, German and Russian)

Covers microbial, biochemical and bioanalogous technology. Includes topical reviews, short communications and book reviews.

Acta Pathologica et Immunologica Scandinavica

Munksgaard International Publishers, 35 Norre Sogade, Copenhagen K, DK-1370, Denmark

6 issues per year
(English summaries)

Divided into three sections: Section A, Pathology; Section B, Microbiology; Section C, Immunology.

Advances in Neuroimmunology

Manchester University Press, Oxford Road, Manchester, M13 9PL, UK
Telephone: 61 273 5539
Fax: 61 274 3346

3 issues per year
(English)

Publishes current reviews on significant research into the interactions between nervous and immune systems. Subjects covered include psychoneuroimmunology, immunopharmacology, immunoautoregulatory mechanisms, neuroimmodulation, neuroimmunoendocrinology.

AgBiotech

CAB International, Wallingford, Oxfordshire, OX10 8DE, UK
Telephone: 491 32111

ISSN: 0954-9897
(English)

Contains both review and abstract sections dealing with most aspects of agricultural biotechnology including policy and planning, genetic engineering applications (isolation of genes, vectors, cloning, transgenic animals and plants, disease and pests, pharmaceuticals and biological products, nuclear and plasmon transfer, targetted mutagenesis, gametoclonal variation, etc), gene sequencing, molecular aspects of gene function, mapping, genetics, immunological techniques, cell and embryo manipulation, gamete and haploid technology, artificial control of reproductive processes, diet and feed, biological control, biosensors and bioengineering, wastes and organic byproduct processing, environmental interactions, economics, etc, as well as social, ethical and legal aspects; details of conferences and book reviews.

Agriculture, Ecosystems & Environment

Elsevier,
Contact: JW Sturrock
Crown House, Linton Road, Barking, Essex, IG11 8JU, UK
Telephone: 81 594 7272

16 issues per year *ISSN: 0167-8809*
(English)

Concerned with the interaction of methods of agricultural production, agroecosystems and the environment. Topics covered include the comparison of different methods of production (intensive, extensive, linear, cyclic) in terms of their ecology; how agricultural production methods affect pollution of soil, water and air, the quality of food, and the use of energy and non-renewable resources; pest control; the effect of industrial pollutants on agriculture; and the policy issues involved in the change and development of agriculture. Publishes original scientific papers, review articles and occasional comment papers.

Agroforestry Today

Int Council for Research in Agroforestry, PO Box 30677, Nairobi, Kenya
Telephone: 332304

Quarterly *ISSN: 1013-9591*
(English or French)

Publishes papers in the agroforestry field.

Agro-Industry Hi-Tech

Teknoscienze, Via V Gioberti 1, Milano, I-20123, Italy
Telephone: 2 481 8011
Fax: 2 481 8070

6 issues per year
(English)

Publishes original papers applicable to all branches of agriculture and related industry. Also contains news items.

Agronomy Journal

American Society of Agronomy, USDA-ARS National Soil Tilth Lab, 2150 Pammel Drive, Ames, IA-50011, USA

6 issues per year *ISSN: 0002-1962*
(English)

Publishes original research findings on: soil-plant relationships; crop science; crop, forage and pasture production and utilisation; turf grass; agroclimatology; organomic modelling.

AIDS Research and Human Retroviruses

Mary Ann Liebert Inc, 1651 Third Avenue, New York, NY-10128, USA
Telephone: 212 289 2300

4 issues per year
(English)

Covers all aspects of acquired immunodeficiency syndrome including aetiology, pathogenesis, prevention and treatment, through original scientific papers. Contents include studies in virology, immunology, epidemiology and the role of opportunistic infections in the natural progress of the disease including aspects of targeted approaches to prevention and therapy, vaccine approaches in animal models and humans and the establishment of targeted antiviral agents.

AIDS Research Reviews

Marcel Dekker Inc, 270 Madison Avenue, New York, NY-10016, USA
Telephone: 212 696 9000

Annual
(English)

Each year covers four to six topics within the areas of human immunodeficiency virus (HIV) molecular biology, host responses, animal models, drug development, and vaccine development.

American Journal of Human Biology

Wiley-Liss Inc, 41E 11th Street, New York, NY-10003, USA
Telephone: 212 475 7700

Bimonthly *ISSN: 1042-0533*
(English)

A journal of the Human Biology Council, provides a medium for the exchange of research results in human biology.

American Journal of Veterinary Research

American Veterinary Medical Association, 930 North Meacham Road, Schaumburg, IL-60196, USA
Telephone: 312 885 8070

12 issues per year *ISSN: 0002-9645*
(English)

Covers all aspects of animal disease and health care in original reports and articles.

American Society for Horticultural Science Journal

American Society for Horticultural Science, 701 North St, Asaph Street, Alexandria, VA-22314, USA

6 issues per year *ISSN: 0003-1062*
(English)

Publishes original research reports concerning fruit, vegetables, food and medicinal crops, including aspects of plant physiology, structure, reproduction, environment, disease disorders and pests, and the action of herbicides, pesticides, plant hormones and growth substances.

Analytical Biochemistry

Academic Press Inc, Journals Division, 1250 Sixth Avenue, San Diego, CA-92101, USA
Telephone: 619 230 1840
Fax: 619 699 6859

16 issues per year *ISSN: 0003-2697*
(English)

Contains original papers and reviews on biochemical techniques, including cell and tissue culture, sequencing and synthesis of nucleic acids, purification methods for enzymes, immunochemical techniques (analytical and preparative) and pharmacological and toxicological research.

Angewandte Chemie

VCH, 220 East 23rd Street Suite 909, New York, NY-10010, USA
Telephone: 6201 602 220
Fax: 6201 602 328

12 issues per month *ISSN: 0570-0833*
(International edition in English)

Contains review articles from all fields of chemistry. Regularly publishes the lectures of the Nobel laureates in chemistry and related fields.

Animal Genetics

Blackwell Scientific publications, Osney Mead, Oxford, Oxon, OX2 0EL, UK
Telephone: 865 240201

Quarterly *ISSN: 0268-9146*
(English)

Official journal of the International Society for Animal Blood Group Research. Reports on current research on genetic variation at the level of single genes, their products and functions in economically important and domestic species of animals.

Animal Reproduction Science

Elsevier Science Publishers, PO Box 330, AH Amsterdam, NL-1000, The Netherlands

12 issues per year *ISSN: 0378-4320*
(English)

Publishes scientific papers dealing with the study of reproduction in all animals which can be regarded as useful to man. The scope includes the oestrus cycle in domesticated animals, cytogenetic selection in relation to reproductive capacity, reproductive physiology, endocrine and artificial control in reproduction, preservation of gametes and embryos, disease and nutrition in relation to reproductive performance, diagnostic and therapeutic techniques, pregnancy and foetal characteristics, sterility, reproduction problems and fertility problems.

Annals of Applied Biology

Portland Press Ltd, PO Box 32, Commerce Way, Colchester, Essex, CO2 8HP, UK
Telephone: 206 46351

6 issues per year *ISSN: 0003-4746*
(English)

Publishes papers, short communications and reviews on entomology, mycology and fungal disease, fungicides, bacterial plant diseases, plant virus diseases, nematology, host plant resistance, plant physiology, plant breeding and other fields of applied biology.

Annals of Clinical Biochemistry

Blackwell Scientific Publications Ltd, Osney Mead, Oxford, Oxon, OX2 0EL, UK
Telephone: 865 240201

6 issues per year *ISSN: 0004-5632*
(English)

Publishes original papers, review articles and book reviews on all aspects of clinical biochemistry, including new methodology.

Annals of Human Biology

Taylor & Francis Ltd, Rankine Road, Basingstoke, Hants, RG24 0PR, UK
Telephone: 256 840366
Fax: 256 479438

Bimonthly *ISSN: 0301-4460*
(English)

International journal publishing papers concerning research into human population biology, reporting investigations on the nature, development and causes of human variation. Covers the disciplines of human genetics, auxology, environmental physiology, ecology, epidemiology and ageing.

Annual Reviews

Annual Reviews Inc, 4139 El Camino Way, Palo Alto, CA-94306, USA
Telephone: 415 493 4400
Fax: 415 855 9815

(English)

Produces annual series of detailed scientific reviews on selected topics in a wide range of scientific areas, including the following separate series of interest to life scientists: Biochemistry, Biophysics and Biophysical Chemistry, Cell Biology, Ecology & Systematics Genetics, Immunology, Medicine, Microbiology, Neuroscience, Pharmacology & Toxicology, Physiology, Plant Physiology.

Annual Reviews in Biodiversity

Annual Reviews Inc,

ISSN: 0066-4154
(English)

Annual Reviews in Cell Biology

Annual Reviews Inc,

ISSN: 0743-4634
(English)

Annual Reviews in Entomology

Annual Reviews Inc,

ISSN: 0066-4170
(English)

Annual Reviews in Genetics

Annual Reviews Inc,

ISSN: 0066-4197
(English)

Annual Reviews in Immunology

Annual Reviews Inc,

ISSN: 0732-00582
(English)

Annual Reviews in Microbiology

Annual Reviews Inc,

ISSN: 0066-4227
(English)

Annual Reviews in Neuroscience

Annual Reviews Inc,

ISSN: 0147-006X
(English)

Annual Reviews in Plant Physiology and Plant Microbial Biology

Annual Reviews Inc,

ISSN: 1040-2519
(English)

Antibiotics and Chemotherapy

S Karger AG, Allschwilerstrasse 10,
Basel, CH-4009, Switzerland
Telephone: 61 306 1111
Fax: 61 306 1234

ISSN: 0066-4758
(English)

Publishes bound volumes devoted to specific topics covered by comprehensive reviews. Example: volume 31 covers Pharmacokinetics.

Antimicrobial Agents and Chemotherapy

American Society for Microbiology,
1325 Massachusetts Avenue NW,
Washington, DC-20005, USA
Telephone: 202 737 3600

12 issues per year *ISSN: 0066-4804*
(English)

Publishes papers on biosynthesis, mechanism of action and resistance, susceptibility, analytical procedures, experimental and clinical therapeutics, and pharmacology of antimicrobial and anticancer drugs.

Antiviral Research

Elsevier Biomedical Press,
Postbus 211, AE Amsterdam,
NL-1000, The Netherlands
Telephone: 20 5803 911

6 issues per year *ISSN: 0166-3542*
(English)

Publishes full papers, short communications and reviews on virus infections in man, plants and lower organisms. Topics covered include host resistance, interferons, chemotherapeutics, vaccine development and preparation of antiviral immunoglobulins.

Applied Agricultural Research

Springer-Verlag New York Inc,
Heidelberger Platz 3, Berlin 33,
D-1000, Germany
Telephone: 30 82 07 1

5 issues per year *ISSN: 0179-0374*
(English)

Publishes articles and brief research reports on the production and protection of crops, livestock, poultry and horticultural products as well as aspects of the management of soil and water resources, computer applications and technology transfer.

Applied Biochemistry and Biotechnology

Humana Press Inc, Crescent Manor
Box 2148, Clifton, NJ-07015, USA
Telephone: 201 773 4389
Fax: 201 773 5235

12 issues per year *ISSN: 0273-2289*
(English)

Publishes papers on genetic engineering, enzyme technology, monoclonal and tissue culture, immobilized biochemicals, cells, cell organelles and bacteria, and their applications in fermentation technology, bioenergy, solid-phase synthesis, assay and characterization of antibodies, receptors, RNA, DNA, proteins and other biologically important compounds.

Applied Biochemistry and Microbiology

Plenum Publishing Corporation, 223 Spring Street, New York, NY-10013, USA
Telephone: 212 620 8468
Fax: 212 463 0742

6 issues per volume *ISSN: 0003-6838*
(English)

English translation of the Russian Journal, Prikladnaya Biokhimiya i Mikrobiologiya.

Applied and Environmental Microbiology

American Society for Microbiology, 1325 Massachusetts Avenue NW, Washington, DC, DC-20005, USA
Telephone: 202 737 3600

12 issues per year *ISSN: 0099-2240*
(English)

Publishes papers on industrial microbiology, food aspects and toxicology, mycotoxins, public health, environmental aspects and general microbial ecology.

Applied and Theoretical Electrophoresis

Macmillan Press Ltd, Houndmills, Basingstoke, Hampshire, RG21 2XS, UK
Telephone: 256 29242
Fax: 256 479476

6 issues per volume *ISSN: 0954-6642*
(English)

Publishes original papers on use of electrophoresis in biotechnical and biomolecular engineering applications.

Aquaculture

Elsevier Science Publishers, PO Box 330, AH Amsterdam, NL-1000, The Netherlands
Telephone: 20 5803 911

32 issues per year *ISSN: 0044-8486*
(English)

Publishes papers on improvement and management of all aquatic food resources including flora and fauna related directly or indirectly to human consumption.

Aquaculture Engineering

Elsevier Applied Science Publishers, Crown House, Linton Road, Barking, Essex, IG11 8JU, UK
Telephone: 81 594 7272

4 issues per year *ISSN: 0144-8609*
(English)

Publishes papers related to engineering and design of aquacultural facilities, engineering based research studies, construction experience and techniques, commissioning and operation of aquaculture facilities, materials selection and their uses as well as quantification of biological data and constraints.

Aquatic Conservation

John Wiley and Sons Inc, 605 Third Avenue, New York, NY-10158, USA
Telephone: 212 850 6000

(English)

Publishes original papers that relate specifically to freshwater, brackish or marine habitats. Also publishes short communications, review articles, discussions and book reviews.

Aquatic Toxicology

Elsevier,
Contact: DC Malins
Pacific Northwest Research Foundation, Labs for Medical Research, Seattle, Washington, USA

12 issues per year *ISSN: 0166-445X*
(English)

Publishes original papers dealing with: mechanisms of toxicity in aquatic environments and the understanding of responses to toxic agents at all levels; toxicant induced alterations in organisms; development of procedures and techniques; identification of toxicants related to the understanding of perturbations in life processes. Includes reports of laboratory and field investigations.

Arboricultural Journal

A B Academic Publishers, PO Box 42, Bicester, Oxon, OX6 7NW, UK
Telephone: 869 320949

4 issues per volume *ISSN: 0307-1375*
(English)

Publishes original papers, reviews and survey articles as well as book reviews, items of news and comments in areas of arboriculture, amenity forestry, landscape design, ecology and horticulture in an arboricultural context.

Archives of Biochemistry and Biophysics

Academic Press Inc, Journals Division, 1250 Sixth Avenue, San Diego, CA-92101, USA
Telephone: 619 230 1840
Fax: 619 699 6859

16 issues per year *ISSN: 0003-9861*
(English)

Publishes original papers on viruses, nucleic acids, bioenergetics, enzyme mechanisms, membranes, hormones, photosynthesis and plant biology, protein structure and function and biochemical regulation.

Archives of Microbiology

Springer-Verlag, Heidelberger Platz 3, Berlin 33, D-1000, Germany
Telephone: 30 82 07 1

12 issues per year *ISSN: 0302-8933*
(English)

Contains reports on basic research on bacteria and other microorganisms.

Arzneimittel-Forschung/Drug Research

Editio Cantor GmbH, PO Box 1255, Aulendorf, D-7960, Germany
Telephone: 7525 431 433

12 issues per year *ISSN: 0004-4172*
(English)

Biotechnology Section: Original papers on all aspects of genetic engineering, biochemical pharmacology, endocrinology, immunology etc.

Australian Journal of Biological Sciences

CSIRO, 314 Albert Street, East Melbourne, Victoria 3002, Australia

4 issues per year
(English)

Publishes with the cooperation of the Australia Academy of Sciences, results of fundamental research in biochemistry, molecular biology, cell biology, microbiology, genetics, physiology, endocrinology and reproductive biology. Also available on microfiche.

Australian Journal of Biotechnology

Australian Industrial Publishers, 19 North Street, Adelaide, SA-6000, Australia
Telephone: 8 231 0220
Fax: 8 212 3939

Published quarterly *ISSN: 0819-3355*
(English)

Reports on scientific, financial and business matters, corporate management, commercialization of research, stock exchange performance and news in the Australian biotechnology industry as well as covering international developments.

Autoimmunity

Harwood Academic Publishers, STBS, PO Box 197, London, WC2E 9PX, UK

8 issues per year *ISSN: 0891-6934*
(English)

Publishes full-length original articles, invited reviews, hypotheses and correspondence on the pathogenesis, genetics, immunology and clinical expression of autoimmunity.

Biocatalysis

Harwood Academic Publishers, STBS, PO Box 197, London, WC2E 9PX, UK

4 issues per volume
(English)

International journal covering the industrial exploitation, both actual and potential, of biological catalysis and the mechanistic principles derived from these catalysts for the interconversion of chemical species, in the form of full-length research papers and reviews as well as occasional shorter communications.

Biochemical and Biophysical Research Communications

Academic Press Inc, Journals Division, 1250 Sixth Avenue, San Diego, CA-92101, USA
Telephone: 619 230 1840
Fax: 619 699 6859

24 issues per year *ISSN: 0006-291X*
(English)

Rapidly publishes papers on experimental biology with emphasis on molecular aspects, protein synthesis, nucleic acids, ion transport, bioenergetics and biochemistry.

The Biochemical Engineering Journal

Elsevier Sequoia SA, PO Box 564, 1001 Lausanne 1, Switzerland
Telephone: 21 20 73 81
Fax: 21 235 444

(English)

Provides international coverage of research and development in the area of biochemical engineering, especially processes associated with nutrition, health care, raw materials, chemical synthesis, fuel and waste water treatment industries.

Biochemical Genetics

Plenum Press, 233 Spring Street, New York, NY-10013, USA
Telephone: 212 620 8468

12 issues per year *ISSN: 0006-2928*
(English)

Covers biochemical genetics and deals with molecular aspects of genetic variation and evolution, mutation, gene action and regulation, immunogenetics, somatic cell genetics and nucleic acid function in heredity and development. Also covers the biochemical aspects of genetic defects.

Biochemical Journal

Biochemical Society, Portland Press, Box 32, Commerce Way, Colchester, Essex, CO2 8HP, UK
Telephone: 206 46351
Fax: 206 549331

24 issues per year *ISSN: 0264-6021*
(English)

Publishes original papers on all aspects of biochemistry; alternate issues cover molecular aspects or cellular aspects. Also available as microfiche or microfilm.

Biochemical Medicine

Academic Press Inc, Journals Division, 1250 Sixth Avenue, San Diego, CA-92101, USA
Telephone: 619 230 1840
Fax: 619 699 6859

6 issues per year *ISSN: 0885-4505*
(English)

An international journal covering original research in biochemistry relevant to physiological problems. Also contains letters to the Editor.

Biochemical Society Transactions

The Biochemical Society, Portland Press, Box 32, Commerce Way, Colchester, Essex, CO2 8HP, UK
Telephone: 206 46351
Fax: 206 549331

6 issues per year
(English)

Contains scientific reports of the Biochemical Society's colloquia and members' communications at meetings held throughout the UK and Eire.

Biochemical Systematics and Ecology

Pergamon Press Inc,
395 Saw Mill River Road, Elmsford, NY-10523, USA
Telephone: 914 592 7700
Fax: 914 592 3625

8 issues per year *ISSN: 0305-1978*
(English)

Biochemistry

American Chemical Society, 1155 Sixteenth St NW, Washington, DC-20036, USA
Telephone: 202 872 4363
Fax: 202 872 4615

24 issues per year *ISSN: 0006-2960*
(English)

Publishes original papers covering all aspects of biochemistry: enzymology, viral studies, amino acid sequences, cloning, bioenergetics, antibodies, membrane phenomena etc. Also available in microfiche form.

Biochemistry and Cell Biology

National Research Council of Canada, Research Journals, Ottawa, Ontario, K1A 0R6, Canada
Telephone: 613 993 9084
Fax: 613 952 7656

ISSN: 0829-8211
(English and French)

Biochimica et Biophysica Acta

Elsevier Biomedical Press,
Postbus 211, AE Amsterdam,
NL-1000, The Netherlands
Telephone: 20 5803 911

126 issues per year *ISSN: 0006-3002*
(English)

Massive publication of original work in biochemistry and biophysics. Divided into sections as follows: Bioenergetics; Biomembranes; Gene Structure and

Expression; Molecular Cell Research; Protein Structure and Molecular Enzymology; General Subjects; Lipids and Lipid Metabolism. Also BBA reviews in Bioenergetics and Biomembranes.

Biochimica et Biophysica Acta - Bioenergetics

ISSN: 0005-2728

Biochimica et Biophysica Acta - Biomembranes

ISSN: 0005-2736

Biochimica et Biophysica Acta - Gene Structure and Expression

ISSN: 0167-4781

Biochimica et Biophysica Acta - General Subjects

ISSN: 0304-4165

Biochimica et Biophysica Acta - Lipids and Lipid Metabolism

ISSN: 0005-2760

Biochimica et Biophysica Acta - Molecular Cell Research

ISSN: 0167-4889

Biochimica et Biophysica Acta - Reviews on Bioenergetics

ISSN: 0304-4173

Biochimica et Biophysica Acta - Reviews on Biomembranes

ISSN: 0304-4157

Biochimie

Societe de Chimie Biologique, Editions Scientifique Elsevier, 29 rue Buffon, Paris, F-75005, France
Telephone: 1 47 07 11 22
Fax: 1 43 36 80 93

12 issues per year *ISSN: 0300-9084*
(English or French)

Publishes details of original work on enzymology, genetics, immunology, microbiology and macromolecular structure. Also publishes reviews and surveys, book reviews and details of forthcoming meetings.

Biocycle

The JG Press Inc, Box 351 Emmaus, PA-18049, US
Telephone: 215 967 4135

8 issues per year *ISSN: 0276-5055*
(English)

Publishes papers on the composting of waste and on landfill techniques, as well as abstracts of relevant papers appearing in other journals. Contains a large number of advertisements for equipment.

Biodegradation

Kluwer Academic Publishers, Distribution Centre, PO Box 322, AH Dordrecht, NL-3300, The Netherlands
Telephone: 78 172 811
Fax: 78 183 273

4 issues per year *ISSN: 0923-9820*
(English)

Publishes papers on all aspects of science pertaining to the detoxification, recycling, amelioration or treatment of waste materials and pollutants by naturally-occurring microbial strains or associations or recombinant organisms.

Biodiversity and Conservation

Chapman & Hall, 2-6 Boundary Row, London, SE1 8HN, UK
Telephone: 71 865 0066
Fax: 71 522 9623

Quarterly *ISSN: 0960-3115*
(English)

International journal devoted to the publication of papers on biological diversity in all its aspects, its description, analysis and conservation, and its controlled and rational use by man. Publishes research papers and incisive articles dealing with the practicalities of conservation management, economic, social and political issues and case studies. Topics include: assessment and monitoring of biodiversity; captive breeding and relocations; marine biota; environmental planning and management; social and economic constraints on conservation; policy; genetic diversity; restoration ecology; new biotechnological product research and development and responses to environmental change.

Bioelectrochemistry and Bioenergetics

Elsevier Sequoia SA, Box 564, Lausanne 1, CH-1001, Switzerland
Telephone: 21 207 381
Fax: 21 235 444

6 issues per year *ISSN: 0302-4598*
(English)

Bioelectromagnetics

John Wiley and Sons Ltd, Baffins Lane, Chichester, West Sussex, PO19 1UD, UK
Telephone: 243 770328
Fax: 243 775878

6 issues per year
(English)

Bioelectronics and Biosensors

Sheffield University Biomedical Information, Service (SUBIS), The University, Sheffield, S10 2TN, UK
Telephone: 742 768 555
Fax: 742 768 555

24 issues per year *ISSN: 0952-0384*
(English)

Covers bioelectrochemistry, biochips, enzyme electrodes and *in vivo* NMR.

BioEngineering - Forschung & Praxis

Redaktion BioEngineering, Klenzestrasse 59, Munchen 5, D-8000, Germany
Telephone: 89 201 41 79

4 issues per year
(Mostly in German with some English summaries)

Covers all aspects of biotechnology and bioengineering including genetics, fermentation, microbiology, downstream processing and legislation in illustrated feature articles with some diagrams and photographs in full colour. Includes editorial comment and sections covering news, products, trends and patents. Also contains a large amount of advertising material from companies producing analytical equipment, diagnostics, chemicals and laboratory supplies as well as job advertisements and notices and reports on conferences and meetings.

BioEssays

Company of Biologists Ltd, Box 32, Commerce Way, Colchester, Essex, CO2 8HP2, UK
Telephone: 206 46351
Fax: 206 549331

ISSN: 0265-9247
(English)

Reports on advances in basic research as well as applied aspects presented as reviews and comments in areas of molecular, cellular and developmental biology, and their impacts on science and society. Published for the ICSU Press (the publishing house of the International Council of Scientific Unions).

BioFactors

Oxford Journals, Walton Street, Oxford, Oxon, OX2 6DP, UK
Telephone: 865 791373

4 issues per year *ISSN: 0951-6433*
(English)

Accepts articles explaining the roles of active trace substances in a given organism or cell line.

Biofouling

Harwood Academic Publishers, STBS, PO Box 197, London, WC2E 9PX, UK

4 issues per volume
(English)

Publishes pure and applied work of all kinds on microbial, plant or animal fouling on natural and man-made surfaces in aquatic and aerial environments as well as medical, dental, industrial and agricultural biofouling.

Biofutur

Editions Scientifique Elsevier, 29 rue Buffon, Paris, F-75005, France
Telephone: 1 47 07 11 22
Fax: 1 43 36 80 93

ISSN: 0294-3506
(French)

Biological Agriculture and Horticulture

A B Academic Publishers, PO Box 42, Bicester, Oxon, OX6 7NW, UK
Telephone: 869 320949

ISSN: 0144-8765
(English)

Publishes original articles and reviews concerning alternatives to traditional methods of plant and animal husbandry with an emphasis on the 'biological' approach.

Biological Conservation

Elsevier Science Publishers, Crown House, Linton Road, Barking, Essex, IG11 8JU, UK
Telephone: 81 594 7272

16 issues per year *ISSN: 0006-3207*
(English)

Publishes papers dealing with the preservation of wildlife and the conservation of biological and natural resources. The journal's main basis is ecological. Publishes analytical studies and reports of discoveries or important original information.

Biological Mass Spectrometry

John Wiley and Sons Inc, 605 Third Avenue, New York, NY-10158, USA
Telephone: 212 850 6000

Monthly
(English)

Publishes papers on the use of mass spectrometry in the study of compounds in the biological, environmental and medical fields. Examines new techniques and instrumentation as well as applications.

Biological Membranes

Harwood Academic Publishers, STBS, PO Box 197, London, WC2E 9PX, UK

ISSN: 0748-8653
(English)

English translation of the Russian journal accepting original papers in the following areas: the structure, composition and physical and chemical properties of membranes; membrane proteins and lipids, bioenergetics, membrane transport, reception and regulation at the membrane level, cell contacts and interactions, membranes and immunology, membranes and medicine, model systems and interface phenomena related to membrane problems.

Biologist

Institute of Biology, 20 Queensberry Place, London, SW7 2DZ, UK
Telephone: 71 581 8333
Fax: 71 823 9409

5 issues per year *ISSN: 0006-3347*
(English)

Journal of the Institute of Biology (UK). Contains editorial (personal comment) matter, correspondance and scientific and technical articles covering general biological subjects of interest to members, as well as a news section detailing institute and national activities and events. Also carries book reviews.

Biology Digest

Plexus Publishing Inc, 143 Old Marlton Pike, Medford, NJ-08055, USA
Telephone: 609 654 6500
Fax: 609 654 4309

Monthly *ISSN: 0095-2958*
(English)

Contains summaries of current science articles and research reports for use in secondary school education - incorporates Environmental Quality Abstracts.

Biology and Fertility of Soils

Springer-Verlag, Heidelberger Platz 3, Berlin 33, D-1000, Germany
Telephone: 30 82 07 1

8 issues per year *ISSN: 0178-2762*
(English)

Publishes original papers, reviews and short communications in English on all fundamental and applied aspects of biology and fertility of soils particularly in relation to understanding of aspects of increasing demands of agriculture, forestry and industrialization. In general, topics fall within the following areas: soil biology, microflora and fauna; soil fertility, biological processes and their effect on nutrient availability; soil hygiene, effects of constraints such as heavy metals, acid rain and pesticide residues; soil biotechnology, waste handling and biological control.

Biomass - an International Journal

Elsevier Applied Science Publishers, Crown House, Linton Road, Barking, Essex, IG11 8JU, UK
Telephone: 81 594 7272

Ceased publication when incorporated into Bioresource Technology.

Biomass & Bioenergy

Pergamon Press, Headington Hill Hall, Oxford, OX3 0BW, UK
Telephone: 865 794141
Fax: 865 743952

12 issues per year *ISSN: 0961-9534*
(English)

Pergamon Press Inc, 395 Saw Mill River Road, Elmsford, NY-10523, USA
Telephone: 914 592 7700
Fax: 914 592 3625

Publishes original papers and reviews covering production and use of plants, microorganisms and enzymes related to energy, fuels and chemicals, including anaerobic digestion and fermentation as well as use of biological residues, energy crops and other organic resources by direct combustion, gasification and chemical processing. The journal also covers environmental, management and economic aspects of biomass and bioenergy and features book reviews, conference reports and letters to the Editor. Editing team was previously associated with the Biomass Journal.

Biomedical Chromatography

Heyden and Son Ltd, Spectrum House, Hillview Gardens, London, NW4 2JQ, UK
Telephone: 71 203 5171
Fax: 71 203 1027

Bimonthly *ISSN: 0269-3879*
(English)

International journal devoted to research on chromatographic methodologies and their applications in the biosciences. Concerned with the application of chromatography and allied technologies in the biological and medical sciences.

Bioorganic and Medicinal Chemistry Letters

Pergamon Press Inc, 395 Saw Mill River Road, Elmsford, NY-10523, USA
Telephone: 914 592 7700
Fax: 914 592 3625

12 issues per year *ISSN: 0960-894X*
(English)

Provides rapid dissemination of preliminary communications on all aspects of bioorganic chemistry, medicinal chemistry and related disciplines.

BioPharm

Aster Publishing, 195 Main Street, Metuchen, NJ-08840, USA
Telephone: 201 549 3000
Fax: 201 549 8927

10 issues per year *ISSN: 1040-8304*
(English)

Publishes features, articles, and columns related to product and process development, scale-up, and manufacture of biotechnologically derived pharmaceuticals, vaccines, and diagnostics.

Biopharmaceutics and Drug Disposition

John Wiley and Sons Ltd, Baffins Lane, Chichester, West Sussex, PO19 1UD, UKA
Telephone: 243 770 328
Fax: 243 775 878

9 issues per year *ISSN: 0142-2782*
(English)

Publishes original reports of studies in biopharmaceutics, drug disposition and pharmacokinetics, especially those which have a direct relation to the therapeutic use of drugs. This includes human pharmacological studies, and therapeutic response and toxicity related to plasma and tissue concentrations of drugs and their metabolites.

Biophysical Chemistry

Elsevier Science Publishers, PO Box 330, AH Amsterdam, NL-1000, The Netherlands
Telephone: 20 5803 911

9 issues per year *ISSN: 0301-4622*
(English)

Publishes original research papers and review articles covering physical aspects of the biological functions of individual macromolecules and of supramolecular structure which these molecules can form with each other as well as papers which deal with systems serving as models for these.

Biopolymers

John Wiley & Sons Inc, 605 Third Avenue, New York, NY-10158, USA
Telephone: 212 850 6000

12 issues per year *ISSN: 0006-3525*
(English)

Contains original research reports on biopolymers - proteins, nucleic acids and polysaccharides - including model systems. Rapid publications appear as research communications.

Bioprocess Engineering

Springer-Verlag, Heidelberger Platz 3, Berlin 33, D-1000, Germany
Telephone: 30 82 07 1

4 issues per year *ISSN: 0178-515X*
(English)

Bioreactors, upstream development and optimization of technical bioprocesses, details of new equipment adapted to such processes, bioprocess unit or plants, computer models and details of bioreactors as well as aspects of measurement, control and automation.

Biorecovery

A B Academic Publishers, PO Box 42, Bicester, Oxon, OX6 7NW, UK
Telephone: 869 320 949

4 issues per volume
(English)

Topics covered include: microbial mining; petroleum microbiology; bioaccumulation and separation of elements; bioaccumulation applied to pollution control and effluent treatment; microbial coal desulphuration; biorecovery of energy from waste materials.

Bioresource Technology

Elsevier Applied Science Publishers, Crown House, Linton Road, Barking, Essex, IG11 8JU, UK
Telephone: 81 594 7272

12 issues *ISSN: 0906 8524*
(English)

Formerly the journals Biological Wastes and Biomass.
Publishes original papers concerning aspects of waste treatment, including anaerobic digestion or fermentation for production of gaseous or liquid fuels, animal feed and fertilizer as well as various aspects of recovery of energy and materials from purpose-grown plant material.

BIOS-Conditionnement Brasserie-Malterie

Pole Technologique, 7 rue du Bois de la Champelle, BP 268, Vandouvre Cedex, France
Telephone: 83 44 25 32
Fax: 83 44 12 90

12 issues per year *ISSN: 0366-2284*
(Mainly in French)

Contains news, articles and techniques for the brewing industry as well as production statistics and other information relating to grain and its use in brewing. Carries trade advertising and classified advertisements.

Bioscience, Biotechnology and Biochemistry

Agricultural Chemical Society of Japan, 2-4-16 Yagoi Bunkyo-ku, Tokyo 1130, Japan

12 issues per year *ISSN: 0916-8451*
(English)

Subscriptions, PO Box 5030, Tokyo International, Tokyo, Japan.
Formerly Agricultural and Biological Chemistry. Publishes original papers on subjects including analytical chemistry, organic chemistry, biological chemistry, food & nutrition, microbiology & fermentation and pesticide chemistry.

Bioscience Reports

Plenum Publishing Corp, 223 Spring Street, New York, NY-10013, USA
Telephone: 212 620 8468

12 issues per year *ISSN: 0144-8463*
(English)

Publishes short reports of original research of significance that merits rapid publication. Covers all areas of biochemistry including biochemical genetics, biochemical pharmacology, endocrinology, immunology, etc. Reports new methods, applications of methods and treatment of existing data, as well as conventional studies in molecular and cellular biology. Formerly published by the Biochemical Society.

Biosensors and Bioelectronics

Elsevier Advanced Technology, Crown House, Linton Road, Barking, Essex, IG11 8JU, UK
Telephone: 81 594 7272

8 issues per year *ISSN: 0956-5663*
(English)

International journal providing information on the latest developments in the research, application and development of biosensors and other related devices that exploit biochemicals. Also includes recent patents and patent applications, information on the development of products-applications, regulatory issues and markets.

Biosystems

Elsevier
Contact: AW Schwartz, Dept of Exobiology, University of Nijmegen, Nijmegen, The Netherlands

8 issues per year *ISSN: 0303-2647*
(English)

Publishes articles on origins, evolution, dynamics, assembly and control of biological systems.

Biotech Forum Europe

Dr Alfred Huethig Verlag GmbH, Im Weiher 10, Heidelberg 1, D-6900, Germany
Telephone: 6221 489 421
Fax: 6221 489 242

10 issues per year *ISSN: 0938-7501*
(Mainly English)

Contains editorials, market analysis, business reviews and technical and scientific articles covering all aspects of biotechnology with an emphasis on Europe. Includes commercial advertisements, product reviews and details of conferences.

Biotech Knowledge Sources

BioCommerce Data Ltd, 95 High Street, Slough, Berkshire, SL1 1DH, UK
Telephone: 753 511777
Fax: 753 512239

Monthly
(English)

Provides listings of new publications (books, journals, market research reports, videos, software, databases) and forthcoming meetings, conferences, courses and exhibitions concerned with biotechnology.

Bio/Technology

Nature Publishing Company,
65 Bleecker Street, New York,
NY-10012-2467, USA
Telephone: 212 477 9600
Fax: 212 505 1364

12 issues per year *ISSN: 0733-222X*
(English)

Publishes news, financial analyses and original papers with industrial implications covering basic discoveries in cell biology, breeding, industrial fermentation and separation techniques relevant to health care, chemicals, energy, food, waste management and the natural product industries.

Biotechnology Advances

Pergamon Press Inc, 395 Saw Mill River Road, Elmsford, NY-10523, USA
Telephone: 914 592 7700
Fax: 914 592 3625

2 issues per year *ISSN: 0734-9750*
(English)

Covers all aspects of biotechnology, applied biology, technical biochemistry, genetic engineering and biochemical engineering. Each issue contains about eight mini-reviews on selected subjects.

Biotechnology and Applied Biochemistry

Academic Press Inc, Journals Division, 1250 Sixth Avenue, San Diego, CA-92101, USA
Telephone: 619 699 6859

6 issues per year *ISSN: 0885-4513*
(English)

Publishes original papers and review articles in the field of applied biochemistry with special reference to biotechnology, with fundamental studies directed at the design of new biotechnology, improvement of existing biochemical processes and the preparation, utilization, conversion and application of biological materials. Publication of the International Union of Biochemistry.

Biotechnology and Bioengineering

John Wiley and Sons Inc, 605 Third Avenue, New York, NY-10158, USA
Telephone: 212 850 6000

22 issues per year *ISSN: 0006-3592*
(English)

John Wiley and Sons Ltd, Baffins Lane, Chichester, West Sussex, PO19 1UD, UK
Telephone: 243 770328
Fax: 243 775878

Contains basic research reports on biological, physical and chemical principles related to technical use of microorganisms and enzymes.

Biotechnology Chemonomics

Economics of Technology, 49 E 41st Street, New York, NY-10165, USA
Telephone: 212 988 9678
Fax: 201 763 5057

ISSN: 0734-5151
(English)

Covers economic aspects of process technology, including costs relating to biotechnology based manufacturing systems.

Biotechnology and Genetic Engineering Reviews

Intercept Ltd, PO Box 716, Andover, Hants, SP10 1YG, UK
Telephone: 264 334748
Fax: 264 334058

Annual *ISSN: 0264-8725*
(English)

Contains specialist reviews covering all aspects of biotechnology and genetic engineering in the widest sense. Fast publication procedure results in up-to-date comment.

Biotechnology Insight

Science and Technology Letters,
PO Box 81, Northwood, Middlesex,
HA6 3DN, UK
Telephone: 9274 20732
Fax: 9274 25066

ISSN: 0268-7291
(English)

Biotechnology Letters

Science Reviews Inc, 707 Foulk Road
Suite 102, Wilmington, DE-19803,
USA

12 issues per year *ISSN: 0141-5492*
(English)

Rapidly communicates experimental results on all aspects of biotechnology: applications of microbiology and molecular biology in combination with chemical systems and process engineering, related to use of living cells or enzymes and the practical applications of biological reactions. Includes relevant aspects of genetics and cell biochemistry, process and reactor design, and pre- and post-treatment steps.

Biotechnology Progress

American Chemical Society, 1155
Sixteenth St NW, Washington,
DC-20036, USA
Telephone: 202 872 4363
Fax: 202 872 4615

Bimonthly *ISSN: 8756-7938*
(English)

Features research reports, reviews, and news of the development and design of new processes, products, or devices for the biotechnology/bioprocess industries. Emphasizing practical applications of research, the journal reports advances in bioseparation, bioconversion, bioreactors, biomedical engineering, cell culture, downstream processing, hybridoma applications, metabolic control, biomodelling, genetic engineering, biomolecules, bioanalytics, bioinstrumentation, and other new technologies from molecular biology.

Biotechnology Techniques

Science and Technology Letters,
PO Box 81, Northwood, Middlesex,
HA6 3DN, UK
Telephone: 9274 20732
Fax: 9274 25066

ISSN: 0951 208X
(English)

Publishes short accounts of new methods and techniques for research and development in biotechnology.

Biotechnology Therapeutics

Marcell Dekker Journals, 270 Madison
Avenue, New York, NY-10016, USA
Telephone: 212 969 9000
Fax: 212 685 4540

(English)

Covers preclinical experiments and clinical trials of therapeutic substances produced using genetic engineering and other bioprocesses.

Botanica Acta

Georg Thieme Verlag, Rudigerstrasse
14, Postfach 732, Stuttgart 1, D-7000,
Germany
Telephone: 711 89 31 0

4 issues per year *ISSN: 0932-8629*
(German or English)

Publishes full-length original papers, rapid communications and concise review articles. The emphasis is on physiology and biochemistry, cytology and molecular biology of both higher and lower plants. An official journal of the German Botanical Society.

British Medical Journal

British Medical Association, Tavistock
Square, London, WC1H 9JR, UK
Telephone: 71 397 4499

52 issues per year *ISSN: 0007-1447*
(English)

Contains leading articles and papers, as well as short reports, covering clinical research and medical practice. Also publishes news and notes.

British Mycological Society Transactions

Cambridge University Press, Edinburgh Buildings, Shaftsbury Road,
Cambridge, CB2 2RU, UK
Telephone: 223 312393

6 issues per year
(English)

Subscriptions include British Mycological Society Bulletin (published twice yearly).

British Phycological Journal

Academic Press Inc (London) Ltd,
24-28 Oval Road, London,
NW1 7DX, UK
Telephone: 71 267 4466
Fax: 71 482 2293

4 issues per year *ISSN: 0007-1617*
(English)

Publishes original papers on current research in algae as well as brief reports of lectures to the British Phycological Society and reviews of books related to phycology. Topics include physiology, ultrastructure, ecology, systematics, effects of pollutants and use of algae as bioindicators.

Canadian Journal of Biochemistry and Cell Biology

National Research Council of Canada,
Research Journals, Ottawa, Ontario,
K1A 0R6, Canada
Telephone: 613 993 9084
Fax: 613 952 7656

(English)

Original publications in the indicated areas of biological science with French and English summaries.

Canadian Journal of Plant Pathology

Canadian Phytopathological Society,
Department of Environmental Biology,
University of Guelph, Ontario,
N1G 2W1, Canada
Telephone: 519 824 4120

ISSN: 0706 0661
(Mainly English, some French)

Canadian Journal of Plant Science

Agricultural Institute of Canada, Suite 907, 151 Slater Street, Ottawa,
Ontario, K1P 5H4, Canada
Telephone: 613 232 9459
Fax: 613 594 5190

ISSN: 0008-4220
(English and French)

Cancer Drug Delivery

Mary Ann Liebert Inc, 1651 Third Avenue, New York, NY-10128, USA
Telephone: 212 289 2300

4 issues per year *ISSN: 0732-9482*
(English)

Reports new and innovative methods of delivery of drugs and biologicals of potential to improve the efficacy, decrease the toxicity or increase the convenience of cancer therapy. Covers both clinical investigations and basic research.

Cancer Research

Waverly Press Inc, 428 East Preston Street, Baltimore, MD-21202, USA
Telephone: 301 528 4000

12 issues per year *ISSN: 0008-5472*
(English)

Publishes original research in biochemical and physiological aspects of cancer, chemical and physical carcinogenesis and mutagenesis, clinical investigations, endocrinology, epidemiology and biostatics, immunology, molecular and cell biology, preclinical pharmacology and experimental therapeutics, radiobiology and virology. Official journal of the American Associations for Cancer Research Inc.

Carbohydrate Polymers

Elsevier Applied Science Publishers Ltd, Crown House, Linton Road, Barking, Essex, IG11 8JU, UK
Telephone: 81 594 7272

8 issues per year *ISSN: 0144-8617*
(English)

Covers the study and exploitation of industrial applications of carbohydrate polymers in areas such as food, textiles, paper, wood, adhesives, pharmaceuticals, oil field applications and industrial chemistry.

Carbohydrate Research

Elsevier Science Publishers, PO Box 330, AE Amsterdam, NL-1000, The Netherlands
Telephone: 20 5803 911

ISSN: 0008-6215
(English)

Carlsberg Research Communications

Springer-Verlag, Heidelberger Platz 3, Berlin 33, D-1000, Germany
Telephone: 30 82 07 1

6 issues per year *ISSN: 0105-1938*
(English)

Publishes details of work at Carlsberg research establishments, including Departments of Chemistry and Physiology, Brewing Chemistry, Biotechnology and Brewing Technology. Subjects covered include fundamental and applied research in biochemistry, cell molecular biology, genetics, microbiology and plant and animal physiology.

Catering and Health

A B Academic Publishers, PO Box 42, Bicester, Oxon, OX6 7NW, UK
Telephone: 869 320949

ISSN: 0627-3851
(English)

Publishes reports and review articles as well as book reviews and reports of relevant meetings of national and international organizations such as the World Health Organization working party on health aspects of catering as well as notes on changes in Regulations and Labelling.

Cell

Cell Press, 50 Church Street, Cambridge, MA-02138, USA
Telephone: 617 253 5242

13 issues per year *ISSN: 0092-8674*
(English)

Publishes reports of research of exceptional significance in any area of biology.

Cell Adhesion and Communication

Harwood Academic Publishers, STBS, PO Box 197, London, WC2E 9PX, UK

4 issues per year
(English)

Provides a central forum for the rapid publication of full-length manuscripts, short communications and reviews of high quality covering all aspects of receptor-based cell recognition and signalling.

Cell Biochemistry and Function

Butterworth Scientific Ltd,
Linacre House, Jordan Hill, Oxford,
Oxon, OX2 8DG, UK
Telephone: 865 310366
Fax: 865 310898

4 issues per year *ISSN: 0263-6484*
(English)

An international journal covering the function and interaction of cells and organelles and the influence of disease, drugs and other chemicals. Contains full papers, communications, book reviews and a calendar of courses, conferences, symposia etc.

Cell Biology: International Reports

Academic Press Inc (London) Ltd,
24-28 Oval Road, London, NW1 7DX, UK
Telephone: 71 267 4466

12 issues per year *ISSN: 0309-1651*
(English)

Rapid publication journal of the International Federation of Cell Biology. Each January issue contains details of Cell Biology Society meetings and details of societies worldwide.

Cell Motility and the Cytoskeleton

Alan R Liss Inc, 41 East 11th Street,
New York, NY-10003, USA
Telephone: 212 475 7700

12 issues per year
(English)

Publishes full-length articles and brief rapid communications as well as invited reviews and book reviews.

Cellular Biochemistry

John Wiley, Baffins Lane, Chichester,
West Sussex, PO19 1UD, UK
Telephone: 243 770328
Fax: 243 775878

Monthly *ISSN: 0730-2312*
(English)

Publishes reviews, commentaries, and descriptions of original research in areas where complex cellular, pathogenic, clinical, or animal model systems are studied by biochemical, genetic or quantitative ultrastructural approaches.

Cellular Immunology

Academic Press Inc, Journals Division,
1250 Sixth Avenue, San Diego,
CA-92101, USA
Telephone: 619 230 1840
Fax: 619 699 6859

ISSN: 0008-8749
(English)

Cellular and Molecular Biology

Pergamon Press Inc, 395 Saw Mill
River Road, Elmsford, NY-10523,
USA
Telephone: 914 592 7700
Fax: 914 592 3625

8 issues per year *ISSN: 0145-5680*
(English)

Publishes original articles in the fields of cellular and molecular biology and enzymology, applied to human tissues as well as to animals and plants. Contributions to biological, medical, dental, veterinary, pharmacological and biological research are published provided that they demonstrate the link between structure and biochemical function.

Cellular Signalling

Pergamon Press Ltd, Headington Hill Hall, Oxford, OX3 0BW, UK
Telephone: 865 794141
Fax: 865 60285

6 issues per volume
(English)

Publishes original papers covering all aspects of mechanisms, actions and structural components of cellular signalling systems as well as short reviews.

Chemico-Biological Interactions

Elsevier, Crown House, Linton Road, Barking, Essex, IG11 8JU, UK
Telephone: 81 594 7272

12 issues per year *ISSN: 0009-2797*
(English)

Includes: research reports, rapid communications, review articles and commentaries covering: molecular aspects of cytotoxicity, carcinogenesis, mutagenesis and teratogenesis; molecular mechanisms by which drugs exert their therapeutic or toxic effects.

Chemistry and Industry

Society of Chemical Industry, 14 Belgrave Square, London, SE1X 8PS, UK
Telephone: 71 235 3681

26 issues per year *ISSN: 0009-3068*
(English)

Contains news, articles and scientific papers with issues related to biotechnology resulting from meetings sponsored by the Society of Chemical Industry. Contains both commercial and job advertisements.

Chimica oggi

Teknoscienze Srl, Via V Gioberti 1, Milano, I-20123, Italy
Telephone: 2 481 8011
Fax: 2 481 8070

10 issues per year
(Italian with many English articles)

Contains full-length articles on the production, application and development of organic chemical products. Incorporates Biotecnologie oggi with full-length feature articles, many in English, including reports on activities within the biotechnology programmes of the European Community. Also contains editorial comment, short paragraphs on developments in the Italian chemical industry and lists suppliers of chemical and biotechnical products, equipment and services as well as carrying trade and industry advertisements, mainly relating to Italy.

Chinese Journal of Physiological Sciences

Taylor & Francis Ltd, Rankine Road, Basingstoke, Hants, RG24 0PR, UK
Telephone: 256 840366
Fax: 256 479438

Quarterly *ISSN: 0258-6428*
(English)

Publishes original papers of physiology as well as functionally oriented research contributions from biochemistry, biophysics, pharmacology, morphology and other related biomedical disciplines.

Chromosoma

Springer-Verlag, Heidelberger Platz 3, Berlin 33, D-1000, Germany
Telephone: 30 82 07 1

10 issues per year *ISSN: 0009-5915*
(English)

Publishes details of nuclear and chromosome research including general aspects of cytotaxonomy, karyotype descriptions and mammalian and human cytogenetics.

Clinical Biochemistry

Pergamon Press Inc,
395 Saw Mill River Road, Elmsford,
NY-10523, USA
Telephone: 914 592 7700
Fax: 914 592 3625

6 issues per year *ISSN: 0009-9120*
(English)

Clinical Chemistry

American Assoc of Clinical Chemistry,
2029 K Street NW, Washington,
DC-20006, USA
Telephone: 919 659 03000
Fax: 919 651 1221

12 issues per year *ISSN: 0009-9147*
(English)

Publishes full papers, scientific notes, case reports and letters on all aspects of clinical chemistry. Numerous advertisements for equipment, test kits, chemicals etc. Official journal of the American Association for Clinical Chemistry.

Clinical Chemistry and Enzymology Communications

Harwood Academic Publishers, STBS,
PO Box 197, London, WC2E 9PX, UK

6 issues per volume
(English)

Rapid publication of new clinically based research in chemistry and enzymology relating to the diagnosis of human disease in the form of research papers, short communications and reviews as well as information on related software.

Clinical and Experimental Immunology

Blackwell Scientific Publications Ltd,
Osney Mead, Oxford, Oxon,
OX2 0EL, UK
Telephone: 865 240201

ISSN: 0009-9104
(English)

Publishes papers on the role of immunology in the diagnosis and pathogenesis of disease, including allergy. Official journal of the British Society for Immunology.

Clinical Microbiology Reviews

American Society for Microbiology,
1325 Massachusetts Avenue NW,
Washington, DC-20005, USA
Telephone: 202 737 3600

ISSN: 0893-8512
(English)

Covers all aspects of clinical microbiology and immunology.

Clinical Science

The Biochemical Society, Portland
Press, Box 32, Commerce Way,
Colchester, Essex, CO2 8HP, UK
Telephone: 206 46351
Fax: 206 549331

12 issues per year *ISSN: 0143-5221*
(English)

Covers biochemical, physiological and immunological aspects of clinical investigation. Published jointly by the Medical Research Society and the Biochemical Society.

Comments on Molecular and Cellular Biophysics

Gordon and Breach Science Publishers,
C/O STBS, PO Box 90 Reading,
Berkshire, RG1 8JL, UK
Telephone: 734 560080
Fax: 734 568211

ISSN: 0143 8123
(English)

Focuses on significant developments in molecular and cellular biosciences, and facilitates dissemination and critical discussion of emerging ideas.

Comments on Theoretical Biology

Gordon and Breach Science Publishers, C/O STBS, PO Box 90 Reading, Berkshire, RG1 8JL, UK
Telephone: 734 560080
Fax: 734 568211

6 issues per volume *ISSN: 0894-8550*
(English)

Facilitates dissemination and critical evaluation of recent theoretical analysis of biological systems ranging from the population to the subcellular level.

Communications in Soil Science

Marcel Dekker Inc, 270 Madison Avenue, New York, NY-10016, USA
Telephone: 212 696 9000

ISSN: 0010-3624
(English)

Rapid publication of articles on soil science and crop production, particularly in the areas of mineral content of soil and plants and plant nutrition.

Comparative Biochemistry and Physiology

Pergamon Press Inc, 395 Saw Mill River Road, Elmsford, NY-10523, USA
Telephone: 914 592 7700
Fax: 914 592 3625

6 issues per year
(English)

Comptes Rendus de l'Academie des Sciences

Centrale des Revues, 11 rue Gossin, Montrouge Cedex, F-92543, France

40 issues per year
(Mainly French)

Reports original discoveries in the form of short notes on topics which include theoretical biology, molecular biology, genetics, marine biology, ecology, animal nutrition and botany. Mainly in French but with some contributions in English or other languages with French summaries.

Computer Applications in the Biosciences - CABIOS

Oxford Journals, Walton Street, Oxford, Oxon, OX2 6DP, UK
Telephone: 865 791373

4 issues per year *ISSN: 0266-7061*
(English)

Specializes in the biological applications of computing. Publishes peer-reviewed papers, reviews programmes and publishes new applications and developments.

CRC Critical Reviews

CRC Inc, 2000 NW 24th Street, Boca Raton, FL-33431, USA

4 issues per year
(English)

Publishes numerous detailed reviews: Analytical Chemistry, Biochemistry, Biomedical Engineering, Clinical Laboratory Sciences, Clinical Neurobiology, Diagnostic Imaging, Environmental Control, Food Science, Nutrition, Microbiology, Plant Sciences, Toxicology.

Critical Reviews in Biochemistry and Molecular Biology

CRC Inc, 2000 NW 24th Street,
Boca Raton, FL-33431, USA

6 issues per year *ISSN: 1040-9238*
(English)

Includes several critical surveys of specific topics of current interest.

Critical Reviews in Biotechnology

CRC Inc, 2000 NW 24th Street,
Boca Raton, FL-33431, USA

6 issues per year
(English)

Presents in-depth review articles covering a broad spectrum of technologies involving the use of living organisms and biological processes in industrial, agricultural and medical applications.

Critical Reviews in Eukaryotic Gene Expression

CRC Inc, 2000 NW 24th Street,
Boca Raton, FL-33431, USA

4 issues per year *ISSN: 1045-4403*
(English)

Assimilates and presents timely concepts and experimental approaches that are contributing to understanding of gene regulation, organization and structure. Has relevance to basic scientists, clinical investigators and other applied scientists working in the biotechnology field.

Critical Reviews in Food Science and Nutrition

CRC Inc, 2000 NW 24th Street,
Boca Raton, FL-33431, USA

8 issues per year *ISSN: 1040-8398*
(English)

Presents critical viewpoints of current technology, food science and human nutrition. Addresses the application of scientific discoveries and acquisition of knowledge related to nutrition.

Critical Reviews in Immunology

CRC Inc, 2000 NW 24th Street,
Boca Raton, FL-33431, USA

4 issues per year *ISSN: 1040-8401*
(English)

Presents an overview of contemporary immunology, bringing together molecular immunology and immunobiology. Encompasses developments ranging from the highly clinical to the highly molecular, and provides insight into the cellular events of the immune response for a better understanding of the molecular mechanisms of immunological disorders.

Critical Reviews in Microbiology

CRC Inc, 2000 NW 24th Street,
Boca Raton, FL-33431, USA

4 issues per year *ISSN: 1040-841X*
(English)

Gives reviews encompassing the many diverse fields of microbiology: viruses, rickettsiae, mycoplasmas, bacteria, fungi, algae, and protozoa.

Critical Reviews in Plant Sciences

CRC Inc, 2000 NW 24th Street,
Boca Raton, FL-33431, USA

4 issues per year
(English)

Presents reviews of timely subjects in areas such as shortages in world food supply, energy resources, the role of research in increasing crop productivity; also basic research areas in photosynthesis, nitrogen fixation, gene regulation and genetic engineering.

Crop Protection

Butterworth Scientific Ltd, Linacre House, Jordan Hill, Oxford, Oxon, OX2 8DG, UK
Telephone: 865 310366
Fax: 865 310898

4 issues per year *ISSN: 0261-2194*
(English)

Covers all practical aspects of pest, disease and weed control, including the development of pesticides, pollution control and effects of plant growth regulators.

Crop Science

Crop Science Society of America Inc, 677 South Segoe Road, Madison, WI-53711, USA
Telephone: 608 273 8080
Fax: 608 273 2021

6 issues per year *ISSN: 0011-183X*
(English)

Contains original reports and reviews covering all aspects of agricultural crops, with the emphasis on the US. Includes aspects of plant genetics, cytology, plant breeding, physiology, ecology, weed science, crop quality and utilization of crop products.

Current Advances in Biochemistry

Pergamon Press Inc,
395 Saw Mill River Road, Elmsford, NY-10523, USA
Telephone: 914 592 7700
Fax: 914 592 3625

12 issues per year *ISSN: 0741-6118*
(English)

Current Advances in Cell and Developmental Biology

Pergamon Press Inc,
395 Saw Mill River Road, Elmsford, NY-10523, USA
Telephone: 914 592 7700
Fax: 914 592 3625

12 issues per year *ISSN: 0741-1626*
(English)

Current Advances in Immunology

Pergamon Press Inc,
395 Saw Mill River Road, Elmsford, NY-10523, USA
Telephone: 914 592 7700
Fax: 914 592 3625

12 issues per year *ISSN: 0741-1650*
(English)

Current Advances in Microbiology

Pergamon Press Inc,
395 Saw Mill River Road, Elmsford, NY-10523, USA
Telephone: 914 592 7700
Fax: 914 592 3625

12 issues per year *ISSN: 0741-1669*
(English)

Current Advances In Neuroscience

Pergamon Press Inc,
395 Saw Mill River Road, Elmsford, NY-10523, USA
Telephone: 914 592 7700
Fax: 914 592 3625

12 issues per year *ISSN: 0741-1677*
(English)

Current Biology

Current Biology Ltd, 34-42 Cleveland Street, London, W1P 5FB, UK

6 issues per year *ISSN: 0960-9822*
(English)

Contains brief reviews of developments in biological research.

Current Genetics

Springer-Verlag, Heidelberger Platz 3, Berlin 33, D-1000, Germany
Telephone: 30 82 07 1

6 issues per year *ISSN: 0172-8083*
(English)

Rapid publication of original articles of immediate importance on genetics of eukaryotes.

Current Methods in Molecular Biology

John Wiley & Sons Inc, Baffins Lane, Chichester, West Sussex, PO19 1UD, UK
Telephone: 243 770328
Fax: 243 775878

(English)

Publishes original articles describing protocols and strains related to the molecular and cellular biology of prokaryotes and eukaryotes as well as comments on established protocols, the quality of commercial reagents, improvements on existing methods, new products for molecular biology and reviews of advantages and disadvantages of current approaches.

Current Microbiology

Springer-Verlag, Heidelberger Platz 3, Berlin 33, D-1000, Germany
Telephone: 30 82 07 1

12 issues per year *ISSN: 0343-8651*
(English)

Covers all areas of microbiology, including general and applied, medical and nonmedical aspects.

Current Opinion in Biotechnology

Current Biology Ltd, Middlesex House, 34-42 Cleveland Street, London, W1P 5FB, UK
Telephone: 71 323 0323
Fax: 71 580 1938

(English)

Provides systematic coverage of all the latest advances, reviews from acknowledged world authorities, highlights of key papers and a comprehensive bibliographic listings of papers and patents. Other titles in the same series include: Current Opinion in Cell Biology; Current Opinion in Genetics & Development; Current Opinion in Immunology; Current Opinion in Genetics & Development; Current Opinion in Neurobiology. The series aims to provide a comprehensive and critical appraisal of all new developments in modern biology as they are reflected in the primary literature.

Current Opinion in Cell Biology

Current Biology Ltd, 34-42 Cleveland Street, London, W1P 5FB, UK
Telephone: 71 323 0323
Fax: 71 580 1938

6 issues per year *ISSN: 0955-0674*
(English)

Each year, the field of cell biology is divided into the following sections: Feb: Cytoplasm and cell motility; Apr: Cell multiplication, cell signalling; Jun: Nucleus and gene expression; Aug: Membranes, membrane permeability; Oct: Cell-to-cell contact, extracellular matrix; Dec: Cell differentiation, post-transcriptional processes.

Current Opinion in Genetics & Development

Current Biology Ltd, 34-42 Cleveland Street, London, W1P 5FB, UK
Telephone: 71 323 0323
Fax: 71 580 1938

6 issues per year *ISSN: 0959-437X*
(English)

Each year, the field of genetics and development is divided into the following sections: Feb: Viral genetics, Oncogenes and cell proliferation; Apr: Gene expression and differentiation; Jun: Gene mapping, Genetics of disease; Aug: Pattern formation and developmental mechanisms; Oct: Prokaryotes and their viruses, Lower eukaryotes; Dec: Gene organization and evolution.

Current Opinion in Immunology

Current Biology Ltd, 34-42 Cleveland Street, London, W1P 5FB, UK
Telephone: 71 323 0323
Fax: 71 580 1938

6 issues per year *ISSN: 0952-7915*
(English)

Each year, the field of immunology is divided into the following sections: Feb: Innate immunity, antigen recognition; Apr: Lymphocyte development, immunological techniques; Jun: Lymphocyte activation and effector functions; Aug: Immunity to infection, immunodeficiency; Oct: Reproduction, cancer, transplantation; Dec: Atopic allergy, autoimmunity.

Current Opinion in Neurobiology

Current Biology Ltd, 34-42 Cleveland Street, London, W1P 5FB, UK
Telephone: 71 323 0323
Fax: 71 580 1938

6 issues per year *ISSN: 0959-4388*
(English)

Each year, the field of neurobiology is divided into the following sections: Feb: Development; Apr: Cognitive neuroscience; Jun: Signalling mechanisms; Aug: Sensory systems; Oct: Neuronal and glial cell biology, disease, transplantation and regeneration; Dec: Neural control.

Current Opinion in Structural Biology

Current Biology Ltd, 34-42 Cleveland Street, London, W1P 5FB, UK
Telephone: 71 323 0323
Fax: 71 580 1938

6 issues per year *ISSN: 0959-440X*
(English)

Each year, the field of structural biology is divided into the following sections: Feb: Protein-nucleic acid interactions, folding and binding; Apr: Macromolecular assemblages, theory and simulation; Jun: Nucleic acids, sequences and topology; Aug: Lipids, membrane proteins, engineering and design; Oct: Carbohydrates and glycosylation, catalysis and regulation; Dec: Proteins, biophysical methods.

Cytobios

Faculty Press, 88 Regent Street, Cambridge, UK

ISSN: 0011-4529
(English, French or German)

International journal of cell biology, publishing original papers on cell organization, including behaviour, structure and function of plant and animal cells, extracellular products and subcellular organelles. Also contains book reviews and details of forthcoming conferences.

Cytokine

Saunders Scientific Publications,
WB Saunders Co, Curtis Center,
Independence Square West,
Philadelphia, PA-19106, USA
Telephone: 215 238 7800
Fax: 215 238 7883

(English)

International journal devoted exclusively to cytokines.

Cytometry

Wiley-Liss Inc, 41 East 11th Street,
New York, NY-10003, USA
Telephone: 212 475 7700

8 issues per year
(English)

John Wiley and Sons Ltd, Baffins Lane, Chichester, West Sussex,
PO19 1UD, UK
Telephone: 243 770328
Fax: 243 775878

Covers all aspects of analytical cytology in the form of reviews, articles, brief reports, letters to the editor, technical notes and tutorials.

Cytotechnology

Kluwer Academic Publishers,
Distribution Centre, PO Box 322,
AH Dordrecht, NL-3300,
The Netherlands
Telephone: 78 172 811

4 issues per year *ISSN: 0920-9069*
(English)

International Journal of Cell Culture and Biotechnology. Publishes reviews, original papers, short communications and technical reports on new and novel developments, methods and techniques in fundamental cell culture research and their applications in vertebrate and invertebrate in vitro cell culture.

Developmental Genetics

John Wiley, Baffins Lane, Chichester,
West Sussex, PO19 1UD, UK
Telephone: 243 770328
Fax: 243 775878

Bimonthly *ISSN: 0192-253X*
(English)

Devoted to biology dealing with gene regulation and expression during development and differentiation in biological systems. Emphasis is on molecular biochemical and physiological aspects of gene action and regulation during development.

Diagnostic Immunology

Wiley-Liss Inc, 41 East 11th Street,
New York, NY-10003, USA
Telephone: 212 475 7700

4 issues per year
(English)

Covers immunoserology, immunochemistry, cellular immunology and immunopathology, with papers on laboratory diagnostic techniques as well as case reports and methodological reports.

Diagnostic Microbiology and Infectious Disease

Elsevier Science Publishing Co Inc,
655 Avenue of the Americas,
New York, NY-10017, USA
Telephone: 212 989 5800
Fax: 212 633 3990

4 issues per year *ISSN: 0732-8893*
(English)

Publishes full papers, critical reviews, case studies, communications, editorial and book reviews on the latest developments in clinical microbiology and diagnosis and treatment of infectious disease.

DNA A Journal of Molecular Biology

Mary Ann Liebert Inc, 1651 Third Avenue, New York, NY-10128, USA
Telephone: 212 289 2300

4 issues per year
(English)

A journal of molecular biology, covering the applications of recombinant DNA technology to problems of gene structure and regulation of gene expression, cell biology, developmental biology, gene transfer and protein synthesis.

DNA Probes

Sheffield University, Biomedical Information Service - SUBIS, The University, Sheffield, S10 2TN, UK
Telephone: 742 768555
Fax: 742 768555

28 issues per annum *ISSN: 0266-6308*
(English)

Publishes information on probes for all types of organisms as well as details and procedures for restriction fragment length polymorphisms (RFLPs), DNA fingerprinting, polymerase chain reactions (PCR) and related bioanalytical techniques.

DNA and Protein Engineering Techniques

Alan R Liss Inc, 41 East 11th Street, New York, NY-10003, USA
Telephone: 212 475 7700

ISSN: 0894-7937
(English)

Serves as an outlet for the rapid exchange of information on the latest experimental methods in molecular genetics and protein engineering featuring step-by-step experimental protocols.

DNA Sequence

Harwood Academic Publishers, STBS, PO Box 197, London, WC2E 9PX, UK

6 issues per volume *ISSN: 1042-5179*
(English)

Publishes papers on DNA sequencing and mapping.

Ecotoxicology

Chapman & Hall, 2-6 Boundary Row, London, SE1 8HN, UK
Telephone: 71 865 0066
Fax: 71 522 9623

Quarterly *ISSN: 0963-9292*
(English)

International journal publishing fundamental research in ecotoxicology. Aimed at understanding the mechanisms and processes whereby chemicals exert their effects on ecosystems, and the impact of the chemicals at the population or community level. Includes technical notes and review articles.

Electroanalysis

VCH, 220 East 23rd Street Suite 909, New York, NY-10010, USA
Telephone: 6201 602 220
Fax: 6201 602 328

6 issues per year *ISSN: 1040-0397*
(English)

Reports on the development and applications of electroanalytical techniques in the biomedical, environmental, industrial, pharmaceutical and food fields.

Electrophoresis

VCH, 220 East 23rd Street Suite 909, New York, NY-10010, USA
Telephone: 6201 602 220
Fax: 6201 602 328

12 issues per year *ISSN: 0173-0835*
(Predominantly English)

The official journal of the Electrophoresis Society and published under the auspices of the Japanese Society of Electrophoresis. Presents innovative applications of electrophoretic methods for researchers working in biochemistry, clinical chemistry, genetics, immunology, microbiology, and food science. Publishes original papers, short communications, review articles, book reviews and important conference dates. Available on microfilm.

EMBO Journal

IRL Press, Pinkhill House, Southfield Road, Eynsham, Oxon, OX8 1JJ, UK
Telephone: 865 882 56767
Fax: 865 882 890

12 issues per year *ISSN: 0261-4189*
(English)

Provides international forum for rapid publication of research results in molecular biology and related areas. Published on behalf of the European Molecular Biology Organization.

Environmental Conservation

Elsevier Science Publishers,
PO Box 330, AH Amsterdam,
NL-1000, The Netherlands

4 issues per year *ISSN: 0376-8929*
(English)

Published for the Foundation for Environmental Conservation with the collaboration of the United Nations Environment Programme (UNEP), International Union for Conservation of Nature and Natural Resources (IUCN), International Association for Ecology (INTECOL), International Conferences on Environmental Future (ICEFs), World Council For the Biosphere-International Society for Environmental Education (WCB-ISEE), World Wide Fund for Nature (WWF). Includes specialist contributions on: enlightened environmental policy and management; imperatives of human population moderation; anti-pollution measures; avoidance of ecodisasters; land-use planning; minimum-waste technology; conservation of biotic populations and ecosystems; development and use of renewable resources.

Environmental and Molecular Mutagenesis

John Wiley, Baffins Lane, Chichester, West Sussex, PO19 1UD, UK
Telephone: 243 770328
Fax: 243 775878

8 issues per year *ISSN: 0893-6692*
(English)

Provides a forum for research on basic mechanisms of mutation, the detection of mutagens, and the implications of environmental mutagens for human health. Also covers all aspects of mutation research and genetic toxicology.

Environmental Pollution

Elsevier,
Contact: JP Dempster
Crown House, Linton Road, Barking, Essex, IG11 8JU, UK
Telephone: 81 594 7272

24 issues per year *ISSN: 0269-7491*
(English)

International journal covering the biological, chemical and physical aspects of environmental pollution and pollution control, publishing research and review articles on the distribution and ecological effects of environmental pollutants and on new techniques for their study and measurement.

Environmental Science & Technology

American Chemical Society, 1155 Sixteenth St NW, Washington, DC-20036, USA
Telephone: 202 872 4363
Fax: 202 872 4615

Monthly *ISSN: 0013 936X*
(English)

Provides news, views and features on environmental science.

Environmental Toxicology and Water Quality

John Wiley and Sons Inc, 605 Third Avenue, New York, NY-10158, USA
Telephone: 212 850 6000

Quarterly *ISSN: 0884-8181*
(English)

Publishes original research papers on all aspects of environmental toxicology and water quality research, monitoring, criteria/standards and policies.

Enzyme

S Karger AG, Allschwilerstrass 10, PO Box, Basel, CH-4009, Switzerland
Telephone: 61 306 1111
Fax: 61 306 1234

Quarterly *ISSN: 0013-9432*

Publishes papers on metabolism and experimental enzymology. Official journal of the International Society for Clinical Enzymology.

Associate Editor: TA MacMurray, HJ Heinz Co, USA.

Enzyme and Fermentation Biotechnology

Academic Press Inc, Journals Division, 1250 Sixth Avenue, San Diego, CA-92101, USA
Telephone: 619 230 1840
Fax: 619 699 6859

(English)

Official journal of the International Society of Cell Biology.

Enzyme and Microbial Technology: Biotechnology Research and Reviews

Butterworth-Heinemann Ltd, Linacre House, Jordan Hill, Oxford, OX2 8DG, UK
Telephone: 865 310366
Fax: 865 310898

12 issues per year *ISSN: 0141-0229*
(English)

Publishes original papers and review articles relevant to the use of enzymes and microorganisms. Also includes literature survey of molecular immobilization and bioaffinity phenomena, patent reports and news of meetings.

UK Editor: Raymond E. Spier, University of Surrey, UK.

Euphytica

Kluwer Academic Publishers, Distribution Centre, PO Box 322, AH Dordrecht, NL-3300, The Netherlands
Telephone: 78 172 811

3 issues per year *ISSN: 0014-2336*
(English)

International Journal of Plant Breeding. Publishes original research, critical reviews and short communications on aspects related to plant breeding research.

European Journal of Biochemistry

Springer-Verlag, Heidelberger Platz 3, Berlin 33, D-1000, Germany
Telephone: 30 82 07 1

24 issues per year *ISSN: 0014-2956*
(English)

Publishes original papers on fundamental aspects of biochemistry, molecular biology and cell biology, nucleic acids, protein synthesis, molecular genetics, enzymology, protein chemistry, natural products, membranes and bioenergetics, immunology and developmental biochemistry. Produced by the Federation of European Biochemical Societies. Available as microfilm on 6mm, 35mm and microfiche.

European Journal of Cell Biology

Wissenschefliche Verlagsgesellschaft mbH, Birkenwaldstrasse 44, Postfach 40, Stuttgart 1, D-7000, Germany
Telephone: 711 25 82 0

6 issues per year *ISSN: 0171-9335*
(English, French or German with English summaries)

Cosponsored by German Society of Electron Microscopy and German Society of Cell Biology, which form the European Cell Biology Organization.

European Journal of Immunology

VCH, 220 East 23rd Street Suite 909, New York, NY-10010, USA
Telephone: 6201 602 220
Fax: 6201 602 328

12 issues per year *ISSN: 0014-2980*
(English)

Publishes papers on immunochemistry, immunobiology, immunopathology, immunogenetics and clinical immunology, as well as on methods and techniques of experimental

The European Journal of Neuroscience

Oxford University Press, Walton Street, Oxford, OX2 6DP, UK
Telephone: 865 56767

Monthly *ISSN: 0953 816X*
(English)

Published on behalf of the European Neuroscience Association.

Evolutionary Ecology

Chapman & Hall, 2-6 Boundary Row, London, SE1 8HN, UK
Telephone: 71 865 0066
Fax: 71 522 9623

Bimonthly *ISSN: 0269-7653*
(English)

Publishes papers on subjects including genetic, behavioural, palaeobiological, physiological, population-level and community studies, covering all types of living organisms.

Experimental Cell Research

Academic Press Inc, Journals Division, 1250 Sixth Avenue, San Diego, CA-92101, USA
Telephone: 619 230 1840
Fax: 619 699 6859

14 issues per year *ISSN: 0014-4827*
(English)

Publishes original papers on physical and chemical aspects of cellular and intracellular structure, biosynthesis with reference to cell growth, reproduction and differentiation, mechanisms of cell division and cell cycles, membrane functions, function of subcellular particles, cell energetics and regulation of cell function. Includes a special article or review paper in each issue.

Experimental and Clinical Immunogenetics

Karger, Allschwilerstrass 10, PO Box, Basel, CH-4009, Switzerland
Telephone: 61 306 1111
Fax: 61 306 1234

Quarterly *ISSN: 0254-9670*
(English)

Explores all aspects of human and animal immunogenetics including histocompatibility, blood group relevance, the evolution and ontogeny of the immunoglobulin superfamily, and the genetics of the immune response.

Farm Chemicals International

Meister Publishing Company, 37733 Euclid Avenue, Willoughby, OH-44094, USA
Telephone: 216 942 2000

Quarterly *ISSN: 0092-0053*
(English)

Publishes reports and articles serving manufacturers, formulators, mixers, dealers and commercial applicators of fertilizers and pesticides.

The FASEB Journal

Federation of American Societies for, Experimental Biology, 9650 Rockville Pike, Bethesda, MD-20814, USA
Telephone: 301 530 7027
Fax: 301 571 1855

15 issues per year *ISSN: 0892-6638*
(English)

Publishes scientific papers and reports and abstracts from meetings of the Federation of American Societies for Experimental Biology.

FEBS Letters

Elsevier Science Publishers, PO Box 330, AH Amsterdam, NL-1000, The Netherlands

28 issues per year *ISSN: 0014-5793*
(English, French or German with English summaries)

Rapidly publishes short reports in biochemistry, biophysics and molecular biology of significant importance to justify this approach. Includes protein chemistry, enzymology, nucleic acids, biochemical genetics, immunochemistry and bioenergetics. Official journal of Federation of European Biochemical Societies.

FEMS Microbiology Ecology

Elsevier Biomedical Press, Postbus 211, NE Amsterdam, NL-1000, The Netherlands

8 issues per year *ISSN: 0168-6496*
(English)

FEMS Microbiology Letters

Elsevier Biomedical Press, Postbus 211, AE Amsterdam, NL-1000, The Netherlands

25 issues per year *ISSN: 0378-1097*
(English)

Rapidly publishes short research reports on all aspects of microbiology. Published on behalf of the Federation of European Microbiology Societies.

Fertilizer Research

Kluwer Academic Publishers Group, Distribution Centre, PO Box 322, AH Dordrecht, NL-3300, The Netherlands
Telephone: 78 172 811

6 issues per year *ISSN: 0167-1731*
(English)

Provides a broad coverage of all aspects of plant fertilizers and their use, including soils, soil fertility, soil chemistry and crop and animal production as affected by fertilizers. Studies are concerned with agricultural, horticultural and plantation crops.

Flavour and Fragrance Journal

John Wiley and Sons Ltd, Baffins Lane, Chichester, PO19 1UD, UK
Telephone: 234 784531

5 issues per year *ISSN: 0882-5734*
(English)

Publishes scientific and technical papers on essential oils and related products as well as aspects of flavours, colours of foodstuffs.

Folia Microbiologica

Academia, Academy of Science, Vodickova 40, 112 20 Prague 1, Czechoslovakia

6 issues per year *ISSN: 0015-5632*
(English)

Contains original papers, reviews and brief reports. Sponsored by the Czechoslovakian Microbiological Society.

Food and Chemical Toxicology

Pergamon Press Inc,
395 Saw Mill River Road, Elmsford, NY-10523, USA
Telephone: 914 592 7700
Fax: 914 592 3625

12 issues per year *ISSN: 0278-6915*
(English)

Food Chemistry

Elsevier Applied Science Publishers Ltd, Crown House, Linton Road, Barking, Essex, IG11 8JU, UK
Telephone: 81 594 7272

ISSN: 0308-8146
(English)

Concerned with the chemistry and biochemistry of foods and their sensory and nutritional properties.

Food Control

Butterworth-Heinemann Ltd,
Linacre House, Jordan Hill, Oxford, OX2 8DG, UK
Telephone: 865 310366
Fax: 865 310898

Quarterly
(English)

Covers current research, technology, legislation and good practices in the food industry. Carries original research papers, notes and comments on new developments, a literature survey, patent reports, a calendar of forthcoming meetings, book reviews and reports of recent conferences.

Associate Editor: TA MacMurray, HJ Heinz Company, USA.

Food Microbiology

Academic Press Inc (London) Ltd, 24-28 Oval Road, London, NW1 7DX, UK
Telephone: 71 267 4466

4 issues per year *ISSN: 0740-0020*
(English)

Research areas include: application of chemical and physical approaches to food microbiology; biotechnological aspects of established processes; food spoilage and safety; predictive microbiology; production of fermented foods; rapid methodology; use of novel microbial processes to produce flavours and food-related enzymes.

Food Research Quarterly

CSIRO, 314 Albert Street, East Melbourne, Victoria 3002, Australia

Quarterly *ISSN: 0310-9070*
(English)

Publishes papers on food research.

Food Sciences and Nutrition

Macmillan Press, Houndmills,
Basingstoke, Hampshire,
RG21 2XS, UK

Quarterly *ISSN: 0963-7486*
(English)

Publishes research articles and critical reviews in the fields of food sciences and nutrition, with special emphasis on the emerging interface between the two sciences.

Food Science and Technology Today

Institute of Food Science and
Technology, 5 Cambridge Court,
210 Shepherds Bush, Road, London,
W6 7NL, UK
Telephone: 71 603 6316

Quarterly *ISSN: 0950-9623*
(English)

Journal of the Institute of Food Science and Technology (UK) containing articles of interest to food scientists and technologists in manufacturing, retailing, research establishments, educational establishments, consultants and government.

Gene

Elsevier Science Publishers,
PO Box 330, AH Amsterdam,
NL-1000, The Netherlands

13 volumes per year *ISSN: 0378-1119*
(English)

International journal devoted to gene cloning and recombinant nucleic acids. Covers purification techniques, analysis, genetic recombination mechanisms, reconstruction and characterization of complex genomes, applications and future developments.

General and Comparative Endocrinology

Academic Press Inc (London) Ltd,
24-28 Oval Road, London,
NW1 7DX, UK
Telephone: 71 267 4466

12 issues per year *ISSN: 0016-6480*
(English)

Publishes papers on fundamental research on cellular mechanisms of hormone action as well as functional, developmental and evolutionary aspects of vertebrate and invertebrate endocrine systems.

Genes, Chromosomes and Cancer

John Wiley and Sons Ltd,
Baffins Lane, Chichester, West Sussex,
PO19 1UD, UK
Telephone: 243 770328
Fax: 243 775878

8 issues per year *ISSN: 1045-2257*
(English)

Covers all aspects of genomic of abnormalities related to meoplasia. Emphasis is placed on work combining molecular and cytogenetic analyses of the acquired genetic alterations of cancer cells. Areas covered include: chromosomal abnormalities in all benign and malignant disorders, including changes that occur during neoplastic progression.

Genes and Development

Cold Spring Harbor Laboratory Press,
POB 100, 1 Bungtown Road, Cold
Spring Harbor, NY-11724-2203, USA
Telephone: 516 367 8325
Fax: 516 367 8432

Monthly *ISSN: 0891-9369*
(English)

Published in association with the Genetic Society of Great Britain. Publishes original articles and papers.

The Genetic Engineer and Biotechnologist

GB Biotechnology Ltd, 4 Beaconsfield Court, Sketty, Swansea, West Glamorgan, SA2 9JU, UK
Telephone: 792 208190
Fax: 792 297402

24 issues per month *ISSN: 0959-020X*
(English)

Journal/magazine incorporating the former International Industrial Biotechnology. Reports on all areas of biotechnology from business and finance to research and development. Includes BioBusiness which covers patents, international news, details of biotechnology in different countries, etc and Biopapers Journal with mainly original communications as well as reviews on special areas of biotechnology. There is also a supplement, European Biopatent Watch, which surveys, classifies and comments on current patents.

Genetic Engineering

Academic Press Inc, Journals Division, 24-28 Oval Road, 1250 Sixth Avenue, San Diego, CA-92101, USA
Telephone: 619 230 1840
Fax: 619 699 6859

1 volume per year *ISSN: 0196-3716*
(English)

Publishes review articles on recombinant DNA techniques, including preparation of genomic libraries, host/vector systems, recombinant gene expression and new techniques.

Genetica

Kluwer Academic Publishers, Postbus 17, AA Dordrecht, NL-3300, The Netherlands
Telephone: 78 334911
Fax: 78 334254

9 issues per year *ISSN: 0016-6707*
(English, French or German with English summaries)

Publishes original papers in cytogenetics, cytotaxonomics, ecological and population genetics.

Genetical Research

Cambridge University Press, Edinburgh Building, Shaftsbury Road, Cambridge, CB2 2RU, UK
Telephone: 223 312393
Fax: 223 315052

6 issues per year. *ISSN: 0016-6723*
(English)

Publishes original papers on all aspects of genetics or any field of research which has an important bearing on genetics as well as analysing the implications of such findings.

Genetics

Genetics Society of America, 9650 Rockville Pike, Bethesda, MD-21814, USA

12 issues per year *ISSN: 0016-6731*
(English)

Provides a periodical record of investigations into hereditary variations. Also provides details of situations vacant. Official journal of the Genetics Society of America.

Genome

Genetics Society of Canada, Research Journals, National Research Council, Ottawa, K1A OR6, Canada
Telephone: 613 993 9084
Fax: 613 952 7656

ISSN: 0831-2796
(English)

Genomics

Academic Press Inc, Journals Division, 1250 Sixth Avenue, San Diego, CA-92101, USA
Telephone: 619 230 1840
Fax: 619 699 6859

8 issues per year *ISSN: 0888-7543*
(English)

International Journal of Gene Mapping and Nucleotide Sequencing Emphasizing Analyses of the Human and Other Complex Genomes.

Geoderma

Elsevier, Crown House, Linton Road, Barking, Essex, IG11 8JU, UK
Telephone: 81 594 7272

Monthly *ISSN: 0016-7061*
(English)

Publishes papers on soil research.

Geomicrobiology Journal

Taylor & Francis Ltd, Rankine Road, Basingstoke, Hants, RG24 0PR, UK
Telephone: 256 840366
Fax: 256 479438

Quarterly *ISSN: 0149-0451*
(English)

Publishes research and review articles on microbial transformations of materials comprising the earth's crust and their geologic impact as well as applied topics including fossil fuel microbiology, bioleaching or ores, geomicrobiological prospecting and ground water microbiology.

Glycoconjugate Journal

Chapman & Hall, 2-6 Boundary Row, London, SE1 8HN, UK
Telephone: 71 865 0066
Fax: 71 522 9623

Bimonthly *ISSN: 0282-0080*
(English)

Official journal of the International Glycoconjugate Organization. Publishes papers and reviews on all aspects of glycoconjugate research and glycobiology. Emphasis is on the metabolism and function of glycolipids, glycoproteins, oligo- and polysaccharides and proteoglycans. Also publishes relevant information on the structure and synthesis of glycoconjugates and developments in methodology.

Growth Factors

Harwood Academic Publishers, STBS, PO Box 197, London, WC2E 9PX, UK

4 issues per volume *ISSN: 0897-7194*
(English)

The Histochemical Journal

Chapman & Hall, 2-6 Boundary Row, London, SE1 8HN, UK
Telephone: 71 865 0066
Fax: 71 522 9623

Monthly *ISSN: 0018-2214*
(English)

Publishes international papers, reviews and rapid communications in histochemistry and cytochemistry. Covers areas including biochemistry, histochemistry, pathology, immunocytochemistry and cell biology.

Histochemistry

Springer Verlag, Heidelberger Platz, Berlin 33, D-1000, Germany
Telephone: 30 8207-1

(English)

Publishes original papers in all areas of cytochemistry and histochemistry, especially those relating to theory and practice.

Human Genetics

Springer-Verlag, Heidelberger Platz 3,
Berlin 33, D-1000, Germany
Telephone: 30 82 07 1

6 issues per year *ISSN: 0340-6717*
(English)

Human Molecular Genetics

IRL, PO Box 1, Eynsham, Oxon,
OX8 1JJ, UK
Telephone: 865 882 2873

9 issues per year *ISSN: 0964-6906*
(English)

Publishes original papers and articles in the field of human molecular genetics.

Hybridoma

Mary Ann Liebert Inc, 1651 Third
Avenue, New York, NY-10128, USA
Telephone: 212 289 2300

4 issues per year *ISSN: 0272-457X*
(English)

Publishes papers on monoclonal antibody-defined antigens as well as their use in diagnostics and therapy. Particularly important for papers on hybridoma research, including T-cell hybridomas.

Immunogenetics

Springer-Verlag, Heidelberger Platz 3,
Berlin 33, D-1000, Germany
Telephone: 30 82 07 1

12 issues per year *ISSN: 0093-7711*
(English)

Covers the immunogenetics of cell interactions and of tissue differentiation and development, phylogeny of alloantigens and of the immune response, genetic control of immune response and disease susceptibility and genetics and biochemistry of alloantigens.

Immunology

Blackwell Scientific Publications Ltd,
Osney Mead, Oxford, Oxon,
OX2 0EL, UK
Telephone: 865 240201

12 issues per year *ISSN: 0019-2805*
(English)

Contains papers on cellular immunology, immunochemistry, allergy, transplant immunology, cancer and chemical aspects. Published for the British Society for Immunology.

Immunology Letters

Elsevier Biomedical Press,
Postbus 211, AE Amsterdam,
NL-1000, The Netherlands

12 issues per year *ISSN: 0165-2478*
(English)

An international journal providing rapid publication short reports in immunology.

Immunology Today

Elsevier Publications, 68 Hills Road,
Cambridge, CB2 1LA, UK
Telephone: 223 315961
Fax: 223 321410

12 issues per year *ISSN: 0167-4919*
(English)

Publishes articles and comments on developments in all aspects of basic and applied immunology as well as book reviews, highlights from workshops, details of conferences and discussions on techniques.

IMPACT AgBioIndustry

CAB International, Wallingford,
Oxfordshire, OX10 8DE, UK
Telephone: 491 32111

4 issues per year *ISSN: 0964-069X*
(English)

Aimed specifically at business men and women responsible for planning, making and implementing policy including those responsible for business development, marketing, R&D and registration. It focuses on new commercial opportunities hidden in the thousands of research papers published annually.

In Vitro

Tissue Culture Association Inc, 19110 Mongomary Village Avenue 300, Gaithersburg, MD-20879, UK
Telephone: 301 869 2900

13 issues per year *ISSN: 0883-8364*
(English)

Covers *in vitro* cultivation of cells, tissues, organs and tumours from multicellular animals and plants.

Indian Journal of Genetics and Plant Breeding

Indian Agricultural Research Institute, New Delhi, 110012, India

3 issues per year *ISSN: 0019-5200*

Publishes original papers and review articles.

Indian Journal of Microbiology

Antibiotics Research Centre, Pimpri, Poona 411018, India

4 issues per year

Publishes original papers on medical, veterinary, agricultural, environmental and economic microbiology. Official journal of the Association of Microbiologists of India.

Industrial Crops and Products

Elsevier, USDA-ARS, US Water Conservation Lab, 4331 East Broadway Road, Phoenix, AZ-85040, USA

4 issues per year *ISSN: 0926-6690*
(English)

Outside Americas, contact:
Agrotechnological Research Institute (ATO-DLO), PO Box 17, AA Wageningen, NL-6700, The Netherlands

Publishes articles covering all aspects of industrial crops and products.

Infection and Immunity

American Society for Microbiology, 1325 Massachusetts Avenue NW, Washington, DC-20005, USA
Telephone: 202 737 3600

12 issues per year *ISSN: 0019-9567*
(English)

Publishes original papers covering infections caused by pathogenic microorganisms, virulence factors, such as toxins and microbial surface structures, host susceptibility and resistance to infection and immunology of infection.

Insect Biochemistry and Molecular Biology

Pergamon Press Inc, 395 Saw Mill River Road, Elmsford, NY-10523, USA
Telephone: 914 592 7700
Fax: 914 592 3625

8 issues per year *ISSN: 0965-1748*
(English)

Formerly Insect Biochemistry. Publishes original contributions and mini-reviews in the fields of insect biochemistry and insect molecular biology.

Interferon Y Biotecnologia

Palacio de las Convenciones, Calle 146 entre 11 y 13, Playa, Apartado 16046, La Habana, Cuba

Publishes reviews, original papers, letters to the Editor, details of techniques, research in Latin American centres, comments on events, interviews, etc, as well as details of conferences, symposia and congresses covering most aspects of the new biotechnology as related to human and animal disease in particular.

Interferons

Academic Press Inc, Journals Division, 1250 Sixth Avenue, San Diego, CA-92101, USA
Telephone: 619 230 1840
Fax: 619 699 6859

(English)

Intern Zeitschrift fur Lebensmittle Technologie und Verfahrentechnik

Fachverlag Oskar Ohler, Stockackerstrasse 6, Coburg, D-8630, Germany
Telephone: 9561 66 122

4 issues per year
(German with some short English abstracts)

Publishes technical articles covering food processing as well as new products and technical innovations. Contains commercial advertisements for relevant equipment and food ingredients and lists suppliers and distributors.

International Biodeterioration

Elsevier Applied Science Publishers Ltd, Crown House, Linton Road, Barking, Essex, IG11 8JU, UK
Telephone: 81 594 7272

6 issues per year *ISSN: 0263-3036*
(English)

Publishes original research papers and reviews on the biological causes of deterioration or advantageous biological upgrading of all types of materials.

International Biotechnology Laboratory

International Scientific Communications Inc, 30 Controls Drive, PO Box 870, Shelton, CT-06484-0870, USA
Telephone: 203 926 9300

4 issues per year *ISSN: 0888-7225*
(English)

Provides information on commercial developments in biotechnology in relation to laboratory aspects and instrumentation in particular. Includes product information and commercial advertisements.

International Immunology

Oxford Journals, Walton Street, Oxford, Oxon, OX2 6DP, UK
Telephone: 865 791373

Monthly *ISSN: 0953-8178*
(English)

International journal publishing a broad range of experimental and theoretical studies in molecular and cellular immunology conducted in laboratories throughout the world.

International Journal of Biochemistry

Pergamon Press Inc, 395 Saw Mill River Road, Elmsford, NY-10523, USA
Telephone: 914 592 7700
Fax: 914 592 3625

12 issues per year *ISSN: 0020-711X*
(English)

Contains general papers, reviews and minireviews on theoretical, plant, bacterial, animal and fungal biochemistry, with applied clinical aspects.

International Journal of Bioethics

Editions Alexandre Lacassagne,
162 avenue Lacassagne, Lyon,
F-69003, France
Telephone: 72 33 40 40
Fax: 72 34 16 74

Quarterly *ISSN: 1145-0762*
(English and French)

Publishes articles on the ethics of biotechnology. Articles predominantly discuss attitudes of churches, the public and governments to human genetic research. A news section covers topics including abortion, contraception, organ transplants.

International Journal of Biological Macromolecules

Butterworth-Heinemann Ltd,
Linacre House, Jordan Hill, Oxford,
OX2 8DG, UK
Telephone: 865 310366
Fax: 865 310898

Bimonthly
(English)

Structure, Function and Interactions. Publishes latest findings from studies of the molecular structure of proteins, carbohydrates, nucleic acids, viruses and membranes. The scope includes conformational studies, molecular associations, interactions and functional properties along with related model systems, new techniques and theoretical developments.

US Associate Editor: Ian Wilson, Scripps Clinic and Research Foundation, California, USA.

International Journal of Clinical Monitoring and Computing

Martinus Nijhoff Publishers,
Postbus 211, AE Amsterdam,
NL-1000, The Netherlands

4 issues per year *ISSN: 0167-9945*
(English)

Publishes original papers and communications concerning clinical monitoring and computing as well as reviews and letters to the Editor.

International Journal of Food Microbiology

Elsevier Science Publishers,
PO Box 330, AH Amsterdam,
NL-1000, The Netherlands

ISSN: 0168-1605
(English)

Published full-length original research papers, short communications, review articles and book reviews covering the microbiological aspects of the safety, quality and acceptability of foods. Official journal of the International Union of Microbiological Societies and the International Committee on Food Microbiology and Hygiene.

International Journal of Insect Morphology and Embryology

Pergamon Press Inc,
395 Saw Mill River Road, Elmsford,
NY-10523, USA
Telephone: 914 592 7700
Fax: 914 592 3625

6 issues per year
(English)

International Journal of Peptide and Protein Research

Munksgaard International Publishers,
35 Norre Sogade, Copenhagen K,
DK-1370, Denmark
Telephone: 33 127030
Fax: 33 129387

12 issues per year *ISSN: 0367-8377*
(English)

Publishes original papers, short communications and letters to the

Editor on the development of peptide and protein research.

International Journal of Radiation Biology

Taylor & Francis Ltd, Rankine Road, Basingstoke, Hants, RG24 0PR, UK
Telephone: 256 840366
Fax: 256 479438

12 issues per year *ISSN: 0020-7616*
(English)

Includes related studies in physics, chemistry and medicine.

International Research Communications Systems

IRCS, St Leonards, St Leonardsgate, Lancs, UK

(English)

Publishes 32 specialist journals covering the whole field of medical and biomedical science, available in electronic or paper form. Titles include Biochemistry, Cell and Membrane Biology, and Immunology and Allergy. A member of Elsevier Science Publishers.

International Reviews of Cytology

Harwood Academic Publishers, STBS, PO Box 197, London, WC2E 9PX, UK

(English)

International Reviews of Immunology

Harwood Academic Publishers, STBS, PO Box 197, London, WC2E 9PX, UK

8 issues per year *ISSN: 0883-0185*
(English)

Each issue is devoted to a single topic, exploring the core areas and boundaries of the field using recent research from laboratories and research centres.

Journal of Agricultural Engineering Research

Academic Press, Journals Division, 1250 Sixth Avenue, San Diego, CA-92101, USA
Telephone: 619 230 1840
Fax: 619 699 6859

Monthly *ISSN: 0021-8634*
(English)

Publishes original papers of agricultural engineering research.

Journal of Agricultural Science

Cambridge University Press, Edinburgh Building, Shaftsbury Road, Cambridge, CB2 2RU, UK
Telephone: 223 312393

6 issues per year *ISSN: 0021-8596*
(English)

Publishes original papers on all aspects of agricultural science.

Journal of the American Medical Association

American Medical Association, 525 North Dearborn Street, Chicago, IL-60610, USA
Telephone: 312 715 6000

52 issues per year
(English)

Publishes original contributions, preliminary and special communications, review articles, international abstracts and book reviews. Large number of full-page advertisements for US drug company products.

Journal of the American Veterinary Medical Association

American Veterinary Medical Association, 930 North Meacham Road, Schaumburg, IL-60196, USA
Telephone: 312 885 8070

24 issues per year
(English)

Publishes scientific articles, clinical reports, letters on aspects of animal disease and health care. Also news and reports on activities of the association, products descriptions, announcements of meetings as well as trade and classified advertisements.

Journal of Animal Ecology

Blackwell Scientific Publications,
Osney Mead, Oxford, Oxon,
OX2 0EL, UK
Telephone: 865 240201

3 issues per year *ISSN: 0021-8790*
(English)

Publishes papers dealing with aspects of animal ecology in the British Isles, such as population and migration studies, and also with the methodology of ecological research with particular reference to animals.

Journal of Antibiotics

Japan Antibiotics Research Association,
20-8 Kamiosaki-2-chome,
Shinagawa-ku, Tokyo 41, Japan

12 issues per year *ISSN: 0021-8820*
(English)

An international journal publishing research papers on antibiotics and other microbial products.

Journal of Antimicrobial Chemotherapy

Academic Press Inc (London) Ltd,
24-28 Oval Road, London,
NW1 7DX, UK
Telephone: 71 267 4466

12 issues per year *ISSN: 0305-7453*
(English)

Contains leading articles, correspondence and book reviews. Official journal of the British Society of Antimicrobial Chemotherapy.

Journal of Applied Bacteriology

Blackwell Scientific Publications Ltd,
Osney Mead, Oxford, Oxon,
OX2 0EL, UK
Telephone: 865 240201

12 issues per year *ISSN: 0021-8847*
(English)

Covers a wide range of topics with original papers and reviews. The official journal of the Society of Applied Bacteriology in UK.

Journal of Applied Toxicology

John Wiley and Sons Ltd, Baffins
Lane, Chichester, West Sussex,
PO19 1UD, UK
Telephone: 234 770328
Fax: 234 775878

Bimonthly *ISSN: 0260-437X*
(English)

Covers the toxic effect of chemicals in industry and the environment. Emphasis is placed on practical applications. Short reports and correspondence are included for rapid communication.

Journal of Bacteriology

American Society for Microbiology,
1325 Massachusetts Avenue NW,
Washington, DC-20005, USA
Telephone: 202 737 3600

12 issues per year *ISSN: 0021-9193*
(English)

Publishes original papers on basic aspects of genetic and molecular biology: plasmids, transposons, membranes, enzymes, physiology and metabolism.

Journal of Biochemical and Biophysical Methods

Elsevier Biomedical Press,
Postbus 211, AE Amsterdam,
NL-1000, The Netherlands

6 issues per year *ISSN: 0165-022X*
(English)

Publishes full papers describing nw methods or significant developments to existing ones. Also contains short notes and concise critical reviews.

Journal of Biochemistry

Japan Academic Societies Centre,
4-16 Yoyoi-2-chome, Bunkyo-ku,
Tokyo 113, Japan

12 issues per year *ISSN: 0021-924X*
(English)

Publishes regular papers and short communications on all biochemical topics. Official journal of the Japanese Biochemical Society.

Journal of Biological Chemistry

American Society of Biological
Chemists Inc, 9650 Rockville Pike,
Bethesda, MD-20814, USA
Telephone: 301 530 7150

24 issues per year *ISSN: 0021-9258*
(English)

Contains original publications concerned with carbohydrates, lipids and other natural products, enzymology, membranes and bioenergetics, nucleic acids, protein synthesis and molecular genetics, and protein chemistry and structure.

Journal of Biological Response Modifiers

Raven Press, 1140 Avenue of the
Americas, New York, NY-10036, USA

6 issues per year *ISSN: 0732-6580*
(English)

Contains original papers on the experimental, theoretical and clinical aspects of biological response modifiers.

Journal of Bioluminescence and Chemiluminescence

John Wiley and Sons Ltd,
Baffins Lane, Chichester, West Sussex,
PO19 1UD, UK
Telephone: 243 770328
Fax: 243 775878

4 issues per year *ISSN: 0884-3996*
(English)

Publishes original scientific papers, short communications, technical notes and review articles on fundamental and applied aspects of both chemical and biological light emitting reactions as well as news and details of forthcoming meetings, information on new products and book reviews. Areas of interest include instrumentation, applications, luminescent immunoassay, luminescence in microbiology and low level luminescence.

Journal of Biotechnology

Elsevier Science Publishers,
PO Box 330, AH Amsterdam,
NL-1000, The Netherlands

6 issues per year *ISSN: 0168-1656*
(English)

Provides a medium for the rapid publication of both scientific articles and short communications on most aspects of biotechnology. In particular the following areas are considered for publication: molecular biology: genetics; biochemical aspects of metabolism and its control in viral, microbial, plant and animal cell systems; process development and its control including computer applications, automation and the development of biosensors as applied to biosystems.

Journal of Cell Biology

The Rockefeller University Press,
1230 York Avenue, New York,
NY-10021, USA

12 issues per year *ISSN: 0021-9525*
(English)

Concerned with structure and function of cells, organelles and macromolecules. Publishes original papers arising from the applications of molecular, biophysical, genetic, immunological and ultrastructural techniques. Official journal of the American Society for Cell Biology.

Journal of Cell Science

The Company of Biologists Ltd,
Box 32, Commerce Way, Colchester,
Essex, CO2 8HP, UK
Telephone: 206 46351
Fax: 206 549331

6 issues per year *ISSN: 0021-9533*
(English)

Publishes original papers on structure and function of both plant and animal cells, microorganisms and viruses, and their extracellular products.

Journal of Cellular Biochemistry

John Wiley and Sons, 605 Third
Avenue, New York, NY-10158, USA
Telephone: 212 850 6000

16 issues per year *ISSN: 0730-2312*
(English)

John Wiley and Sons Ltd,
Baffins Lane, Chichester, West Sussex,
PO19 1UD, UK
Telephone: 243 770328
Fax: 243 775878

Publishes original papers on conditions, agents or differentiation states that influence structure, structure-function relationships or assembly mechanisms in cells, viruses or supramolecular structures.

Journal of Cellular Physiology

John Wiley and Sons, 605 Third
Avenue, New York, NY-10158, USA
Telephone: 212 850 6000

Monthly *ISSN: 0021-9541*
(English)

Covers physiology and pathology at the cellular level, the biochemical and biophysical mechanisms concerned in the regulation of cellular growth, differentiation and function.

Journal of Chemical Technology and Biotechnology

Elsevier Science Publishers,
PO Box 330, AH Amsterdam,
NL-1000, The Netherlands

12 issues per year *ISSN: 0268-2575*
(English)

Contains original publications concerning all aspects of chemical engineering, with a greater emphasis on those aspects related to biotechnology. Relevant areas for biotechnology include: fermentation and downstream processing; separation technology; process engineering; computer control; immobilised biocatalysts and bioreactors; biodegradation and bioconversion; microbial, algal, animal and plant cell biotechnology; diagnostics, recombinant DNA technology; biosensors; cell physiology; biochemistry; environmental and safety aspects. Published on behalf of the Society of Chemical Industry.

Journal of Chromatography

Elsevier Scientific Publishing Co,
PO Box 330, AH Amsterdam,
NL-1000, The Netherlands

57 issues per year *ISSN: 0021-9673*
(English, French or German with English summaries)

Publishes original papers on all chromatography, electrophoresis and related methods. Includes Biochemical Applications and Chromatographic Reviews.

Journal of Clinical Immunology

Plenum Press, 233 Spring Street, New York, NY-10013, USA
Telephone: 212 620 8000

4 issues per year *ISSN: 0271-9142*
(English)

Contains a 'special article', original papers, short communications, book reviews and abstracts of significant papers from other journals. Also contains advertisements for relevant products.

Journal of Clinical Microbiology

American Society for Microbiology, 1325 Massachusetts Avenue NW, Washington, DC-20005, USA
Telephone: 202 737 3600

12 issues per year *ISSN: 0095-1137*
(English)

Journal of Economic Entomology

Entomological Society of America, 9301 Annapolis Road, Lanham, MD-20706, USA

Bimonthly *ISSN: 0022-0493*
(English)

Publishes papers on: insecticide resistance and resistance management, physiology, toxicology and biochemistry, apiculture and social insects, biological and microbial control, ecology, veterinary entomology.

Journal of Environmental Management

Academic Press Inc, Journals Division, 1250 Sixth Avenue, San Diego, CA-92101, USA
Telephone: 619 230 1840
Fax: 619 699 6859

12 issues per year *ISSN: 0301 4797*
(English)

Publishes papers on all aspects of management and use of the environment. Tries to publish examples of the use of modern mathematical and computer techniques and encourages contributions from the developing countries in the Third World.

Journal of Environmental Pathology, Toxicology, and Oncology

CRC Inc, 2000 NW 24th Street, Boca Raton, FL-33431, USA

6 issues per year *ISSN: 0731-8898*
(English)

Provides an international forum for works that relate to toxicology and pathology. Publishes the refereed results of both preliminary and more exhaustive investigations into the harmful components within the environment, and the effects of these substances in humans and animals.

Journal of Enzyme Inhibition

Harwood Academic Publishers, STBS, PO Box 197, London, WC2E 9PX, UK

4 issues per volume *ISSN: 8755-5093*
(English)

Publishes papers on enzyme inhibition.

Journal of Experimental Biology

The Company of Biologists Ltd,
Cambridge Commercial Park,
140 Cowley Road, Cambridge,
CB4 4DL, UK
Telephone: 223 420482
Fax: 223 353980

7 issues per year *ISSN: 0022-0949*
(English)

Covers all aspects of comparative physiology, ranging from cellular neurophysiology to active transport and excretion. 7 issues per year including a review volume. Also produce two journals, Bioessays and Development plus one or two supplements (books) per year.

Journal of Experimental Botany

Oxford University Press, Walton
Street, Oxford, OX2 6DP, UK
Telephone: 865 56767
Fax: 865 882890

12 issues per year *ISSN: 0022-0957*
(English)

Publishes papers on functional anatomy, ultrastructure, plant biochemistry and biophysics of plants. An official publication of the Society for Experimental Biology and of the Federation of European Societies for Plant Physiology.

Journal of Experimental Marine Biology and Ecology

Elsevier, Institute for Environmental
Research, Plymouth, Devon, UK

30 issues per year *ISSN: 0022-0981*
(English)

Provides a forum for work in the biochemistry, physiology, behaviour, and genetics of marine plants and animals in relation to their ecology. Main emphasis on experimental work.

Journal of Experimental Pathology

Mary Ann Liebert Inc, 1651 Third
Avenue, New York, NY-10128, USA
Telephone: 212 289 2300

4 issues per year *6X(WSSTN0-8485*
(English)

Publishes original papers which report on ways in which modern biochemical, biophysical and physiological approaches provide insight into disease processes covering the cellular and molecular biology of disease and response to disease in particular.

Journal of Fermentation Technology

Society of Fermentation Technology,
Engineering, Osaka University,
2-1 Yamadaoka Suita, Osaka 565,
Japan

6 issues per year
(English)

Contains original papers on production of fermentation products.

Journal of General and Applied Microbiology

Microbiology Research Foundation,
c/o Japan Academic Societies Centre,
4-16 Yayoi-2-chome, Bunkyo-ku,
Tokyo, Japan

6 issues per year *ISSN: 0022-1260*
(English, French and German)

Contains original papers concerning general and applied microbiology.

Journal of General Microbiology

Society of General Microbiology,
14 Belgrave Square, London,
SE1X 8PS, UK
Telephone: 71 235 3681

12 issues per year *ISSN: 0022-1287*
(English)

Contains original papers on fundamental aspects of bacteria, viruses, microscopic fungi, protozoa and microscopic algae.

Journal of General Virology

Society for General Microbiology, Journal Sales, 62 London Road, Reading, Berkshire, RG1 5AS, UK

12 issues per year *ISSN: 0022-1317*
(English)

Publishes full papers and short communications giving details of original research on viruses that infect hosts of any kind. Also features review articles.

Journal of High Resolution Chromatography

Dr Alfred Huthig Verlag GmbH, Im Weiher 10, Heidelberg, D-6900, Germany
Telephone: 6221 489 0
Fax: 6221 489 279

12 issues per year
(English)

Deals with all aspects of chromatographic science. Places special emphasis on high resolution techniques and covers recent developments. Includes reviews, book reviews and conference reports.

Journal of Horticultural Science

East Malling Research Station, East Malling, Maidstone, Kent, ME19 6BJ, UK

4 issues per year *ISSN: 0022-1589*
(English)

Reports original research on fruit and other perennial crops, vegetables and flowers, covering both temperate and tropical species. Includes aspects of genetics, plant disease, breeding etc.

Journal of Immunoassay

Marcel Dekker Journals, 270 Madison Avenue, New York, NY-10016, USA
Telephone: 212 696 9000

4 issues per year *ISSN: 0197-1522*
(English)

Publishes original papers on immunoassay, receptor assay and ligand assay using tracers (radioisotopes, enzymes, fluor etc) or immunological markers of ligand-receptor interactions.

Journal of Immunogenetics

Blackwell Scientific Publications Ltd, Osney Mead, Oxford, Oxon, OX2 0EL, UK
Telephone: 865 240201

6 issues per year *ISSN: 0305-1811*
(English)

Covers immunogenetics of antigens in man and animals including proteins, histocompatibility, blood group and other cell surface antigens.

Journal of Immunological Methods

Elsevier Science Publishers, PO Box 330, AH Amsterdam, NL-1000, The Netherlands

ISSN: 0022-1759
(English)

Publishes original reports covering the following techniques: detection of antibodies or antigens, fractionation and purification of immunoglobulins, isolation of antigens and other substances with properties which relate to immunological processes, labelling of antigens and antibodies with radioactive and other markers, localizing antigens and antibodies in tissues, detecting, enumerating and fractionating immunocompetent cells, assays for cellular immunity and detection of cell-surface antigens and initiation of immune unresponsiveness. Also contains information on potentially valuable monoclonal antibodies.

Journal of Immunology

Marcel Dekker Journals, 270 Madison Avenue, New York, NY-10016, USA
Telephone: 212 696 9000

4 issues per year *ISSN: 0022-1767*
(English)

Covers the use of drugs and other agents in the modulation of the immune system.

Journal of Immunopharmacology

Marcel Dekker Journals, 270 Madison Avenue, New York, NY-10016, USA
Telephone: 212 696 9000

4 issues per year
(English)

Covers the use of drugs and other agents in the modulation of the immune system.

Journal of Industrial Microbiology

Elsevier Science Publishers,
PO Box 330, AH Amsterdam,
NL-1000, The Netherlands

6 issues per year *ISSN: 0169-4146*
(English)

Publishes original research letters, short communications, critical reviews and letters to the Editor in the fields of biotechnology, fermentation, environmental microbiology, biodegradation, biodeterioration, quality control and other aspects of industrial microbiology. Published on behalf of the Society for Industrial Microbiology.

Journal of Interferon Research

Mary Ann Liebert Inc, 1651 Third Avenue, New York, NY-10128, USA
Telephone: 212 289 2300

4 issues per year *ISSN: 0197-8357*
(English)

Publishes original papers on all aspects of interferon research.

Journal of Labelled Compounds and Radiopharmaceuticals

John Wiley and Sons Ltd,
Baffins Lane, Chichester, West Sussex,
PO19 1UD, UK
Telephone: 243 770328
Fax: 243 775878

12 issues per year
(English)

Journal of Leukocyte Biology

John Wiley and Sons, 605 Third Avenue, New York, NY-10158, USA
Telephone: 212 850 6000

Monthly *ISSN: 0741-5400*
(English)

Focuses on the origins, the developmental biology and the functions of granulocytes, lymphocytes and mononuclear phagocytes.

Journal of Medical Microbiology

Longman Group Ltd, Longman House,
Burnt Hill, Harlow, Essex,
CM20 2JE, UK
Telephone: 279 426721
Fax: 279 431059

4 issues per year *ISSN: 0022-2615*
(English)

Contains papers on microbiology, bacteriology, protozoology, virology, mycology and immunology as the subjects apply to human or veterinary medicine. Journal of the Pathological Society of Great Britain and Ireland.

Journal of Medical Virology

Alan R Liss Inc, 41 East 11th Street, New York, NY-10003, USA
Telephone: 212 475 7700

8 issues per year *ISSN: 0146-6615*
(English)

Publishes original papers and details of conferences and courses.

Journal of Membrane Science

Elsevier Science Publishers, PO Box 330, AH Amsterdam, NL-1000, The Netherlands

(English)

Publishes papers related to the structure and function of membranes including theory of membrane transport, membrane permeation, etc as well as applications in non-biological processes such as separation, biomedical packaging, membrane electrodes, membrane bound enzymes and biomembrane analogues.

Journal of Microbial Biotechnology

c/o Subodh Saxana, Harsh Naga, Kampur, 208012, India

2 issues per year

Covers fermentation, immobilized biocatalysts, genetic engineering and related aspects of biotechnology.

Journal of Microbiological Methods

Elsevier Science Publishers, PO Box 330, AH Amsterdam, NL-1000, The Netherlands

6 issues per year *ISSN: 0167-7012*
(English)

Publishes original papers and invited reviews in all microbiological methods except virology and immunology.

Journal of Molecular and Applied Genetics

Raven Press, 1140 Avenue of the Americas, New York, NY-10036, USA

12 issues per year
(English)

Publishes original papers on genetic techniques, especially recombinant DNA methods, and their use in the study of basic and applied problems in genetics, biochemistry, medicine and agriculture.

Journal of Molecular and Cellular Immunology

Springer-Verlag, Heidelberger Platz 3, Berlin 33, D-1000, Germany
Telephone: 30 82 07 1

6 issues per year
(English)

Publishes research articles of general interest in immunology.

Journal of Molecular Biology

Academic Press Inc, 1 Vincent Square, London, SW1P 2PN, UK
Telephone: 71 630 7881

24 issues per year *ISSN: 0022-8508*
(English)

Publishes papers, reviews and short letters on subcellular organization, structure and function of cells, molecular genetics, structure and replication of viruses, structure of proteins and nucleic acids and energy transfer. Also contains communications on results obtained using special techniques.

Journal of Molecular Catalysis

Elsevier Sequoia SA, PO Box 564,
1001 Lausanne 1, Switzerland
Telephone: 21 20 73 81
Fax: 21 235 444

ISSN: 0304-5102
(English)

Publishes papers on all aspects of molecular activation in homogenous and heterogenous catalysis divided into three sections including one specifically dealing with aspects of biochemical catalysis such as enzyme activity, binding between ligands and enzymes, enzyme analogues, photobiocatalysis, multienzyme systems, kinetics and catalytic aspects of photosynthesis.

Journal of Molecular Graphics

Butterwoth-Heinemann Ltd,
Linacre House, Jordan Hill, Oxford,
OX2 8DG, UK
Telephone: 865 310366
Fax: 865 310898

Quarterly *ISSN: 0263-7855*
(English)

Update of the latest developments in computer-aided molecular design. Provides detailed appraisal of the techniques and applications of computer graphics to the investigation of molecular structure. Contains original research papers, short communications, reports from conferences, book reviews, news, a calendar of events and colour illustrations.
US Associate Editor: Vivian Cody, Med Foundation of Buffalo Inc, USA.

Journal of Molecular Recognition

John Wiley and Sons, 605 Third
Avenue, New York, NY-10158, USA
Telephone: 212 850 6000

4 issues per year *ISSN: 0952-3499*
(English)

John Wiley and Sons Ltd,
Baffins Lane, Chichester, West Sussex,
PO19 1UD, UK
Telephone: 243 770328
Fax: 243 775878

International journal devoted to research on specific molecular recognition in chemistry, biology, biotechnology and medicine, the basic principles, characterization and application of specific molecular interactions in chemistry, biology, biotechnology and medicine.

Journal of Neurocytology

Chapman & Hall, 2-6 Boundary Row,
London, SE1 8HN, UK
Telephone: 71 865 0066
Fax: 71 522 9623

Bimonthly *ISSN: 0300-4864*
(English)

Neurobiology, cell biology, biochemistry, physiology, pharmacology, pathology and cytochemistry. Publishes studies of neurons, receptors, synapses, neuroeffector junctions, glia and other elements of the peripheral and central nervous systems.

The Journal of Nutritional Biochemistry

Butterworth-Heinemann,
Linacre House, Jordan Hill, Oxford,
OX2 8DG, UK
Telephone: 865 310366
Fax: 865 310898

12 issues per year *ISSN: 0955-2863*
(English)

Includes submissions of work in basic biochemical and molecular biology with nutritional orientation, experimental nutrition at the animal or human level, and review articles in the following areas: biochemistry, neurochemistry, toxicology, molecular biology, physiology and pharmacology.

Journal of Planar Chromatography - Modern TLC

Dr Alfred Huthig Verlag GmbH,
Im Weiher 10, Heidelberg, D-6900,
Germany
Telephone: 6221 489 0
Fax: 6221 489 279

6 issues per year *ISSN: 0933-4173*
(English)

Publishes research papers on analytical and preparative planar chromatography. Includes relevant developments in analytical methodology, instrumentation, computation, and interpretation, as well as applications of planar chromatography, the separation, identification, and determination of synthetic and naturally occurring compounds in pharmacy and medicine, biochemistry, biotechnology, environmental analysis, and related disciplines.

Journal of Plant Physiology

Gustav Fischer Verlag,
Wollgrasweg 49, Stuttgart 70,
Hohenheim, D-7000, Germany

20 issues per year *ISSN: 0176-1617*
(English, French and German papers)

Covers results of basic research in all aspects of plant physiology, including photosynthesis, nitrogen metabolism, plant tissue and cell culture, in vitro propagation, nucleic acids and proteins, enzymology and biochemistry, membranes and energy-dependent transport phenomena. Contains papers in English, French, German. Formerly Zeitschrift fur Planzenphysiologie.

Journal of the Science of Food and Agriculture

Elsevier Applied Science Publishers
Ltd, Crown House, Linton Road,
Barking, Essex, IG11 8JU, UK
Telephone: 81 594 7272

16 issues per year *ISSN: 0022-5142*
(English)

Publishes original research and critical reviews with particular emphasis on interdisciplinary studies at the agriculture/food interface. Covers fundamental and applied research, and the development of the production of crops and animals.

Journal of Soil Contamination

Lewis Publishers Inc,
2000 Corporate Blvd NW, Boca Raton,
FL-33431, USA

Quarterly
(English)

Journal of the Association for the Environmental Health of Soils (AEHS). Focuses on scientific and technical information, data, and critical analysis in the areas of analytical chemistry, site assessment, environmental fate, environmental modelling, remediation techniques, risk assessment issues, risk management, regulatory issues, and legal considerations.

Journal of Soil Science

Blackwell Scientific Publications,
Osney Mead, Oxford, Oxon,
OX2 0EL, UK
Telephone: 865 240201

ISSN: 0022-4588
(English)

Published for the British Society of Soil Science. Publishes papers on soil science.

The Journal of Steroid Biochemistry and Molecular Biology

Pergamon Press Inc,
395 Saw Mill River Road, Elmsford,
NY-10523, USA
Telephone: 914 592 7700
Fax: 914 592 3625

24 issues per year *ISSN: 0960-0760*
(English)

Publishes papers dealing with new experimental or theoretical developments in areas related to steroids.

Journal of Structural Biology

Academic Press, Journals Division,
1250 Sixth Avenue, San Diego,
CA-92101, USA
Telephone: 619 230 1840
Fax: 619 699 6850

6 issues per year *ISSN: 1047-8477*
(English)

Publishes papers dealing with the structural analysis of biological matter at all levels of organization by means of light and electron microscopy, X-ray diffraction, nuclear magnetic resonance, as well as other imaging, diffraction, spectroscopic, or scanning probe techniques yielding structural information.

Journal of Tissue Culture Methods

Tissue Culture Association, 1 Bank Street, Suite 210, Gaithersburg,
MD-20878, UK
Telephone: 301 869 2900

4 issues per year *ISSN: 0271-8057*
(English)

Publishes papers on techniques, methods and procedures for the cultivation of cells, tissues, organs and tumours.

The Journal of Trace Elements in Experimental Medicine

Alan R Liss Inc, 41 East 11th Street,
New York, NY-10003, USA
Telephone: 212 475 7700

Quarterly
(English)

Covers the role of trace elements in human health and disease.

Journal of Virological Methods

Elsevier Science Publishers,
PO Box 330, AH Amsterdam,
NL-1000, The Netherlands

ISSN: 0166-0934
(English)

Publishes original papers and invited reviews covering all aspects of virology including aspects of structure, immune response, detection and identification as well as transmission and pathogenicity.

Journal of Virology

American Society for Microbiology,
1325 Massachusetts Avenue NW,
Washington, DC-20005, USA
Telephone: 202 737 3600

12 issues per year *ISSN: 0022-538X*
(English)

Publishes fundamental papers on viruses of bacteria, plants and animals, including biochemistry, genetics, immunology, morphology and physiology.

La Recherche

Societe d'Editions Scientifiques,
57 rue de Seine, Paris, F-75006,
France
Telephone: 1 35 43 28 4

12 issues per year
(French with English summaries)

Publishes news, articles and reviews on advances in medical research. Includes news on medical biotechnology.

Land Management and Environmental Law Report

John Wiley and Sons Inc, 605 Third
Avenue, New York, NY-10158, USA
Telephone: 212 850 6000

6 issues per year
(English)

Covers relevant law and policy issues. Includes articles on applicable law, policy and practice. Gives coverage of legislation, statutory instruments, EEC Directives, Ministerial circulars and cases. Summarises proposals for legal or policy changes. Covers law, policy and practice within the United Kingdom, with regular reports from the USA and other countries.

Life Sciences

Pergamon Press Inc, 395 Saw Mill River Road, Elmsford, NY-10523, USA
Telephone: 914 592 7700
Fax: 914 592 3625

52 issues per year *ISSN: 0024-3205*
(English)

Provides rapid communications in pharmacology and biochemistry with a broad subject coverage.

Lymphokine Research

Mary Ann Liebert Inc, 1651 Third Avenue, New York, NY-10128, USA
Telephone: 212 289 2300

4 issues per year *ISSN: 0277-6766*
(English)

Covers recent advances in lymphokine and monokine research.

Macromolecular Structures

Current Biology Ltd, 34-42 Cleveland Street, London, W1P 5FB, UK

Annual
(English)

A full-colour guide to the biological macromolecular structures reported in the year prior to publication. Provides information on all new atomic-level structures obtained by X-ray crystallography, nuclear magnetic resonance, electron crystallography or fibre diffraction. Each entry comprises a description of the structure with illustrations, technical information and a guide to the biological relevance of the molecule.

Mammalian Genome

Springer-Verlag, Heidelberger Platz 3, Berlin 33, D-1000, Germany
Telephone: 30 82 07 1

Monthly *ISSN: 0938-8990*
(English)

Journal of the International Mammalian Genome Society.

Marine Environmental Research

Elsevier Science Publishers, PO Box 330, AH Amsterdam, NL-1000, The Netherlands

8 issues per year *ISSN: 0141-1136*
(English)

Publishes original research papers on chemical, physical, and biological interactions in the oceans and coastal waters.

Medical Microbiology and Immunology

Springer-Verlag, Heidelberger Platz 3, Berlin 33, D-1000, Germany
Telephone: 30 82 07 1

6 issues per year *ISSN: 0300-8584*
(English)

Contains original papers on virology, bacteriology and immunology, with emphasis on pathogenetic mechanisms.

Membrane Biochemistry

Taylor & Francis Inc, Rankine Road, Basingstoke, Hants, RG24 0PR, UK
Telephone: 256 840366
Fax: 256 479438

Quarterly *ISSN: 0149-046X*
(English)

Publishes original full-length research papers, short communications and reviews of current topics in membrane science.

Methods in Molecular and Cellular Biology

John Wiley and Sons Ltd,
Baffins Lane, Chichester, West Sussex,
PO19 1UD, UK
Telephone: 243 770328
Fax: 243 775878

6 issues per year *ISSN: 0898-7750*
(English)

Incorporating DNA and Protein Engineering Techniques. Publishes reports describing new methods and methodologies, improvements of common protocols, troubleshooting guides, and new simplified protocols related to broad aspects of molecular biology.

Microbial Pathogenesis

Academic Press, Journals Division,
1250 Sixth Avenue, San Diego,
CA-92101, USA
Telephone: 619 230 1840
Fax: 619 699 6859

12 issues per year *ISSN: 0882-4010*
(English)

Molecular and cellular biology of infectious disease. Research areas include: pathogenesis; virulence factors; host susceptibility or resistance; immune mechanisms; identification; cloning, and sequencing of relevant genes; genetic studies; viruses, prokaryotic organisms, and protozoa.

Microbial Releases

Springer for Science, PO Box 503,
AM Ijmuiden, NL-1970,
The Netherlands

ISSN: 0940-9653

Microbiological Reviews

American Society for Microbiology,
1325 Massachusetts Avenue NW,
Washington, DC-20005, USA
Telephone: 202 737 3600

4 issues per year *ISSN: 0146-0749*
(English)

Reviews scientific literature in all areas of microbiology.

Microbiology

Consultants Bureau, 233 Spring Street,
New York, NY-10013, USA

6 issues per year *ISSN: 0026-2617*
(English)

Complete English translation of the Russian journal Mikrobiologiya, published by Akademiya Nauk SSR. Contains short papers on all aspects of microbiology (no abstracts).

Microbiology and Immunology

Centre for Academic Publications
Japan, 4-16 Yayoi-2-chome,
Bunkyo-ku, Tokyo 113, Japan

6 issues per year *ISSN: 0385-5600*
(English)

Contains original articles and reviews. Journal of the Japanese Society of Bacteriology, Society of Japanese Virologists and Japanese Society for Immunology.

Micron and Microscopica Acta

Pergamon Press Inc,
395 Saw Mill River Road, Elmsford,
NY-10523, USA
Telephone: 914 592 7700
Fax: 914 592 3625

4 issues per year *ISSN: 0739-6260*
(English)

Covers technical aspects of electron microscopy, electronprobe microanalysis, image processing etc, including medical and biological applications.

MIRCEN

Oxford University Press, Walton Street, Oxford, OX2 6DP, UK
Telephone: 865 56767

(English)

Journal of Applied Microbiology and Biotechnology
Publishes full-length papers, short communications and review papers on biotechnology in the developing world. Published in association with UNESCO.

Molecular Biology Reports

Kluwer Academic Publishers, Distribution Centre, PO Box 322, AH Dordrecht, NL-3300, The Netherlands
Telephone: 78 172 811

4 issues per year *ISSN: 0301-4851*
(English)

Molecular Biotherapy

Butterworth-Heinemann Ltd, Linacre House, Jordan Hill, Oxford, OX2 8DG, UK
Telephone: 865 310366
Fax: 865 310898

Quarterly *ISSN: 0952-8172*
(English)

Carries original manuscripts about experimental, clinical and epidemiological aspects of therapeutic and prophylactic biologicals. Presents clinical and preclinical results in biotherapy as well as findings from international researchers in medical laboratories and biotechnology companies.

Molecular and Cellular Biochemistry

Martinus Nijhoff Publishers, Postbus 211, AE Amsterdam, NL-1000, The Netherlands

16 issues per year *ISSN: 0300-8177*
(English)

Publishes original articles dealing with physical and chemical structure of proteins, nucleic acids, carbohydrates, lipids and all aspects of enzymology. Also covers cellular aspects: structure, function, hormonal effects, transport and secretion.

Molecular and Cellular Biology

American Society for Microbiology, 1325 Massachusetts Avenue NW, Washington, DC-20005, USA
Telephone: 202 737 3600

12 issues per year *ISSN: 0270-7306*
(English)

Covers the molecular biology of eukaryotic cells, both microbial and higher organisms.

Molecular and Cellular Endocrinology

Elsevier Scientific Publishing Co Inc, PO Box 330, AH Amsterdam, NL-1000, The Netherlands

12 issues per year *ISSN: 0303-7207*
(English)

Covers synthesis and secretion of extracellular signals, hormones, neurotransmitters etc. Provides an understanding of cellular regulation mechanisms involved in hormonal control.

Molecular and Cellular Probes

Academic Press Inc, Journals Division, 1250 Sixth Avenue, San Diego, CA-92101, USA
Telephone: 619 230 1840
Fax: 619 699 6859

6 issues per year *ISSN: 0890-8508*
(English)

The Location, Diagnosis and Monitoring of Diseases by Specific Molecules and Cell Lines. Presents advances in molecular biology, the production of hybridomas and other cloned cell lines, and the designing of new labelling and detection methods as they relate to medical diagnosis.

Molecular and General Genetics

Springer-Verlag, Heidelberger Platz 3, Berlin 33, D-1000, Germany
Telephone: 30 82 07 1

12 issues per year *ISSN: 0026-8925*
(English)

Publishes original papers on bacterial and phage genetics, plasmids, transposons, plastid and mitochondrial aspects, determination of structure and function of ribosomes and molecular genetics. Microfilm editions in 16mm, 35mm and microfiche available.

Molecular Immunology

Pergamon Press Inc,
395 Saw Mill River Road, Elmsford, NY-10523, USA
Telephone: 914 592 7700
Fax: 914 592 3625

12 issues per year
(English)

Publishes papers dealing with the molecular analysis of the cellular components and processes underlying the behaviour of cells in response to the immune phenomena. Includes research reports, preliminary communications, book reviews, review articles, announcements of forthcoming meetings, summaries of meetings, letters to the Editor and brief communications formulating new ideas.

Molecular Microbiology

Blackwell Scientific Publications Ltd, Osney Mead, Oxford, Oxon, OX2 0EL, UK
Telephone: 865 240201

6+ issues per year *ISSN: 0950-382X*
(English)

Publishes research papers on molecular biology, genetics and biochemistry of prokaryotic and eukaryotic microorganisms.

Molecular Phylogenetics and Evolution

Academic Press, Journals Division, 1250 Sixth Avenue, San Diego, CA-92101, USA
Telephone: 619 230 1840
Fax: 619 699 6859

4 issues per year *ISSN: 1055-7903*
(English)

Provides a forum for molecular studies that advance understanding of phylogeny and evolution.

M/S Medicine Sciences

CDR, 11 rue Gossin, Montrouge, F-92543, France

10 issues per year
(English)

Carries international reviews, papers, discussion and news in the fields of biology and medicine. Result of collaboration between the French and Canadian Governments.

Mutation Research

Elsevier, Crown House, Linton Road, Barking, Essex, IG11 8JU, UK
Telephone: 81 594 7272

12 issues per year *ISSN: 0165-1218*
(English)

International journal on mutagenesis, chromosome breakage and related subjects.

The Mycologist

Cambridge University Press, Edinburgh Building, Shaftsbury Road, Cambridge, CB2 2RU, UK

ISSN: 0269-915X
(English)

Contains general interest articles, original reports, news and reviews as well as full reports on the activities of the British Mycological Society on whose behalf it is published. Areas covered include fungal involvement in biotechnology, ecology, conservation, genetics, molecular biology, pathogenicity, physiology and sytematics.

Mycopathologia

Martinus Nijhoff/Dr W Junk Publishers, Postbus 211, AE Amsterdam, NL-1000, The Netherlands

12 issues per year *ISSN: 0301-486X*
(English)

Publishes original research papers and critical reviews in medical and veterinary mycology, plant pathology, crop protection, mycotoxins and industrial mycology as well as aspects of systematics and biology of pathogenic fungi. Published in association with the CAB International Mycological Institute.

Mycorrhiza

Springer for Science, PO Box 503, AM IJmuiden, NL-1970, The Netherlands

ISSN: 0940-6360

Natural Product Reports

Royal Society of Chemistry, Thomas Graham House, Science Park, Milton Road, Cambridge, CB4 4WF, UK
Telephone: 602 507411

6 issues per year *ISSN: 0265-0568*
(English)

Provides critical coverage of recent developments in natural product and bio-organic chemistry in the form of review articles, papers, technical notes and literature reviews.

Nature

Macmillan Journals Ltd, 4 Little Essex Street, London, WC2R 3LF, UK
Telephone: 71 836 6633

52 issues per year *ISSN: 0028-0836*
(English)

Publishes editorial matter, news and views, book reviews, original articles and letters covering all aspects of science. Attracts significant papers in the areas of molecular genetics in particular. Numerous advertisements for products and jobs.

Nature Genetics

Macmillan Journals Ltd, 4 Little Essex Street, London, WC2R 3LF, UK

12 issues per year
(English)

Publishes original research articles of broad interest in all fields relating to human genetics. Subjects include molecular medicine, gene mapping, the genome project and developmental genetics.

The New Biologist

Saunders Scientific Publications, WB Saunders Co, Curtis Center, Independence Square West, Philadelphia, PA-19106, USA
Telephone: 215 238 7800
Fax: 215 238 7883

(English)

Multidisciplinary journal in cell and molecular biology.

New Phytologist

Cambridge University Press, The Edinburgh Building, Shaftsbury Road, Cambridge, CB2 2RU, UK

(English)

Publishes commissioned and non-commissioned reviews as well as research papers and book reviews on all aspects of plant sciences including physiology, biophysics, biochemistry, biotechnology and pathology with an emphasis on pure science.

The New Phytologist

Academic Press Inc, 11 Fifth Avenue, New York, NY-10003, USA
Telephone: 212 741 6800

12 issues per year *ISSN: 0028-646X*
(English)

Publishes full-length research papers, short reports, review articles and book reviews in areas of plant physiology, biophysics, biochemistry, phytochemistry, mycology, phycology, plant pathology, etc.

New Scientist

New Science Publications,
Holborn Publishing Group,
King's Reach Tower, Stamford Street,
London, SE1 9LS, UK
Telephone: 71 261 5000
Fax: 71 261 6464

52 issues per year *ISSN: 0028-6664*
(English)

Publishes features, news, letters and book reviews on all aspects of science, technology and sociology. Also contains advertisements for commercial and public-sector research, information science and administrative posts.

Nucleic Acids Research

IRL Press Ltd, PO Box 1, Eynsham, Oxon, OX8 1JJ, UK
Telephone: 865 882 2873

24 issues per year *ISSN: 0305-1048*
(English)

Publishes surveys, summaries and original papers on molecular biological, enzymological and chemical aspects of nucleic acids, nucleosides and nucleotides, but does not cover practical techniques in synthesis of oligonucleotides or papers dealing with properties and functions of simple purine and pyrimidine derivatives.

Nucleosides and Nucleotides

Marcel Dekker Inc, 270 Madison Avenue, New York, NY-10016, USA
Telephone: 212 696 9000

ISSN: 0732-8311
(English)

A rapid publication journal publishing papers on the chemistry and biology of nucleosides, nucleotides and related compounds.

Oncogene

Macmillan Press Ltd, Houndmills, Basingstoke, Hants, RG21 2XS, UK

12 issues per year *ISSN: 0950-9232*
(English)

Publishes the latest research on molecular biology of malignant change.

Peptides

Pergamon Press Inc,
395 Saw Mill River Road, Elmsford,
NY-10523, USA
Telephone: 914 592 7700
Fax: 914 592 3625

6 issues per year *ISSN: 0196-9781*
(English)

Pesticide Biochemistry and Physiology

Academic Press, Journals Division, 1250 Sixth Avenue, San Diego, CA-92101, USA
Telephone: 619 230 1840
Fax: 619 699 6859

6 issues per year *ISSN: 0048-3575*
(English)

Contains original scientific papers on all aspects of the biochemistry and physiology of insecticides, fungicides, acaricides, nematocides, rodenticides and antifeeding compounds, as well as chemosterilants, the biosynthesis of pheromones, hormones and plant resistance agents, effects of such agents on biochemistry and physiology of both target and non-target organisms, biotransformations of such agents themselves, and mechanisms of resistance, synergism and antagonism.

Pesticide Outlook

Royal Society of Chemistry, Thomas Graham House, Science Park, Milton Road, Cambridge, CB4 4WF, UK
Telephone: 602 507411

ISSN: 0956-1250
(English)

Publishes review articles on pesticides in crop protection, animal and human health, wood preservation, amenity use, etc as well as covering worldwide developments including alternatives to the use of chemicals such as biopesticides. Each issue contains reviews, feature articles, company profiles, book reviews, conference reports, letters, comment, world news and details of forthcoming events.

Pesticide Science

Elsevier Applied Science Publishers Ltd, Crown House, Linton Road, Barking, Essex, IG11 8JU, UK
Telephone: 81 594 7272

12 issues per year *ISSN: 0031-613X*
(English)

Covers all aspects of research and technology on crop protection and pest control. Deals with synthesis, screening and structure/activity studies of new compounds, formulation, application, metabolism and safety in use of new and existing products, synthetic and naturally occurring insecticides, herbicides, fungicides, nematocides, plant and insect hormones, veterinary and public health uses of insecticides and ectoparasites as well as ecological implications of such applications.

Pharmacogenetics

Chapman & Hall, 2-6 Boundary Row, London, SE1 8HN, UK
Telephone: 71 865 0066
Fax: 71 522 9623

Bimonthly *ISSN: 0960-314X*
(English)

Devoted to the rapid publication of research papers, brief review articles and short communications on genetic variation in response to drugs and other chemicals in humans and animals. Emphasis is given to an understanding of the genetic mechanisms of species differences in foreign compound metabolism and responsiveness, and to the regulation of drug metabolizing enzymes and their genes.

Philosophical Transactions of the Royal Society: Series B

The Royal Society, 6 Carlton Terrace, London, SW1Y 5AG, UK
Telephone: 71 839 5561

(English)

Covers all biological science.

Photosynthesis Research

Martinus Nijhoff Publishers,
Postbus 211, AE Amsterdam,
NL-1000, The Netherlands

4 issues per year *ISSN: 0166-8595*
(English)

Publishes original papers on both fundamental and applied aspects of photosynthesis covering a range of topics from primary reactions to the use of photosynthesis as a means of harvesting solar energy in the form of plant biomass. Areas covered include electron transport, photophosphorylation, carbon assimilation, photorespiration, regulation at the genetic and cellular level, and environmental and ecological aspects.

Physiologia Plantarum

Minksgaard International Publishers,
35 Norre Sogade, Copenhagen K,
DK-1370, Denmark

12 issues per year *ISSN: 0031-9317*
(Mainly English, French, German)

Publishes original reports of scientific research on all aspects of plant physiology, including photosynthesis, plant nucleic acids, plant growth substances and hormones, and whole-plant physiology. Larger proportion of papers from Scandinavia and rest of Europe. Official journal of the Federation of European Societies of Plant Physiology.

Physiological and Molecular Plant Pathology

Academic Press Inc (London) Ltd,
24-28 Oval Road, London,
NW1 7DX, UK
Telephone: 71 267 4466

6 issues per year *ISSN: 0885-5765*
(English)

Publishes original papers relating to fungal, viral and bacterial diseases of plants.

Phytochemical Analysis

John Wiley and Sons Ltd,
Baffins Lane, Chichester, West Sussex,
PO19 1UD, UK
Telephone: 243 770328
Fax: 243 775878

6 issues per year *ISSN: 0958-0344*
(English)

An International Journal of Plant Biochemical Techniques. Publishes articles on the application of analytical methodology in plant sciences. Covers methods and techniques relevant to the extraction, separation, purification, identification and quantification of substances in plant biochemistry, plant cellular and molecular biology, plant biotechnology, the food sciences, agriculture and horticulture.

Phytochemistry

Pergamon Press Inc,
395 Saw Mill River Road, Elmsford,
NY-10523, USA
Telephone: 914 592 7700
Fax: 914 592 3625

12 issues per year *ISSN: 0031-9422*
(Mainly English, French and German)

Covers research on all aspects of pure and applied plant biochemistry, especially those related to the chemistry of plant products and an understanding of factors underlying control of growth development and differentiation. Sections cover original papers and short reports in areas of biochemistry, biosynthesis, chemotaxonomy, phytochemistry and the identification of new chemical compounds derived from plants. Official organ of the Phytochemical Society of Europe.

Phytopathology

American Phytopathological Society,
3340 Pilot Knob Road, St Paul,
Minnesota, MN-55121, USA
Telephone: 612 454 7250

12 issues per year *ISSN: 0031-949X*
(English)

Contains scientific reports concerning basic research in all aspects of plant disease and its control, including control by chemicals or by management as well as details of effects of disease or effects of treatment of causative agents.

Plant Cell and Environment

Blackwell Scientific Publications Ltd,
Osney Mead, Oxford, Oxon,
OX2 0EL, UK
Telephone: 865 240201

9 issues per year *ISSN: 0140-7791*
(Mainly English)

Publishes experimental and theoretical papers on physiology of green plants, including biochemistry, biophysics, cell physiology, whole-plant physiology, crop physiology and physiological ecology.

Plant, Cell, Tissue and Organ Culture

Martinus Nijhoff/Dr W Junk
Publishers, Postbus 211,
AE Amsterdam, NL-1000,
The Netherlands

Monthly *ISSN: 0167-6857*
(English)

Publishes original results of studies on in vitro culture of higher plants, including fundamental, methodological and technological aspects relating to anatomy, histology, ultrastructure, physiology, biochemistry, genetics, breeding and phytopathology. Genetics and breeding include aspects of fusion, somatic hybridization, uptake of nuclei, chromosome and DNA transformation, cell screening and sorting, and organelle transfer. Specific biotechnology aspects include biosynthesis of secondary products, biotransformation, low-temperature storage and mass production of plants *in vitro*.

Plant Disease

American Phytopathological Society,
3340 Pilot Knob Road, St Paul,
Minnesota, MN-55121, USA
Telephone: 612 454 7250

12 issues per year *ISSN: 0191-2917*
(English)

Contains editorial comment, research reports, information on new diseases and epidemics (in the US), new chemicals (fungicides, pesticides, nematocides) and treatments for plant disease. Also covers serological identification of plant disease and applied plant pathology.

Plant Growth Regulation

Kluwer Academic Publishers,
Distribution Centre, PO Box 322,
AH Dordrecht, NL-3300,
The Netherlands
Telephone: 78 172 811

6 issues per year *ISSN: 0167-6903*
(English)

The Plant Journal

Blackwell Scientific Publications,
Osney Mead, Oxford, Oxon,
OX2 0EL, UK
Telephone: 865 240201

Bimonthly
(English)

Publishes original research papers in all key areas of plant molecular sciences.

Plant Molecular Biology

Kluwer Academic Publishers,
Distribution Centre, PO Box 322,
AH Dordrecht, NL-3300,
The Netherlands
Telephone: 78 172 811

12 issues per year *ISSN: 0167-4412*
(English)

A rapid publication outlet for research on plant molecular biology and genetics. Special emphasis is given to aspects of plant genomes and genes; the regulation of expression at various levels; cloning and sequencing of plant genes; construction of molecular cloning vectors; controlling DNA elements; protein biosynthesis; molecular basis of the action of plant hormones and the molecular biology of nitrogen fixation.

Plant Pathology

Blackwell Scientific Publications Ltd,
Osney Mead, Oxford, Oxon,
OX2 0EL, UK
Telephone: 865 240201

4 issues per year *ISSN: 0032-0862*
(English)

Covers all aspects of research in plant disease and its control by chemical and experimental studies and trials. Publication of the British Society for Plant Pathology.

Plant Physiology

American Society of Plant Physiologists, USDA-ARS National Soil Tilth Laborator, 2150 Pammel Drive, Ames, IA-50011, USA

12 issues per year *ISSN: 0032-0889*
(English)

Contains reports of original research on all aspects of plant physiology, including plant tissue culture, enzymology, photosynthesis and energy metabolism, nitrogen fixation, plant nucleic acids and protein synthesis.

Plant Physiology and Biochemistry

Centrale des Revues, 11 rue Gossin,
Montrouge Cedex, F-92543, France

6 issues per year *ISSN: 0981-9428*
(English or French)

Publishes manuscripts in English or French as well as short communications, reviews and letters to the Editor and papers devoted to computer methods. Main topics are physiology, biochemistry, molecular biology, biophysics, structure and genetics at different levels from the molecular basis to whole plant and environmental considerations. Publishes occasional thematic issues on topics such as photosynthesis, herbicides and storage of compounds. An official journal of the Federation of Societies of Plant Physiology.

Plant Science

Elsevier Science Publishers, Institute for Photobiology, Brandeis University, Waltham, MA-02154, USA

12 issues per year *ISSN: 0168-9452*
(English)

An international journal covering all areas of experimental plant biology including enzymology, cell biology, genetics, molecular biology and physiology.

Plant Science Letters

Elsevier Biomedical Press,
Postbus 211, AE Amsterdam,
NL-1000, The Netherlands

15 issues per year
(English)

Rapidly publishes original papers in plant biology: cells and tissue culture, cell organelles (including development of mitochondria and chloroplasts), cell walls, cellular and molecular genetics, enzymology and intermediate metabolism, nucleic acids and proteins, photosynthesis, respiration and nitrogen metabolism, phytochemistry and plant products, regulation of growth and viruses.

Planta

Springer-Verlag, Heidelberger Platz 3, Berlin 33, D-1000, Germany
Telephone: 30 82 07 1

18 issues per year *ISSN: 0032-0935*
(English)

Reports original scientific research on all aspects of plant physiology, biochemistry, structure and function: including cell and tissue culture; cytogenetics; phytopathology; photosynthesis and nitrogen metabolism; synthesis of proteins and nucleic acids; membrane phenomena and biotransformations.

Plants Today

Blackwell Scientific Publications, Osney Mead, Oxford, Oxon, OX2 0EL, UK
Telephone: 865 240201

6 issues per year *ISSN: 0952-3855*
(English)

Publishes news and reviews of current developments in plant sciences.

Plasmid

Academic Press Inc, 11 Fifth Avenue, New York, NY-10003, USA
Telephone: 212 741 6800

6 issues per year *ISSN: 0147-619X*
(English)

An international journal devoted to extrachromosomal gene systems in prokaryotes and eukaryotes. Publishes review articles, original full papers and short communications.

Pollution Prevention

Pollution Prevention Magazine, 90 Long Lane, Rickmansworth, Hertfordshire, WD3 2YG, UK
Telephone: 923 776245
Fax: 923 776245

6 issues per year
(English)

Contains features on various aspects of pollution prevention and the protection of the environment.

Proceedings of the National Academy of Sciences

PO Box 97, Berkhamsted, Herts, HP4 2PX, UK

24 issues per year
(English)

Publishes original papers on all aspects of biological science including: biochemistry, biophysics, cell biology developmental biology, immunology, medical sciences, microbiology, neurobiology and physiological sciences.

Proceedings of the Society for Experimental Biology and Medicine

Academic Press Inc, 11 Fifth Avenue, New York, NY-10003, USA
Telephone: 212 741 6800

11 issues per year
(English)

Includes original papers on endocrinology, haematology, microbiology, immunology, oncology, pharmacology and physiology.

Process Biochemistry

Morgan Grampian Publishers Ltd, 30 Calderwood Street, London, SE18 6HQ, UK

6 issues per year *ISSN: 0032-9592*
(English)

Publishes original research reports on biochemical aspects of food, pharmaceuticals, distilling, effluent and waste treatment industries. Supplements detail products, processes and information sources featuring various areas of biotechnology.

Progress in Biophysics and Molecular Biology

Pergamon Press Inc,
395 Saw Mill River Road, Elmsford,
NY-10523, USA
Telephone: 914 592 7700
Fax: 914 592 3625

6 issues per year *ISSN: 0079-6107*
(English)

Protein Expression and Purification

Academic Press, Journals Division,
1250 Sixth Avenue, San Diego,
CA-92101, USA
Telephone: 619 230 1840
Fax: 619 699 6859

6 issues per year *ISSN: 1046-5928*
(English)

Provides a forum for the presentation of protein isolations employing the variety of molecular biological procedures used in the expression of proteins as well as those based on conventional fractionation techniques.

Proteins: Structure, Function, and Genetics

John Wiley and Sons Inc, 605 Third Avenue, New York, NY-10158, USA
Telephone: 212 850 6000

12 issues per year *ISSN: 0887-3585*
(English)

John Wiley and Sons Ltd,
Baffins Lane, Chichester, West Sussex,
PO19 1UD, UK
Telephone: 243 770328
Fax: 243 775878

Publishes original research articles, reviews and brief research communications on advances in all areas of protein biochemistry.

Protoplasma

Springer-Verlag, Heidelberger Platz 3,
Berlin 33, D-1000, Germany
Telephone: 30 82 07 1

24 issues per year *ISSN: 0033-183X*
(English)

Publishes original manuscripts in the area of biology of cell-animal, plant, algal, protozoan, fungal, including descriptive and experimental ultrastructure, differentiation, pathology and pathogenesis, cytoskeletal elements, membranes and membrane biogenesis, cell fractionation (organelle isolation), analysis and intracellular distributions of cell constituents, cytochemistry, environmental cell biology, and papers dealing with new methodologies in these areas.

Pure and Applied Chemistry

Pergamon Press Inc,
395 Saw Mill River Road, Elmsford,
NY-10523, USA
Telephone: 914 592 7700
Fax: 914 592 3625

12 issues per year *ISSN: 0033-4545*
(English)

Official journal of the International Union of Pure and Applied Chemistry (IUPAC).

Quantitative Structure-Activity Relationships

VCH, 220 East 23rd Street Suite 909,
New York, NY-10010, USA
Telephone: 6201 602 220
Fax: 6201 602 328

4 issues per year *ISSN: 0931-8771*
(English)

Interdisciplinary forum for contributions on QSAR, molecular modelling, computer graphics and other computer-assisted methods helpful in drug design and development. Publishes original papers and provides an abstracting service. The journal covers: biology, medicinal-organic-analytical and inorganic chemistry, pharmacology, toxicology, and environmental protection research.

Receptors and Channels

Harwood Academic Publishers, STBS, PO Box 197, London, WC2E 9PX, UK

(English)

Publishes manuscripts covering all aspects of original research on receptors and ion channels. Brings together studies on these and the related signalling elements at the levels of molecular biology and genetics, biochemistry, physical analysis, molecular modelling, immunology, molecular pharmacology and electrophysiology.

Regulatory Peptides

Elsevier Science Publishers, PO Box 330, AII Amsterdam, NL-1000, The Netherlands

12 issues per year
(English)

A rapid publication journal covering the physiology and pathology of peptides of the gut, endocrine and nervous systems involved in regulation of cell or tissue function.

Renewable Energy

Pergamon Press Ltd,
Contact: Ms C Morgan
Headington Hill Hall, Oxford, OX3 0BW, UK
Telephone: 865 74341

6 issues per year *ISSN: 0960-1481*
(English)

Reproduction Nutrition Development

CDR Gauthier Villars, 11 rue Gossin, Montrouge, F-92543, France

6 issues per year
(French with English summaries)

Publishes original papers concerning all aspects of reproduction and development in animals ranging from invertebrates, through vertebrates of all types to primates.

Research in Veterinary Science

British Veterinary Association, 7 Mansfield Street, London, W1M 0AT, UK

6 issues per year
(English)

Publishes original scientific contributions in all aspects of veterinary medicine and animal health, including bacterial, fungal and viral diseases and their treatment, immunoassays and immunization. Official publication of the British Veterinary Association.

Resources, Conservation and Recycling

Elsevier Science Publishers, PO Box 330, AH Amsterdam, NL-1000, The Netherlands

4 issues per year *ISSN: 0921-3449*
(English)

Incorporates the journals: Resources and Conservation (formerly published by Elsevier Science Publishers BV) and Conservation & Recycling (formerly published by Pergamon Press plc). Gives investigations, analyses and reviews of renewable and nonrenewable resource management, particularly for their conservation. Topics include: technological management and change for reduced consumption or substitution; recovery of materials and energy from established and new sources; management of natural resources.

Review of Agricultural Entomology

CAB International, Wallingford, Oxfordshire, OX10 8DE, UK
Telephone: 491 32111

Monthly
(English)

Review of Medical and Veterinary Entomology

CAB International, Wallingford, Oxfordshire, OX10 8DE, UK
Telephone: 491 32111

Monthly
(English)

Review of Medical and Veterinary Mycology

CAB International, Wallingford, Oxfordshire, OX10 8DE, UK
Telephone: 491 32111

Quarterly
(English)

Reviews in Fish Biology and Fisheries

Chapman & Hall, 2-6 Boundary Row, London, SE1 8HN, UK
Telephone: 71 865 0066
Fax: 71 522 9623

Quarterly *ISSN: 0960-3166*
(English)

International journal publishing important review articles on any aspect of fish and fisheries biology. All fields of fish biology where the emphasis is on the whole organism. Subjects covered include physiology, evolutionary biology, taxonomy, zoogeography, behaviour, ecology and exploitation. Reviews provide an up-to-date synopsis for research workers in the chosen field with an outline of the next problems which should be tackled. Provide no-specialist fish biologists with an awareness and understanding of the field. Caters for all with an interest in fish biology and fisheries.

Revue d'Ecologie et de Biologie du Sol

Centrale des Revues, 11 rue Gossin, Montrouge Cedex, F-92543, France

4 issues per year
(Abstracts in French and English)

Covers all aspects of the biology and ecology of the soil as original papers in English, Spanish, German or French, with abstracts in English and French.

RIA (Revue de l'Industrie Agro-Alimentaire)

Groupe SEPAIC, 42 Rue Louvre, Paris, F-75001, France
Telephone: 1 42 21 84 00
Fax: 1 42 21 16 23

24 issues per year
(French)

Publishes articles and technical and financial analyses of the French agroindustry. Includes commercial and classified advertising. Published twice a month in French.

Scandinavian Journal of Immunology

Blackwell Scientific Publications Ltd, Osney Mead, Oxford, Oxon, OX2 0EL, UK
Telephone: 865 240201

12 issues per year *ISSN: 0300-9475*
(Mostly in English)

Contains papers on cellular and molecular immunology. Official journal of the Scandinavian Society of Immunology.

Science

American Assoc for the Advancement of Science, 1333 H Street NW, Washington, DC-20005, USA

52 issues per year *ISSN: 0036-8075*
(English)

In addition to original reports covering all aspects of science, publishes letters, editorial comment, leading articles, news and comments, research news and book reviews. Also publishes special issues and supplements on biotechnology and a 'Guide to Biotechnology Products and Instruments'.

Scientific American

415 Madison Avenue, New York, NY-10017, USA
Telephone: 212 754 0550

12 issues per year *ISSN: 0036-8733*
(English)

Popular science magazine. Often has articles on genetic engineering and cell biology. Well illustrated with many colour illustrations and plates.

Seed Science Research

CAB International, Wallingford, Oxfordshire, OX10 8DE, UK
Telephone: 491 32111

ISSN: 0960-2585
(English)

Publishes original articles on the physiology, biochemistry and molecular biology of seeds.

Seed Science and Technology

International Seed Testing Association, Reckenholz, PO Box 412, Zurich, CH-8046, Switzerland
Telephone: 1 371 31 33
Fax: 1 377 72 01

3 issues per year *ISSN: 0251-0952*
(English)

Proceedings of the International Seed Testing Association.

Soil Biology and Biochemistry

Pergamon Press Inc, 395 Saw Mill River Road, Elmsford, NY-10523, USA
Telephone: 914 592 7700
Fax: 914 592 3625

6 issues per year *ISSN: 0038-0717*
(English)

Publishes original papers, short communications and book reviews.

Soil Science

Williams and Wilkins, 428E Preston Street, Baltimore, MD-21202, USA

Monthly *ISSN: 0038-075X*
(English)

Publishes papers and book reviews on soil science.

Somatic Cell and Molecular Genetics

Plenum Press, 233 Spring Street, New York, NY-10013, USA
Telephone: 212 620 8000

6 issues per year
(English)

Publishes original papers on somatic cell genetics and molecular genetics of higher eukaryotes, with particular emphasis in gene expression and regulation, gene transfer in cultured cells and embryos, gene isolation, genetic recombination, mutation, chromosome replication, gene mapping and genetics of subcellular organelles.

Soviet Biotechnology (Biotekhnologiya)

Allerton Press Inc, 150 Fifth Avenue,
New York, NY-10011, UK
Telephone: 212 924 3950

6 issues per year
(Full English translation)

Russian journal devoted to theoretical and applied aspects of biotechnology covering biochemistry, microbiology, molecular genetics, genetic engineering, medicine and agriculture as well as aspects of immunology, biological control agents and bioregulators, biocatalysts, biomembranes and receptors, applications of biopolymers, use of animal cultures in production of chemicals, bioconversion of lignocellulose, leaching of minerals and mineral enrichment and the use of biological methods in pollution control and waste management.

Spectra Biologie

Pole technologique, 7 rue du Bois de la Champelle, BP 268, Vandoevre Cedex,
F-54512, France
Telephone: 83 44 25 32
Fax: 83 44 12 90

6 issues per year
(Mainly in French)

Articles, interviews, press reports and diary of events as well as advertisements and classified section covering commercial, instrumentation and laboratory techniques in clinical instrumentation with an emphasis on immunodiagnostics.

Spectroscopy Europe

VCH, 220 East 23rd Street Suite 909,
New York, NY-10010, USA
Telephone: 6201 602 220
Fax: 6201 602 328

A controlled circulation journal, available free-of-charge to individuals engaged in spectroscopy within Europe.

Starch

VCH, 220 East 23rd Street Suite 909,
New York, NY-10010, USA
Telephone: 6201 602 220
Fax: 6201 602 328

12 issues per year *ISSN: 0038-9056*
(English)

Examines new technologies necessary to exploit carbohydrates and reports on recent developments concerning new starch sources, starch derivatives and starch saccharification products.

Supramolecular Chemistry

Gordon and Breach Science Publishers,
C/O STBS, PO Box 90, Reading,
Berkshire, RG1 8JL, UK
Telephone: 734 560080
Fax: 734 568211

(English)

Publishes papers from the fields and sub-disciplines related to supramolecular chemistry, including the fields of analytical, inorganic, organic, and physical chemistry as well as interdisciplinary sub-disciplines such as bio-inorganic chemistry.

Systema Ascomycetum

CAB International Mycological
Institute, Wallingford, Oxfordshire,
OX10 8DE, UK
Telephone: 491 32111

2 issues per year
(English)

Provides information on the largest division of the fungi, many of which are important plant pathogens, agents of biodeterioration or of biotechnical and industrial importance. Publishes lists of generic names, notes on ascomycetes systematics, lists of genera by families including synonyms, original papers on ascomycetes systematics. Published by the University of Umea (Sweden) in collaboration with the CAB International Mycological Institute.

Technique

Saunders Scientific Publications, WB Saunders Co, Curtis Center, Independence Square West, Philadelphia, PA-19106, USA
Telephone: 215 238 7800
Fax: 215 238 7883

(English)

Journal of methods in cell and molecular biology.

Theoretical and Applied Genetics

Springer-Verlag, Heidelberger Platz 3, Berlin 33, D-1000, Germany
Telephone: 30 82 07 1

8 issues per year *ISSN: 0040-5752*
(English)

Theriogenology - An International Journal of Animal Reproduction

Butterworth-Heinemann
Contact: Mr VM Shille, University of Florida, Florida, USA

12 issues per year
(English)

Covers the field of animal reproduction, including: effects of environment, radiation and nutrition, function of the ovary and testis, control of ovulation and egg laying, cryobiology of spermatozoa and embryos, *in vivo* fertilization, transmission of viruses by embryos.

Topics in Enzyme and Fermentation Biotechnology

Ellis Horwood Ltd, Market Cross House, Cooper Street, Chichester, W Sussex, PO19 1EB, UK

ISSN: 0140-0835
(English)

Bound publication containing review articles.

Topics in Fluorescence Spectroscopy

Plenum Publishing, 223 Spring Street, New York, NY-10013, USA
Telephone: 212 620 8000

(English)

A series that synthesizes recent advances and the current state of the art in fluorescence spectroscopy. Features articles by international experts, discusses emerging areas of the field where recent discoveries have occurred and further breakthroughs are expected, and includes coverage of new methods and applications. Vol 1 - Techniques, Vol 2 - Principles, Vol 3 - Biochemical Applications.

Toxicology

Elsevier
Contact: HP Witschi, Toxicology Substance Programme, LEHR Facility, Davis, California, USA

21 issues per year *ISSN: 0300-483X*
(English)

An international journal for the publication of original scientific papers on the biological effects arising from the administration of chemical compounds, principally to animals, tissues or cells, but also to man.

Toxicology and Applied Pharmacology

Academic Press Inc, Journals Division, 1250 Sixth Avenue, San Diego, CA-92101, USA
Telephone: 619 230 1840
Fax: 619 699 6859

12 issues per year
(English)

Publishes original research papers related to alterations in cell and tissue structure or function resulting from administration of chemicals, drugs or natural products to animals or man. Also presents reviews, communications, announcements and letters to the Editor. Official journal of the Society of Toxicology.

Toxicology Letters

Elsevier
Contact: DB Menzel, Southern Occupational Health Center, University of California, Irvine, California, USA

15 issues per year *ISSN: 0378-4274*
(English)

An international journal for the rapid publication of short reports on biochemical mechanisms of mammalian toxicity. Contains meeting, symposium and workshop dates.

Transgenic Research

Chapman & Hall, 2-6 Boundary Row, London, SE1 8HN, UK
Telephone: 71 865 0066
Fax: 71 522 9623

Bimonthly *ISSN: 0962-8819*
(English)

Dedicated to the rapid publication of research in transgenic higher organisms including their production, properties resulting from the transgenic state, use as experimental tools, exploitation and application, and environmental impact. Publishes studies in animals, plants and fungi in which genetic manipulation has been used to confer on the organism novel metabolic or developmental properties by the insertion of transgenes or by altering the expression of homogenes. Investigations into the physiology, biochemistry, molecular biology, development, genetics, behaviour and exploitation of such transgenic organisms are also covered.

Trees - Structure and Function

Springer For Science, Postbus 503, Ijmuiden, NL-1970, The Netherlands

4 issues per year
(English)

Publishes original articles on the physiology, biochemistry, functional anatomy, structure and ecology of trees and other woody plants as well as papers concerned with tree pathology and technical aspects of forestry where they contribute to the basic understanding of the structure and function of trees. Also publishes reviews on related topics.

Trends in Biochemical Sciences

Elsevier Publications (Cambridge), 68 Hills Road, Cambridge, CB2 1LA, UK
Telephone: 223 315961
Fax: 223 321410

12 issues per year *ISSN: 0376-5067*
(English)

Publishes short reviews and news features on a wide range of biochemical topics. Also includes letters, technical notes, book reviews and a meetings diary. Published in association with the International Union of Biochemistry and Molecular Biology.

Trends in Biotechnology

Elsevier Publications (Cambridge)
Contact: Dr C Robinson, 68 Hills Road, Cambridge, CB2 1LA, UK
Telephone: 223 315961
Fax: 223 321410

12 issues per year *ISSN: 0167 9430*
(English)

Publishes minireviews and technical notes, editorial commentaries, features, letters and worldwide views, taking an integrated approach to the applied biosciences. A diary of events and display advertising are also carried. Includes occasional supplements.

Trends in Cell Biology

Elsevier Trends Journals Ltd, 68 Hills Road, Cambridge, CB2 1LA, UK
Telephone: 223 315961
Fax: 223 321410

Monthly
(English)

Trends in Genetics

Elsevier Publications (Cambridge), 68 Hills Road, Cambridge, CB2 1LA, UK
Telephone: 223 315961
Fax: 223 321410

12 issues per year *ISSN: 0168-9479*
(English)

Publishes reviews, discussions and comment on all aspects of genetics. Includes a journal monitoring section covering the latest research papers.

Tropical Animal Health and Production

Longmans Group Ltd, Longman House, Burnt Hill, Harlow, Essex, CM20 2JE, UK
Telephone: 279 426721
Fax: 279 431059

4 issues per year *ISSN: 0049-4747*
(English)

Publishes original research papers covering investigations and observations in any field of animal health and production. The objective is to improve tropical animal husbandry. Published under the auspices of the Centre for Tropical Veterinary Medicine, UK.

Vaccine

Butterworth-Heinemann Ltd, Linacre House, Jordan Hill, Oxford, Oxon, OX2 8DG, UK
Telephone: 865 310366
Fax: 865 310898

12 issues per year *ISSN: 0264-410X*
(English)

Contains editorial matter, reviews, original papers, patents, reports, reports of conferences and details of future events. Published monthly plus supplement.

Veterinary Immunology and Immunopathology

Elsevier Science Publishers, PO Box 330, AH Amsterdam, NL-1000, The Netherlands

12 issues per year *ISSN: 0165-2427*
(English)

Publishes papers concerned with basic immunology of domestic animals and disease processes that result from immunological defects.

Veterinary Microbiology

Elsevier Science Publishers, PO Box 330, AH Amsterdam, NL-1000, The Netherlands

8 issues per year *ISSN: 0378-1135*
(English)

Publishes papers on microbiological diseases of animals useful to man, including studies of antibiotic resistance and organisms such as bacteria, mycoplasma, rickettsia, fungi and viruses.

Veterinary Parasitology

Elsevier Science Publishers,
PO Box 330, AH Amsterdam,
NL-1000, The Netherlands

16 issues per year *ISSN: 0304-4017*
(English)

Publishes papers on aspects of protozoology, entomology and helminthology of interest to animal health practitioners covering aspects of disease prevention, pathology, treatment, epidemiology and control of parasites in animals useful to man.

Veterinary Record

British Veterinary Association,
7 Mansfield Street, London,
W1M 0AT, UK

52 issues per year *ISSN: 0042-4900*
(English)

Contains editorial comment, news and reports, short communications, committee reports, book news and reviews. Official publication of the British Veterinary Association.

Viral Immunology

American Society for Microbiology,
1325 Massachusetts Avenue NW,
Washington, DC-20005, USA
Telephone: 202 737 3600

4 issues per year
(English)

Publishes original papers on human and animal viral immunology, research and development of viral vaccines, immunological characterization of viral components, virus-based immunological diseases, viral diagnostics, tumour and cancer immunology with virus as the primary factor and related topics.

Virology

Academic Press Inc, Journals Division,
1250 Sixth Avenue, San Diego,
CA-92101, USA
Telephone: 619 230 1840
Fax: 619 699 6859

16 issues per year *ISSN: 0042-6822*
(English)

Publishes the results of basic research in all branches of virology, including the viruses of vertebrates and invertebrates, plants, bacteria, and yeasts/fungi. The journal features articles on the nature of viruses; on the molecular biology of virus multiplication; on molecular pathogenesis; and on molecular aspects of the control and prevention of viral infections.

Waste Management Today

Waste Management Information
Bureau, B7.12 Harwell Laboratory,
Oxon, OX1 0RA, UK
Telephone: 235 432 919
Fax: 235 832 591

12 issues per year *ISSN: 0954-495X*
(English)

Aimed at the waste and environmental management business. Provides information on management of wastes including biological treatment of sewage sludges and domestic refuse. Covers disposal, materials recovery, environmental hazards, legislative and policy aspects as well as case studies, modelling, economics and recycling.

Water Air and Soil Pollution

D Reidel Publishing Co, Postbus 17,
AA Dordrecht, NL-3300,
The Netherlands

8 issues per year *ISSN: 0049-6979*
(English)

An international journal covering all aspects of environmental pollution.

Water Pollution Control Federation Journal

Water Pollution Control Federation,
2626 Pennsylvania Avenue NW,
Washington, DC-20037, USA
Telephone: 202 337 2500

12 issues per year
(English)

Publishes papers on process design, process research, waste water analysis, maintenance of water quality and treatment of industrial wastes. Contains numerous advertisements for professional services and equipment.

Water Research

Pergamon Press Inc,
395 Saw Mill River Road, Elmsford,
NY-10523, USA
Telephone: 914 592 7700
Fax: 914 592 3625

12 issues per year *ISSN: 0043-1354*
(English)

Covers aspects of water treatment.

Water Science Technology

Pergamon Press Inc,
395 Saw Mill River Road, Elmsford,
NY-10523, USA
Telephone: 914 592 7700
Fax: 914 592 3625

12 issues per year
(English)

Official journal of the International Association of Water Pollution Research.

Water Services

Industrial and Marine Publications Ltd,
Queensway House, 2 Queensway,
Redhill, Surrey, RH1 1QS, UK
Telephone: 737 68611

ISSN: 0301-7028
(English)

The Journal for Water Sewage and Industrial Effluent Treatment. Publishes technical articles and short news items dealing with all aspects of the water cycle, including water resources location and exploitation, planning, development of surface and groundwater supplies, treatment, distribution and use for public supply, industrial purposes, irrigation and the generation of electrical power. Also covers aspects of water pollution control including the treatment of sewage and industrial waste waters, hydrology, hydraulics and flood control.

Wood Science and Technology

Springer-Verlag Inc,
Heidelberger Platz 3, Berlin 33,
D-1000, Germany
Telephone: 30 82 07 1

4 issues per year *ISSN: 0043-7719*
(English)

Publishes original papers and reviews covering all aspects of wood and pulp; microbiological degradation of wood; cytology of cambium, xylem and phloem; anatomy and ultrastructure; chemistry of wood and bark and wood physics. Also covers problems related to wood technology, including combustion, drying, impregnation and conversion to pulp and paper. Official journal of the International Academy of Wood Science.

Yeast

John Wiley and Sons Ltd,
Baffins Lane, Chichester, West Sussex,
PO19 1UD, UK
Telephone: 243 770328
Fax: 243 775878

12 issues per year *ISSN: 0749-503X*
(English)

Covers physiology, microbiology, molecular biology and biochemistry of yeasts.

Zeitschrift fur Pflanzenernahrung und Bodenkunde

VCH, 220 East 23rd Street, Suite 909, New York, NY-10010, USA
Telephone: 6201 602 220
Fax: 6201 602 328

6 issues per year *ISSN: 0044-3263*
(English and German summaries)

All papers are written in German or English, with summaries in both languages. Covers entire field of plant nutrition and soil science.
Plant nutrition section includes fertilization, food quality, ecology. Soil science section includes soil physics and the chemistry, biology and genetics of soil. Other reports on soil science, geomorphology and use, nature protection and environmental problems are also published. A society journal of the German Soil Science Society and the German Society for Plant Nutrition.

Zentralblatt fur Mikrobiology

VEB Gustav Fisher Verlag Jena, Villengang 2, Jena, D-6900, Germany

8 issues per volume
(English, French or German; English summaries)

Publishes original papers and reviews concerned with bacteriological, mycological and virological investigations of agricultural interest related to environmental protection. Published irregularly.

ZFL

Dr Alfred Huethig Verlag GmbH, Im Weiher 10, Heidelberg, D-6900, Germany
Telephone: 6221 489 0
Fax: 6221 489 279

Monthly
(German)

Journal of food technology and processing. Covers topics including: colorants and flavours produced by fermentation; the utilization of enzymes for improving processes and quality; use of microorganisms for reducing waste and residue emissions; use of biotechnological methods for preservation.

NEWSLETTERS

ABA Bulletin

Australian Biotechnology Association, PO Box 303, Clayton, Victoria 3168, Australia
Telephone: 3 542 2393
Fax: 3 543 6613

6 issues per year
(English)

Contains articles on biotechnology matters, letters to the Editor, information on new product developments, member companies and conferences.

AFRC News

AFRC, Babraham Hall, Babraham, Cambridge, CB2 4AT, UK
Telephone: 223 832312
Fax: 223 836122

Quarterly
(English)

Newsletter of the Agricultural and Food Research Council.

AgBiotech News and Information

CAB International, Wallingford, Oxfordshire, OX10 8DE, UK
Telephone: 491 32111

24 issues per year
(English)

Contains a worldwide selection of abstracts of literature on agriculture, biotechnology, news items and short reviews plus annual author and subject indexes.

Agribusiness Intelligence

G V Olsen Associates, 170 Broadway, New York, NY-10038, USA
Telephone: 212 866 5034

(English)

Produces summaries and references on articles, studies, legislation, government regulations, plant and technology, including details of source of material. Topics covered are as follows: Alcohol Fuel (methanol, gasohol etc from renewable sources); Cellulose to Alcohol (conversion, acid and enzyme hydrolysis); Biotechnology (genetic engineering and recombinant DNA technology systems); Olsen's Agribusiness Report (agroindustry intelligence).

Agricell Report

Agritech Consultants Inc, PO Box 255, Shrub Oak, New York, NY-10588, USA
Telephone: 914 528 3469

12 issues per year
(English)

Contains short news articles as well as details of techniques, recent patents, reports from meetings, letters to the Editor, book reviews and brief listing of titles of noteworthy papers from the scientific literature. Most items have contact addresses for further information.

Agricultural Genetics Report

Mary Ann Liebert Inc, 1651 Third Avenue, New York, NY-10128, USA
Telephone: 212 289 2300

6 issues per year *ISSN: 0278 9736*
(English)

Covers genetic engineering applications related to agriculture and food.

Agrow

PJB Publications Ltd,
Agrow Head Office, 18-20 Hill Rise,
Richmond, Surrey, TW10 6UA, UK
Telephone: 81 948 3262
Fax: 81 948 5598

24 issues per year *ISSN: 0268-313X*
(English)

World Agricultural News. International newsletter for the crop protection industry. Covers news of the industry in short paragraphs with sections on activities, investments and takeovers of companies, company news, financial results, details of pesticide trade, discussions of regulations and legislation, etc in separate sections for the UK, Europe, the US and the rest of the world. Also contains meetings reports and details of future meetings as well as trademark applications and news of appointments in the agrochemical industry. Also available on-line (see AGROW OnScreen, listed under Databases and Abstracts).

Alcohol Update

Information Resources Inc, 499 South Capitol Street Suite 406, Washington, DC-20003, USA
Telephone: 202 554 0614
Fax: 202 554 0613

26 issues per year
(English)

Covers all aspects of the commercial production of fermentation alcohol for fuel use including legislation, national programmes, economic aspects and current prices of products, raw materials and relevant petroleum products.

Alcohol Week

Inside Washington Publishers,
PO Box 7167, Ben Franklin Station,
Washington, DC-20044, USA
Telephone: 703 892 8516

52 issues per year
(English)

Covers alcohol fuels and feedstocks with articles on programmes, legislation, tax incentives, prices and volumes of products traded.

Animal Pharm

PJB Publications, 18/20 Hill Rise,
Richmond, Surrey, TW10 6UA, UK
Telephone: 81 948 0751
Fax: 81 332 8991

(English)

World Animal Health News. Also available on-line (see Agrow OnScreen, listed under Databases and Abstracts).

Applied Genetics News

Business Communications Co Inc,
PO Box 2070C, Stamford,
CT-06906, USA

12 issues per year *ISSN: 0271-7107*
(English)

Contains abstracts and comments on most aspects of genetic engineering.

At the CENTRE

CCOHS, 259 Main St East, Hamilton,
Ontario, L8N 1H6, Canada
Telephone: 416 572 2981
Fax: 416 572 2206

(English)

Published by the Canadian Centre for Occupational Health and Safety as part of an information service covering occupational health and safety through articles and reviews as well as detailing legislation, statistics, conferences and meetings, etc and providing specific hazard alerts.

BIA Bulletin

BioIndustry Association, 1 Queen Anne's Gate, London, SW1H 9BT, UK
Telephone: 71 222 2809
Fax: 71 222 8876

(English)

The newsletter of the BioIndustry Association (UK) issued to members. Covers activities of the association, gives profiles of members' activities and provides summaries of events and legislation of importance worldwide as well as book reviews and details of courses, seminars and exhibitions. It also provides a diary of worldwide events in biotechnology.

Biobulletin

Science and Engineering Research Council, Polaris House, North Star Avenue, Swindon, SN2 1ET, UK
Telephone: 793 26222
Fax: 793 511181

(English)

The Biotechnology Directorate. Reports briefly on work of the Directorate of the Science and Engineering Research Council. Aims to act as a forum for discussion and information exchange on biocatalysis, plant genetics and biochemistry, large-scale growth of plant and animal cells, fermentation technology and microbial physiology, downstream processing, recombinant DNA technology and bioelectronics.

Biocontrol News and Information

CAB International, Wallingford, Oxfordshire, OX10 8DE, UK
Telephone: 491 32111

4 issues per year
(English)

In addition to news contains relevant abstracts. Produced by the CAB International Institute of Entomology.

Bioenergi

AB Svensk Travarutidning, Midskogsgrand 5, Stockholm, S-11543, Sweden
Telephone: 8 664 34 00
Fax: 8 664 30 05

12 issues per year
(Swedish)

Official newsletter of SVEBIO, Swedish Bioenergy Association. Publishes detailed articles on biomass-based energy systems with the emphasis on wood as raw material. Well illustrated with numerous advertisements for related technology.

BioEngineering News

Deborah J Mysiewicz Publishers Inc, 109 Minna Street, Suite 304, San Francisco, CA-94105, USA

48 issues per year *ISSN: 0275-4207*
(English)

Covers news items on a broad range of topics as well as covering legal issues, new patents and stock price index.

Biofutur FLASH

29 rue Buffon, Paris, F-75005, France
Telephone: 1 47 07 11 22

12 issues per year
(French)

A series of newsletters which cover the United States, Japan and the Agricultural and Food Industries as three separate publications on industrial aspects of biotechnology. Contents include scientific, technical and economic aspects of new discoveries, innovations and commercial activities in agriculture, food, fermentation, energy, environment, genetic engineering, health and medicine.

BioGen

Polycom Verlagsgesellschaft mbH, Feuerbachstrass 15, Braunschweig, D-3300, Germany

22 issues per year
(German)

Biotechnology and genetic technology newsletter concerning company activities and new products and markets.

Bioinvention

OMEC Publishing Company, PO Box 446, McLean, VA-22101, USA

26 issues per year
(English)

Monitors US Patent Office and gives details of both new and old patents. Features include details of relevant patents issued in the previous two weeks, as well as related patents from other sectors and details of legal decisions and rulings in relation to patent law. Lists patents available for licensing and general items of interest in terms of activities of trade associations, scientific societies, etc.

Biojoule

Biomass Energy Institute Inc, 1329 Niakwa Road East, Winnipeg, Manitoba, R2J 3T4, Canada
Telephone: 204 257 3891

6 issues per year
(English)

Covers biomass, bioconversion and production of alcohol, as well as other renewable energies. Newsletter of the Biomass Energy Institute.

Bio: La Lettre Des Biotechnologies

SELF, 47 bis rue du Rocher, Paris, F-75008, France
Telephone: 1 293 6901

12 issues per year *ISSN: 0291-2430*
(English)

Covers all aspects of biotechnology in France and elsewhere, with information on developments and conferences.

Biologics and Biotech Regulatory Report

PAREXEL International Corp, 195 West Street, Waltham, MA-02154, USA
Telephone: 617 487 9900
Fax: 617 487 0525

Provides details on the latest trends in the regulation and manufacture of vaccines, bacterial products, cytokines, blood products, gene therapies and biological diagnostics. Has agents in Japan and Italy.

Biomass Bulletin

Multi-Science Publishing Co Ltd, 43-45 New Broad Street, London, EC2M 1QY, UK

4 issues per year *ISSN: 0262-7183*
(English)

Provides abstracts directly from journals, covering all aspects of production of fuels from plants to waste recycling. Includes an annual index.

BioPharm

Aster Publishing, 859 Willamette Street, PO Box 10460, Eugene, OR-97440-2460, USA
Telephone: 503 343 1200
Fax: 503 344 3514

10 issues per year *ISSN: 1040-8045*
(English)

Publishes features, articles and columns related to product and process development, scale-up and manufacture of biotechnologically-derived pharmaceuticals, vaccines and diagnostics. Authors are people working in the industry. Sponsors annual BioPharm Conference.

Bioprocessing Technology

Technical Insights Inc, 158 Linwood Plaza, PO Box 1304, Fort Lee, NJ-07024, USA
Telephone: 201 944 6204

12 issues per year *ISSN: 0885-6207*
(English)

Publishes information on energy and chemicals from renewable sources, based on journals and meetings, plus news items with contact addresses, patents (available from US Patent Office) and news of future meetings (mainly in US).

Bioscan

Oryx Press, 2214 NC at Encanto, Phoenix, AZ-85004-1483, USA
Telephone: 602 254 6156

12 issues per year *ISSN: 0887-6207*
(English)

Provides profiles of biotechnology companies, listing products under development etc.

Bio Sciences

SELF, 47 bis rue du Rocher, Paris, F-75008, France
Telephone: 1 293 6901

6 issues per year
(English)

Publishes short reports of original research of significance in all aspects of biotechnology including: microbiology, molecular biology, biochemical pharmacology, endocrinology, immunology, cell biology, genetic engineering and biochemical engineering.

Biotech Knowledge Sources

BioCommerce Data Ltd, 95 High Street, Slough, Berkshire, SL1 1DH, UK
Telephone: 753 511777
Fax: 753 512239

(English)

Gives brief details of recently published books, directories and market research reports as well as newly available databases and journals. It also includes details of conferences. Produced jointly with the British Library's Biotechnology Information Service.

Biotech Market News and Strategies

Vertis Ltd, Box 11155, Fort Lauderdale, FL-33339, USA
Telephone: 305 914 6692

(English)

Provides information concerning strategic factors in marketing biological compounds, market applications, competitive products worldwide, current market assessments and future projections. Also deals with key products and marketing methods. Areas include restriction enzymes, cloning vectors, molecular linkers, hybridomas, immunoglobulins, prostaglandins, diagnostics and monoclonal antibodies, as well as many other products.

Biotech News

Microinfo Ltd, PO Box 3, Newman Lane, Alton, Hants, GU34 2PG, UK
Telephone: 420 86848

12 issues per year *ISSN: 0263-8029*
(English)

Produces detailed reports and short jottings directed at pharmaceutical, chemical, food processing, agricultural and brewing industries. Lists reports available from US National Technical Information Service.

Biotechnologia

Apdo Postal 20.147, Madrid,
E-28080, Spain

12 issues per year
(Spanish)

General biotechnology newsletter with comments, details of new products, company activities, patents and future meetings. Articles contain references to original source material and addresses for further contacts.

BioTechnologie

Polycom Verlags GmbH,
Feuerbachstrasse 15, Braunschweig,
D-3300, Germany
Telephone: 531 33 39 28

12 issues per year
(German)

Covers general interest aspects of biotechnology in German. Contents include editorial comment, interviews, details of market trends for biotechnical products, articles on processes and techniques, reports on symposia, descriptions of companies' activities, details of laboratory and process equipment and details of forthcoming meetings. Also contains commercial advertisements.

Biotechnology Bulletin

12 issues per year
(English)

Contains short reports on policies, markets and companies; includes details of future meetings in Europe.

Biotechnology Business News

Financial Times Newsletters,
Tower House, Southampton Street,
London, WC2E 7HA, UK
Telephone: 71 240 9391
Fax: 71 240 7946

24 issues per year
(English)

Commentary, reports and news in areas of commerce and finance, regulatory updates and details of new products as well as company profiles and analysis of markets on a national basis.

Biotechnology Information News

British Library Service,
25 Southampton Buildings, London,
WC2A 1AW, UK
Telephone: 71 323 7293

4 issues per year
(English)

British Library Biotechnology Information Service. Provides news about the Biotechnology Information Service and British Library as well as general reports on industry, science and information systems and new companies.

Biotechnology Insight

Science and Technology Letters,
PO Box 81, Northwood, Middlesex,
HA6 3DN, UK
Telephone: 9274 20732
Fax: 9274 25066

12 issues per year *ISSN: 0268-7291*
(English)

Science & Technology Letters. A commentary on commercial prospects in biotechnology aimed at investors, venture capital managers, corporate strategists, research and development directors and business development executives.

The Biotechnology Instrument Market

Delco Scientific Resources, 2315 Post
Road, Fairfield, CT-06430, USA
Telephone: 203 254 0652

4 issues per year
(English)

Contains competitive information and trends about the instruments, equipment, supplies and reagents that impact the bioanalytical market.

Biotechnology in Japan Newservice

Japan Pacific Associates,
441 California Avenue, Suite 3,
Palo Alto, CA-94306, USA
Telephone: 415 322 8441

(English)

Provides up-date information of developments in biotechnology research and business in Japan. Also provides a customer research and consulting service.

Biotechnology Law Report

Mary Ann Liebert Inc, 1651 Third Avenue, New York, NY-10128, USA
Telephone: 212 289 2300

12 issues per year *ISSN: 0278-9728*
(English)

Analyzes legal issues and litigation, including patents, with the help of a large editorial board of experts.

Biotechnology News

CTB International Publishing Co,
PO Box 579, Summit, NJ-07901, USA

26 issues per year *ISSN: 6273-3226*
(English)

Contains short reports with contact addresses, international activities, at-a-glance summaries and details of recent books as well as worldwide patent coverage with issuing addresses.

Biotechnology Newswatch

McGraw-Hill Inc, 1221 Avenue of the Americas, New York, NY-10020, USA

24 issues per year
(English)

Provides wide coverage in subject matter and geographical area with periodic company reviews. Includes original material not published elsewhere.

Biotechnology Software

Mary Ann Liebert Inc, 1651 Third Avenue, New York, NY-10128, USA
Telephone: 212 289 2300

6 issues per year *ISSN: 0749-0372*
(English)

Covers available software, hardware/software systems and publicly accessible databases related to biotechnology and molecular biology including aspects of service, costs, compatibility, availability, language, speed, format etc as well as data handling analysis, mathematical and statistical treatments, molecular kinetics, literature searching, CAD/CAM graphics, microbial growth kinetics, automated gel analysis, nucleic acid and protein sequence management, homology searches, restriction enzyme mapping, secondary and tertiary structure and clinical microbiology.

Biotechnology Techniques

Science Reviews Inc,
707 Foulk Road, Suite 102,
Wilmington, DE-19803, USA

4 issues per year *ISSN: 0951-208X*
(English)

Publishes short accounts of new methods and techniques for research and development in biotechnology.

BioVenture View

BioVenture Publishing, 300 West 23rd Avenue, San Mateo, CA-94403, USA
Telephone: 415 574 7128
Fax: 415 574 8319

Monthly *ISSN: 0892-1903*
(English)

Reports on international business and research developments in biotechnology.

BioWorld Today

Io Publishing Inc, 217 South B Street, San Mateo, CA-94401, USA
Telephone: 415 696 6555
Fax: 415 696 6590

Daily
(English)

A biotechnology newspaper published every business day. Consists of 2 to 3 pages including coverage of companies, reviews of articles from major science journals and news stories about new and developing products, financial markets, corporate developments, regulatory and policy issues.
The newspaper is transmitted to subscribers' fax machines each morning.

Brewing and Distilling International

5/7 Southwark Street, London, SE1 1RQ, UK
Telephone: 71 407 6981

12 issues per year
(English)

Contains feature articles, news largely concerning the British industry as well as comments on overseas matters, new products, plant applications, crop reports, drinks data and news of people. Carries a significant proportion of trade and commercial advertisements and includes classified advertisements and a trade directory. Incorporates the Brewers' Journal.

Bulletin of the Society for Experimental Biology

Society for Experimental Biology, Burlington House, Piccadilly, London, W1V 0LQ, UK
Telephone: 71 439 8732

(English)

Contains editorial matter, articles and details of activities of the SEB.

CAB International News

CAB International, Wallingford, Oxfordshire, OX10 8DE, UK
Telephone: 491 32111

(English)

Covers general news of activities, services, databanks and research activities both in the UK and overseas.

Canadian Biotech News

Canadian Biotech News Service, PO Box 7131, Station J, Ottawa, Ontario, K2A 4C5, Canada
Telephone: 613 761 1312
Fax: 613 761 9464

ISSN: 1188 455X
(English)

Provides news items related to the Canadian biotechnology industry.

Changing Medical Markets

Theta Corp, Theta Building, Middlefield, CT-06455, USA
Telephone: 203 349 1054
Fax: 203 349 1227

(English)

As well as the newsletter, Theta Corp produce healthcare and biotechnology reports.

Clinica

PJB Publications, 18/20 Hill Rise, Richmond, Surrey, TW10 6UA, UK
Telephone: 81 948 0751
Fax: 81 332 8991

(English)

World Medical Device and Diagnostic News. Also available on-line (see Agrow OnScreen, listed under Databases and Abstracts).

Details

Association of Biotechnology Companies, 1666 Connecticut Ave NW, Suite 330, Washington, DC-20009-1039, USA

6 issues per year *ISSN: 0740-4921*
(English)

Newsletter of the Association of Biotechnology Companies.

The Digest

British Anaerobic and Biomass Association Ltd, PO Box 7, Southend, Reading, RG7 6AZ, UK
Telephone: 635 862131

3 issues per year
(English)

Contains scientific articles, news items, book reviews, details of conferences and reports from member companies of the association interested in energy from biomass with emphasis on UK.

D-J-M Enzyme Report

Deborah J Mysiewicz Publishers Inc, 109 Minna Street, Suite 304, San Francisco, CA-94105, USA

12 issues per year *ISSN: 0731-4027*
(English)

Covers developments in enzyme technology as well as listing relevant patents.

DNA and Protein Engineering Techniques

Alan R Liss Inc, 41 East 11th Street, New York, NY-10003, USA
Telephone: 212 475 7700

6 issues per year *ISSN: 0894-7937*
(English)

Provides information on the latest experimental methods in molecular genetics and protein engineering.

EBIS newsletter

CUBE, DG-XII, 200 Rue de la Loi, Brussels, B-1049, Belgium
Fax: 2 235 53 65

Newsletter produced by members of the Concertation Unit for Biotechnology in Europe (CUBE). EBIS makes suggestions and reports on activities such as research, development and demonstration programmes and regulatory aspects in biotechnology. Also reviews books and reports association news.

EFB Newsletter

Dechema, Postfach 970146, Theodor Hause Allee 25, Frankfurt am Main, D-6000, Germany
Telephone: 69 75 64 235

(English)

Newsletter of the European Federation of Biotechnology. Published in the Journal Biotech Forum Europe in collaboration with the Frankfurt Office of the General Secretariat of the European Federation of Biotechnology.

ENDS

Environmental Data Services Ltd, Unit 24, Finsbury Business Centre, 40 Bowling Green Lane, London, EC1R 0NE, UK
Telephone: 71 278 4745

12 issues per year
(English)

Contains reports on environmental issues within UK and Europe.

Environment Business

Information for Industry Ltd,
521 Old York Road, London,
SW18 1TG, UK
Telephone: 81 877 9130
Fax: 81 877 9938

Fortnightly *ISSN: 0959-7042*
(English)

Contains news on policy, issues, industry and projects related to the state of the environment. Subscribers frequently receive supplement reports on subjects such as waste disposal.

Environmental Impact

Industrial and Environment Associates,
77 Temple Sheen Road, East Sheen,
London, SW14 7RS, UK
Telephone: 81 876 3367

12 issues per year
(English)

Covers environment impact assessment, hazardous materials management, environmental audits, risk management, waste disposal, planning, legislation and medical effects in short articles with some information to help follow up particular interests. Offers an additional information service giving copies of source material to subscribers paying higher rates.

European Biotechnology Newsletter

Biofutur SA, 29 rue Buffon, Paris,
F-75005, France
Telephone: 1 47 07 11 22

22 issues per year
(English)

Covers new developments in European biotechnology with topics ranging from basic research and financing, through technical innovation to marketing. Mostly brief notes and a few brief articles with occasional special reports which deal with topical areas in some depth.

European Clinical Laboratory

International Scientific Communications
Ltd, ISC House, Progress Business
Centre, 5 Whittle Parkway, Slough,
Berkshire, SL1 6DQ, UK
Telephone: 628 668881
Fax: 628 669199

6 issues per year
(English)

Product information relating to clinical laboratory products including as regular features product highlights, literature review, laboratory equipment, calendar of events, company news and advertising index. Available free to certain qualified individuals.

FESPP Newsletter

FESPP, University of Lund, Box 7007,
Lund, S-22007, Sweden
Telephone: 46 10 77 89
Fax: 46 11 94 98

(English)

Newsletter of the Federation of European Societies of Plant Physiology (FESPP).

Food Industry News

Beacon Publishing, 95 Bridger Way,
Crowborough, East Sussex,
TN6 2XD, UK
Telephone: 892 668172
Fax: 892 668173

12 issues per year *ISSN: 0962-7235*
(English)

Contains news and articles of interest to the food industry.

FoodInfo

IFIS Publishing, Lane End House, Shinfield, Reading, RG2 9BB, UK
Telephone: 734 883895
Fax: 734 885065

2 issues per year *ISSN: 0965-6146*
(English)

Formerly the International Food Information Service (IFIS) Newsletter. The newsletter's objectives are to keep existing and potential users informed about the full range of IFIS services and how to make best use of them. See International Food Information Service, listed under International Organizations.

Frontiers in Immunoassay and Biotechnology

Scientific Newsletters Inc, 2421 West Broadway, PO Box 4546, Anaheim, CA-92803, USA
Telephone: 714 838 1371

10 issues per year
(English)

Lists references from journals and books, as well as news and reviews concerning assay methods aimed at the user of immunological techniques.

Genetic Engineering and Biotechnology Monitor

Technology Programme of UNIDO, PO Box 300, Vienna, A-1400, Austria

Contains short reports on technical aspects of biotechnology as well as reports on workshops, activities of the UNIDO/WHO/UNEP working party on development of safety guidelines relating to biotechnology research and industrial processes.

Genetic Engineering Letter

Environews Inc,
1331 Pennsylvania Avenue NW, Washington, DC-20004, USA
Telephone: 202 347 3828

24 issues per year *ISSN: 0276-1882*
(English)

Covers public policy issues of genetic engineering.

Genetic Engineering News

Mary Ann Liebert Inc, 1651 Third Avenue, New York, NY-10128, USA
Telephone: 212 289 2300

6 issues per year
(English)

Provides newspaper approach to the subject with news, reviews, photographs, advertisements and reports on equipment.

Genetic Technology News

Technical Insights Inc,
158 Linwood Plaza, PO Box 1304, Fort Lee, NJ-07024, USA
Telephone: 201 944 6204

12 issues per year *ISSN: 0272-9032*
(English)

Contains news and reviews on genetic engineering.

The Government Chemist

Laboratory of the Government Chemist, Queens Road, Teddington, Middlesex, TW11 OLY, UK
Telephone: 81 943 7000
Fax: 81 943 2767

ISSN: 0954 7975
(English)

Newsletter of the Laboratory of the Government Chemist.

Hazardous Substances

Monitor Press, Rectory Road, Great Waldingfield, Sudbury, Suffolk, CO10 0TL, UK
Telephone: 787 78607
Fax: 787 880201

Monthly
(English)

Newsletter on the use, transport, storage, disposal, monitoring and control of hazardous substances.

Hydro Oakland Research Bulletins

Oakland Consultancy and Publishing Services, 10 Jesus Lane, Cambridge, CB5 8BA, UK
Telephone: 223 300475
Fax: 223 300480

Two issues per year
(English)

A series of four bulletins containing summaries of research proposals in science and engineering which have recently been submitted to the SERC for Government funding. The projects have all been awarded a high rating in SERC's peer merit review system. The bulletins enable industrial colleagues to make direct contact with the academic teams proposing research projects of relevance to them.
The four bulletins cover: chemistry; materials and related technologies; environment, civil engineering and construction; process engineering.

IBSA BioNews

International Biotechnology Suppliers Association, 3097 Moorpark Avenue, Siote 202, San Jose, CA-95128, USA

(English)

A publication of the Non-Profit International Biotechnology Suppliers Association.

ICRAF Newsletter

ICRAF, PO Box 30677, Nairobi, Kenya
Telephone: 2 332304

(English)

International Council for Research in Agroforestry. An occasional publication issued free containing editorial comment, news of ICRAF activities, details of agroforestry networks, details of workshops and news of forthcoming events.

IEA Bioenergy Newsletter

International Energy Agency, WSRG, Dept of Forestry, Aberdeen University, 23 St Machar Drv, Aberdeen, AB9 2UD, UK
Telephone: 224 272 683
Fax: 224 272 685

6 issues per year *ISSN: 0964-9409*

The newsletter of the International Energy Agency covering the International Bioenergy Agreement. It gives details of tasks and activities in areas of production, conversion and use of biological resources in energy generation - including aspects of municipal solid waste treatment and recycling.

Industrial Waste Management

Faversham House Group Ltd, Faversham House, 111 St James Road, Croydon, CR9 2TH, UK
Telephone: 81 684 2660
Fax: 81 684 9729

ISSN: 0961-4710
(English)

Magazine containing news, articles and products related to environmental technology and pollution control.

Irish Biotechnology News

National Board for Science & Technology, Sherbourne House, Sherbourne Road, Dublin 4, Ireland
Telephone: 1 683311

(English)

Four page photo-offset sheet issued at intervals by the Biotechnology Unit of the National Board for Science and Technology giving news of people, company activities and meetings in Ireland.

The Japan Biotech Letter

EGIS, 22-1 Ichibancho, Chiyoda-ku,
Tokyo, Japan
Telephone: 3 3264 1060
Fax: 3 3265 2260

24 issues per year

Provides a number of short items covering Government and public organizations as well as business activities.

Japan Biotech and Medical News

A Fujiwara Scientific Liaison Office,
1059-7 Rueta, Kamakura-shi,
Kanagawa-ken 248, Japan
Telephone: 467 31 7833
Fax: 467 31 5491

32 issues per year
(English summaries)

Contains English summaries of biotechnology and medical articles in major Japanese regular and technical newspapers.

Japan Chemical Week

Chemical Daily Co Ltd,
19-16 Shibarura 3-chome, Minato-ku,
Tokyo 108, Japan
Telephone: 3 437 9530

52 issues per year
(English)

Provides up-to-date information on the Japanese chemical industry in general, with an editorial, news and views, production figures and advertisements.

Japan Report: Biotechnology

Japan Reports, Glen House,
200-208 Tottenham Court Road,
London, W1P 9LA, UK
Telephone: 71 637 4178
Fax: 71 631 3667

12 issues per year
(English)

Collates information on biotechnological research by major corporations and government-sponsored bodies in Japan.

Japanese Biotechnology Industry

Newmedia International Japan,
21A Colville Terrace, London,
W11 2BU, UK

12 issues per year
(English)

Contains a series of short paragraphs describing research or commercial advances relating to products and processes developed by or concerning Japanese companies.

Japanscan

North Oxford Publishing Co Ltd,
242-254 Banbury Road, Oxford,
OX2 7DR, UK
Telephone: 865 511166

12 issues per year
(English)

Bio-Industry bulletin reviews market developments and industrial activities, new products, policies and regulations in Japan. Other bulletins deal with the food industry and the pharmaceutical and cosmetics industry.

LINK - Crops for Industrial Use

The LINK programme, c/o AFRC,
Institute of Engineering Research,
Wrest Park, Silsoe, Bedford,
MK45 4HS, UK

(English)

Newsletter of The LINK programme Crops for Industrial Use.

Microbial Clean-up

Henry Stewart Publications, 2/3 Cornwall Terrace, Regent's Park, London, NW1 4QP, UK
Telephone: 71 935 2382
Fax: 71 486 7083

10 issues per year
(English)

Each issue contains: research reports; informed briefings; sources of funding; register of new and useful microorganisms (listed under agriculture, cellulose and pulp, enzymes, halogenated compounds, inorganics and metals, organic solvents and petroleum) giving details of the organism's name and classification, what it does and whom to contact; relevant and important patent registrations; case studies; background and opinion; meetings, conferences and exhibitions; abstracts of important reviews of relevance in bioremediation, summarized and their implications highlighted.

Microbial Genetics Bulletin

Ohio University College of Biological Science, 484 West 12th Avenue, Columbus, OH-43210, USA
Telephone: 212 289 2300

2 issues per year
(English)

Monoclonal Antibody News

Mary Ann Liebert Inc, 1651 Third Avenue, New York, NY-10128, USA
Telephone: 212 289 2300

4 issues per year
(English)

Provides information on new hybrid cell lines, methodology and information on the use of monoclonal antibodies.

Network News

Editor: Dr Woraphat Arthayukti, 84 Soi Rajakroo, Raholyothin Road, Bangkok, 10400, Thailand
Telephone: 2 278 2798
Fax: 2 541 1949

6 issues per year *ISSN: 0857-8109*
(English)

A newsletter of the Biomass Users Network. Contains articles, questions, critical comments and letters. Correspondance can also be sent to the Biomass Users Network (see International Organizations).

New Bioresources

Winter House Scientific Publications, PO Box 7131, Station J, Ottawa, Ontario, K2A 4C5, Canada
Telephone: 613 761 1312
Fax: 613 761 9464

(English)

Covers aspects of biotechnology related to primary resources.

New Biotech

Winter House Scientific Publications, PO Box 7131, Station J, Ottawa, Ontario, K2A 4C5, Canada
Telephone: 613 761 1312
Fax: 613 761 9464

11 issues per year
(English)

Contains news items, feature articles, application and research notes, product reviews, details of new books and notices of forthcoming meetings, conferences and exhibitions. Periodically publishes special features on specific topics such as automated analysis, DNA synthesizers, diagnostic kits, monoclonal antibodies, etc. Contains some commercial advertisements.

New Biotech Busine$$

9a Maple Ridge Crescent, Nepean, Ontario, K2J 3L3, Canada
Telephone: 613 825 6444
Fax: 613 825 6445

(English)

Provides details of biotechnology business activities of Canadian companies through short articles covering stock performance, joint ventures, market growth and related topics as well as detailing forthcoming events.

Nostrum

Scottish Development Agency, 120 Bothwell Street, Glasgow, G2 7JP, UK
Telephone: 41 248 2700

4 issues per year
(English)

Published by the Health Care and Biotechnology Division of the SDA to provide updates on established and new companies and research groups.

Patents Information News

British Library, Science Reference and Info Service, Holborn Reading Rm, 25 Southampton Bld, London, WC2A 1AW, UK

Quarterly
(English)

Newsletter of the Patents Information Network. Given free of charge.

Pharmaceutical Business News

Financial Times Business Information, Tower House, Southampton Street, London, WC2E 7HA, UK
Telephone: 71 240 9391
Fax: 71 240 7946

(English)

Provides coverage of the worldwide pharmaceutical industry, including new product introductions, joint venture agreements, financial performance analysis, mergers and acquisitions, research and development activities, pre-clinical trials, details of conferences and seminars, regulations, registration and licensing, etc as well as commentary on movements and trends in stock markets.

Pharmaprojects

PJB Publications Ltd, 18-20 Hill Rise, Richmond, Surrey, TW10 6UA, UK
Telephone: 81 948 3262
Fax: 81 948 6866

Monthly updated
(English)

Scientific and commercial information on over 5500 new drugs including biotechnology products being researched and developed by more than 800 companies worldwide. Also available on-line via Data-Star, STN, BRS and Dialog.

Practical Biotechnology

Bailliere Tindall, 1 Vincent Square, London, SW1P 2PN, UK

6 issues per year
(English)

Contains editorial comment, articles and news items presented in magazine format, without advertising or photographs. Worldwide coverage with emphasis on the UK.

Radioassay-Ligand Assay News

Scientific Newsletters Inc, 2421 West Broadway, PO Box 4546, Anaheim, CA-92803, USA
Telephone: 714 838 1371

10 issues per year
(English)

Includes information on developments in radiochemical methods related to clinical diagnostics and immunoassays in particular.

Recombinant DNA Technical Bulletin

US National Inst of General Medical Science, Bethesda, MD-21214, USA
Telephone: 301 496 4000

4 issues per year
(English)

Official publication which reports changes in guidelines for the US and contains extensive bibliographic information.

Recycling and Resource Management

Taylor Marketing Services Ltd, Oakwood House, 3 Moulton Park Office Village, Northampton, NN3 1AP, UK
Telephone: 604 790707
Fax: 604 790708

Quarterly
(English)

Contains news and articles on the recycling of waste.

Renewable Energy Bulletin

Multi-Science Publishing, 43-45 New Broad Street, London, EC2M 1QY, UK

ISSN: 0306-364X
(English)

Presents reports of work in progress on all renewable energies.

Renewable Energy News

CREN Publishing, PO Box 4869, Postal Station E, Ottawa, Ontario, K1S 5J1, Canada

12 issues per year
(English)

Newspaper format with articles on technical and policy aspects of renewable energies. Also contains letters, editorial comment, calendar of events mainly covering North America. Biomass and fuel alcohol information included, plus extensive advertisements for literature, consultancies and process plant.

Resource Recovery Report

5313 38th Street NW, Washington, DC-20015, USA
Telephone: 202 362 6034

12 issues per year
(English)

A Monthly Review of Resource Recovery Progress. Contains short paragraphs summarizing all aspects of resource recovery including water use mainly with reference to the US. Comments cover aspects of legislation and government policy in particular as well as regional news concerning new installations.

REVIEW

Department of Trade and Industry (DTI), Editorial Office: Room 823, Bridge Place, 88-89 Eccleston Square, London, SW1V 1PT, UK

4 issues per year *ISSN: 0955-5188*
(English)

Covers all aspects of renewable energy in the UK including research, legislation, news and UK and overseas events.

Science International

International Council of Scientific Unions, 51, Bd de Montmorency, Paris, F-75016, France
Telephone: 1 45 25 03 29
Fax: 1 42 88 94 31

2 issues per year *ISSN: 1011-6257*
(English)

Newsletter of the International Council of Scientific Unions (ICSU). Prints news items, meeting reports and details of future events of interest to members of ICSU.

Science Reference and Information Service (SRIS) Newsletter

British Library, Science Reference and Info Service, Holborn Reading Rm, 25 Southampton Bld, London, WC2A 1AW, UK
Telephone: 71 323 7494/7496
Fax: 71 323 7495

4 issues per year *ISSN: 0951-4635*
(English)

Free newsletter giving details of publications, services, training courses and personnel associated with the British Library Science Reference and Information Service.

Scrip

PJB Publications Ltd, 18-20 Hill Rise, Richmond, Surrey, TW10 6UA, UK
Telephone: 81 948 3262
Fax: 81 948 6866

52 issues per year
(English)

World Pharmaceutical News. Publishes pharmaceutical news, including commercial aspects of genetic engineering. Also available on-line (see Agrow OnScreen, listed under Databases and Abstracts).

Society for General Microbiology Quarterly

Society for General Microbiology, Journal Sales, 62 London Road, Reading, Berkshire, RG1 5AS, UK

4 issues per year
(English)

Contains articles, programmes and meetings. Details group news and book reviews.

Soft Energy Worldwide

OVE Danish Organization for Renewable Energy, Willemoesgade 14, Copenhagen O, DK-2100, Denmark
Telephone: 3142 9091
Fax: 3142 9095

ISSN: 0906-9054

Non-government organization newsletter on renewable energy and a sustainable energy planning. Contains articles, conference and seminar details, events and short news.

Swiss Biotech

Verlag Dr Felix Wust AG, Freiestrasse 204, Postfach 239, Zurich, CH-8032, Switzerland
Telephone: 1 55 63 11

6 issues per year
(German)

Publishes review articles, news and reports covering all aspects of biotechnology in Switzerland.

Technology Transfer Today

LES, 39 Cloth Fair, London, EC1A 7JQ, UK

(English)

Newsletter of Licensing Executives Society, Britain and Ireland. Members also receive Les Nouvelles, a quarterly report on the latest developments in licensing and technology transfer.

UK Monoclonal Bulletin

Ludwig Institute for Cancer Research, Sutton, Surrey, UK

(English)

Lists makers and developers of monoclonal antibodies.

UNITAR/UNDP Newsletter

Centre on Small Energy Resources,
via Panama 12, I-00198, Italy
Telephone: 845 95 48

(English)

Covers activities of UNITAR in renewable energies including biomass and alcohol fuels. Lists training courses and forthcoming conferences and meetings.

US Regulatory Reporter

PARAXEL International Corp,
195 West Street, Waltham,
MA-02154, USA
Telephone: 617 487 9900
Fax: 617 487 0525

Provides regulatory information for the pharmaceutical, medical, chemical and allied industries. Has agents in Japan and Italy.

Warmer Bulletin

The Warmer Campaign, 83 Mount Ephraim, Tunbridge Wells, Kent,
TN4 8BS, UK
Telephone: 892 24626
Fax: 892 25287

(English)

Newsletter of World Action for Recycling Materials & Energy from Rubbish Campaign. Contains news and features on waste management and resource recovery from waste.

Water Quality International

IAWPRC, 1 Queen Anne's Gate,
London, SW1H 9BT, UK
Telephone: 71 222 3848

Quarterly *ISSN: 0892-211X*
(English)

Published by the International Association on Water Pollution Research and Control. Contains news and articles on all aspects of water pollution including sanitation, wastewater reuse, disease, processes, products, management and legislation.

Weekly Bulletin

Home-Grown Cereals Authority,
Hamlyn House, Highgate Hill, London,
N19 5PR, UK

Weekly
(English)

Newsletter of the Home-Grown Cereals Authority.

Weekly Digest

Home-Grown Cereals Authority,
Hamlyn House, Highgate Hill, London,
N19 5PR, UK

Weekly
(English)

Brief newsletter of the Home-Grown Cereals Authority.

XIII Magazine

Commission of the EC Directorate General XIII
Contact: T Howell, CEC DG-XIII-D,
Rue de la Loi, 200, Brussels, B-1049,
Belgium
Telephone: 2 235 03 94

6 issues per year
(English)

Integrating IES News and I'M News previously published by Directorate-General XIII. Provides information covering the activities and programmes supported by DG-XIII.

DIRECTORIES AND REFERENCE BOOKS

The Academic File - Biosciences

Oakland Consultancy, 10 Jesus Lane, Cambridge, CB5 8BA, UK
Telephone: 223 300475
Fax: 223 300480

Annual *ISBN: 1 873090 05 6*
(English)

Provides industrialists with an insight into the research strengths and supporting facilities of bioscience-based departments, centres and units within government-funded Institutions in the UK. The book includes details of public funding for collaborative research, quick reference guide to polytechnic name changes, location map of institutes, profiles contents list, department profiles, overview of bioscience discipline specialisms, comprehensive list of all bioscience departments in UK universities and subject index. Each profile includes major research areas, core research facilities, research and training services to industry, background information and contact name, address and numbers. Fields covered include aquaculture, biomechanics, biotechnology, computer applications, experimental facilities, forensic science, immunology, microbiology, pathology, physiology, statistics, biochemistry, biomedical sciences, cell biology, environment, food sciences, health sciences, materials, molecular biology, pharmacy, plant science and toxicology.

The Academic File - Chemistry

Oakland Consultancy, 10 Jesus Lane, Cambridge, CB5 8BA, UK
Telephone: 223 300475
Fax: 223 300480

Annual *ISBN: 1 873090 04 8*
(English)

Provides industrialists with an insight into the research strengths and supporting facilities of chemistry-based departments, centres and units within government-funded Institutions in the UK. The book includes details of public funding for collaborative research, quick reference guide to polytechnic name changes, location map of institutes, profiles contents list, department profiles, comprehensive list of all chemistry departments in UK universities and subject index. Each profile includes major research areas, core research facilities, research and training services to industry, background information and contact name, address and numbers. Fields covered include analytical chemistry, atmospheric chemistry, biological chemistry, catalysis, environmental chemistry, pharmaceutical chemistry, physical organic chemistry, proteins and protein engineering, spectroscopy and structural chemistry.

Australian and New Zealand Biotechnology Directory

Australian Industrial Publishers Pty Ltd, 2 Wilford Avenue, Underdale, SA 5032, Australia
Telephone: 8 234 0022
Fax: 8 234 0058

ISSN: 1032-8068
(English)

Includes biotechnology-related companies and businesses in Australia and New Zealand, publicly funded R&D organisations in Australia and New Zealand, R&D organizations in Asia (Hong Kong, Indonesia, Japan, Korea, Malaysia, Philippines,

Singapore, Taiwan, Thailand), government departments - contacts/product registration in Australia and New Zealand, Overseas and Australian associations and societies, training courses in Australia and New Zealand, incentive schemes and grants in biotechnology, recent reports and books, microbial culture collections, molecular biology databases, laboratory and pilot fermenter facilities and a calendar of events.

Beilstein Handbook

The Beilstein Institute,
Springer-Verlag, Heidelberger Platz 3,
Berlin 33, D-1000, Germany
Telephone: 30 8207 1

Produced by the Beilstein Institute and Springer-Verlag Inc. Contains information on around 5 million organic compounds. Also available on-line.

BioScan

The Oryx Press, 4041 North Central,
Phoenix, AZ-85012-3397, USA
Telephone: 312 951 9003/708 392 3022
Fax: 800 279 4663

Bimonthly updates
(English)

The Worldwide Biotech Industry Reporting Service. An annual master directory with bimonthly updates. An information service for biotechnology companies, university offices of technology transfer suppliers, venture capitalists, advertising/public relations firms, management consultants, law firms and executive search firms. Currently covers 565 US companies and 340 foreign companies actively involved in product research and development. Does not include suppliers or vendors. Reports on industry leaders and identifies new and emerging companies. It covers start-ups and spin-offs, announces new products and changes in research and technology, and reveals companies entering into strategic alliances and licensing agreements. Entries include details of products in development/awaiting FDA approval, products on the market, business strategy, investments, parent company (and percentage owned), financial information, research and development activities, history and new facilities.

Biotechnology Guide Japan - Company Directory and Comprehensive Analysis

Macmillan Publishers Ltd, Distributed by Globe Book Services Ltd, Brunel Road, Houndmills, Basingstoke, Hants, RG21 2XS, UK

ISBN: 0 333 51805 5
(English)

Published in the United States and Canada by Stockton Press, 15 East 26th Street, New York, NY-10010, USA

Each entry includes name of the organization in Japanese, address, telephone, fax and telex numbers, president, number of employees, year established, capital, annual sales and operating profit in yen, activities and R&D projects with contact names. The index lists organizations and products with page number references.

Biotechnology Guide USA - Companies, Data and Analysis

Macmillan Publishers Ltd, Distributed by Globe Book Services Lt, Brunel Road, Houndmills, Basingstoke, Hants, RG21 2XS, UK
Telephone: 256 29242
Fax: 256 479 476

(English)

Published in the United States and Canada by Stockton Press, 15 East 26th Street, New York, NY-10010, USA

Includes directory of companies, biotechnology company location and type (location by state, state rankings, public, private and subsidiary companies), the founding of biotechnology companies (by year of founding, companies founded per year, founders), company areas of interest (classification codes, 29 listings from agriculture to testing/analytic services, company focus, ranking by primary focus and by combined primary and secondary focus), personnel, financing, revenues, R&D, patents and products, large US corporations, partnerships, state biotechnology centres, analysis of trends, appendices (classification codes and other abbreviations, companies no longer in business or with names changed, new companies and last-minute additions, references and information sources) and an index with page references.

Biotechnology International 1992

Century Press, Holford Mews,
Cruikshank Street, London,
WC1X 9HD, UK
Telephone: 71 278 3000/333 0050
Fax: 71 278 1674

ISBN: 1 85745 060 4
(English)

Contains articles on policy, strategy, finance, national and international news; genetic engineering; bioprocessing - new techniques; equipment; diagnostics, enzymes and analytical techniques; pharmaceuticals and health care; regulations, safety, containment and cleanroom design. Fast access listings give organizations, addresses, contact names and numbers under the topics: chromatography, spectroscopy, electrophoresis, microscopy, general laboratory, genetic engineering, chemical technology, industrial applications, sterile environment, human therapeutics and diagnostics, processing and regulatory affairs. Contains a large number of advertisements.

Biotechnology Worldwide

CPL Press, CPL Scientific Ltd,
Science House, Winchcombe Road,
Newbury, Berks, RG14 5QX, UK
Telephone: 635 524064
Fax: 635 529322

ISBN: 1 872691 30 7
(English)

Results of the COBIOTECH Survey of the Worldwide Biotechnology Industry. This investigation of the biotechnology industry in more than 50 countries from Australia to Zimbabwe provides a unique insight into the state of this growing international business. The survey was carried out in 1990 by CPL Scientific on behalf of COBIOTECH, a committee formed under the auspices of the International Council of Scientific Unions (ICSU). The book includes national profiles providing detailed information on policy, resources, legislation, current research, commercial developments, future growth potential, education and administration, supplemented by the results of a questionnaire completed by biotechnologists in each country. International organizations, databases and information sources, associations and national bodies are detailed.

Biotechnology Yearbook

polycom Verlagsgesellschaft mbH,
Postfach 37 48, Braunschweig, D-3300,
Germany
Telephone: 531 333928
Fax: 531 336460

Annual *ISSN: 0938-9342*
(German with English Contents List)

BioTechnologie Das Jahr- und Adressbuch. Contains addresses of German, Swiss and Austrian organizations concerned with biotechnology research, industry, consulting, venture capital and technology transfer, as well as information services, associations and societies, authorities and federal offices. Also includes details of culture collections and separate indexes for research, industry production and marketing, and services.

DBIR - Directory of Biotechnology Information and Resources

Contact: Dr K Arnett
Bioinformatics Department, ATCC, 12301 Parklawn Drive, Rockville, MD-20852, USA
Telephone: 301 231 5527

Contains databases, access information, characteristics, index terms, etc as well as details of hardware or software required to access such information sources as appropriate.

Dictionary of Information Technology

The Macmillan Press Ltd, 4 Little Essex Street, London, WC2R 3LF, UK
Telephone: 71 836 6633
Fax: 71 379 4980

Describes the concepts, systems and technologies associated with information technology in over 6,000 entries.

The Directory of Biotechnology in France

Editions Scientifique Elsevier, 29 rue Buffon, Paris, F-75005, France
Telephone: 1 47 07 11 22

Annual
(French)

Annuaire des Biotechnologies et des Bioindustries. Published on behalf of ADEBIO. Text largely in French however it includes a glossary and English translation of biotechnological terms. Includes a general alphabetic index of French biotechnology organizations followed by details of service industries and societies listed by products and services, national and government research centres listed by discipline, professional ministerial and associated organizations, financial organizations with relevant indexes.

Directory of British Associations - DBA

CBD Research Ltd, Chancery House, 15 Wickham Road, Beckenham, Kent, BR3 2JS, UK
Telephone: 81 650 7745
Fax: 81 650 0768

ISBN: 0 900246 57 X

Guide to national associations, societies, institutes and similar organizations which have voluntary membership, in all fields of activity including scientific. Each entry gives full name, abbreviation, year of establishment, address, telephone and fax numbers, secretary and other contacts, category and sphere of interest, groups and branches, activities (eg conferences and meetings), membership data, affiliations to other organizations and publications.

Directory of California Biotechnology

Venture Information Services, 6114 La Salle Ave, Suite 250, Oakland, CA-94611, USA
Telephone: 415 530 5030
Fax: 415 531 4140

Lists Californian organizations involved in biotechnology.

Directory of EC Industry Information Sources

Macmillan Publishers Ltd, Distributed by Globe Book Services Ltd, Brunel Road, Houndmills, Basingstoke, Hants, RG21 2XS, UK
Telephone: 256 292 42
Fax: 256 479 476

ISBN: 0 333 57272 6
(English)

Published in the United States and Canada by Stockton Press, 257 Park Avenue South, New York, NY-10010, USA

ISBN: 1 56159 050 9

This book describes such schemes as the EC Business Cooperation Network (BC-Net) and other pan-European operations, giving industry direct access to named contacts through the twelve member states and from there to local business contacts. Includes a directory of EC industry contacts, EC on-line databases, EC-funded research and development programmes, BC-Net, Euro-Info-Centres, business innovation centres (BICs), economics and social committees and a glossary of abbreviations, acronyms, and commercial scientific and technical terms.

Directory of Environmental Consultants

Environmental Data Services Ltd, Finsbury Business Centre, 40 Bowling Green Lane, London, EC1R 0NE, UK
Telephone: 71 278 4745

ISBN: 0 907673 04 X
(English)

Details 225 organizations offering environmental consulting services in the UK. Its principal objectives are to inform potential clients of the range of environmental consultancy services now on offer, and to provide them with as much assistance as possible in drawing up a short list of consultancies.

Directory of European Agricultural Organizations

Kogan Page, 120 Pentonville Road, London, N1 9JN, UK

(English, German and French Versions)

ISBN: 0 85038 929 1 English
ISBN: 3 7890 0874 5 German
ISBN: 92 830 0033 1 French

Office for Official Publications of the European Communities, L-2985, Luxembourg

ISBN: 92 830 0032 1
Catalogue No: EX-38-83-524-FR-C

Compiled by the Studies and Research Division of the General Secretariat of the Economic and Social Committee of the European Communities. This handbook attempts to implement the 1979 proposal that a survey be made of the most important European and national agricultural organizations, setting out their structures, objectives and activities, with a view to promoting a better understanding of the role of agricultural groups in Europe. It aims at the same time to answer the question of how Europe's farmers have organized themselves in the early 1980s to exert influence wherever decisions are made which substantially affect their standard of living and the development of their farms. The main feature is the examination, one by one, of the European agricultural organizations based in Brussels, and the national organizations of the Community member states and the members of EFTA, making a total of seventeen European countries.

Directory of Nordic Companies in Biotechnology

Bioprint, Sjhagsvgen 51, Huddinge, S-14171, Sweden
Telephone: 8 94 59 00

Annual

Provides selected company profiles on Nordic companies in biotechnology.

Directory of Products and Buyers Guide

Hamlet Information Services Ltd, 59a High Street, West Wratting, Cambs, CB1 5LU, UK
Telephone: 223 290838
Fax: 223 290687

(English)

Chemical Industries Association directory of companies. Gives information on the company, its products, its trade names and contract processing.

EC Research Funding 3rd Edition

CEC DG XIII, Telecommunications, Information Industries and Innovation, Office for EC Official Publications, Luxembourg

ISBN: 92-826 3640 2
(English)

Answers the questions that concern most potential applicants with no previous experience of EC research, who often search in vain for answers in the mass of available technical and legal information. It aims to provide a descriptive outline - intelligible to all, including the layman - of the problems associated with information, consultation, submission of applications, financing, etc; a concentrated and schematic outline of programme content and programme procedures; practical hints for interested parties, ranging from the quest for partners to the completion of a potentially successful project proposal. Includes details of research and technological development (RTD), supporting measures (education and training, practical application of research results, technology transfer), cooperation in research and technology beyond the EC, applying, project selection, EC contracts, Framework Programmes (information and communications technologies, industrial and materials technologies, environment, life sciences and technologies, energy, human capital and mobility), related programmes outside the Framework Programme, sources of information and databases.

European Animal Health Directory

PJB Publications Ltd, 18-20 Hill Rise, Richmond, Surrey, TW10 6UA, UK
Telephone: 30 82 07 1

(English)

Contains names, addresses, telephone, telex and fax numbers of European companies and organizations.

The European Community 1991/2 Third Edition

The Macmillan Press Ltd, 4 Little Essex Street, London, WC2R 3LF, UK
Telephone: 71 836 6633
Fax: 71 379 4980

ISBN: 0 333 39838 6
(English)

A guide to issues and affairs, policy and practice throughout the Community. Its objective is to give those with a need to know, an accessible route to all the information they require, in 500 pages. It covers agriculture, education, the environment, European Monetary Union, law, movement of goods, people, services and capital within the Community. It also lists details of Community institutions including the European Parliament, the Commission (CEC), the Council of Ministers, the Court of Justice and the people who run them.

European Directory of Renewable Energy Suppliers and Services

James & James Science Publishers Ltd,
5 Castle Road, London,
NW1 8PR, UK
Telephone: 71 284 3833
Fax: 71 284 3737

Annual *ISSN: 0266-8041*
(English)

A joint publication with the University of Wales College of Cardiff. Includes an overview of renewable energy in Europe, profiles of renewable energy in Europe (Eastern Europe, Denmark, Finland, The Netherlands, Norway, Spain, France, UK, Portugal, Germany, Italy, Sweden and Switzerland), overview and case studies on energy in buildings and industry, overviews, case studies, glossaries and classified listings on instrumentation and metering, photovoltaics, solar heating, as well as biomass, geothermal, hydro wave and tidal and wind energy. Also includes classified listings on services with a list of organizations active in the field of renewable energy and general information (conferences and courses, journal listing and sources of statistical information, bibliography, A-Z company listing, conversion factors).

FAO Yearbook

Food and Agriculture Organization of the, United Nations, Basic Data Unit, Statistics Division, Rome, I-00100, Italy

Annual *ISSN: 0071-7126*
(English, French and Italian)

Contains FAO regional index numbers of agricultural trade and statistics on trade in agricultural products, trade in agricultural requisites and value of agricultural trade by country. Import and export values of fishery and forestry products are also shown in the countries' value tables. Data on these two items are also included in total merchandise trade but are not included in other totals such as agricultural products total, food and live animals and crude materials, nor are they included in the index numbers of agricultural trade. Detailed statistics on trade in fishery and forest products can be found in the FAO Yearbook of Fishery Statistics and FAO Yearbook of Forest Products.

FAO Yearbook of Fishery Statistics

Food and Agricultural Organization of the, United Nations, Basic Data Unit, Statistics Division, Rome, I-00100, Italy

Annual
(English, French and Italian)

Contains statistical information on worldwide trade in fishery products.

FAO Yearbook of Forest Products

Food and Agriculture Organization of the, United Nations, Basic Data Unit, Statistics Division, Rome, I-00100, Italy

Annual
(English, French and Italian)

Contains statistical information on worldwide trade in forest products.

Farm Chemicals Handbook

Meister Publishing Company, Executive Office: 37733 Euclid Avenue, Willoughby, OH-44094, USA
Telephone: 216 942 2000
Fax: 216 942 0662

Annual
(English)

Includes sine index containing every product, term and regulation mentioned in the Handbook in alphabetical order; fertilizer dictionary (materials and processes, nutrient suppliers, companies/products, tradenames, consumption/production statistics, agronomics/application) and buyers' guide; pesticide dictionary, toxicity charts, biocontrols dictionary and buyers' guide; regulatory file (abbreviations and acronyms, state control officials, EPA regional offices, quick-glance chart of state regulations); safety/application and buyers guide; alphabetical list of farm chemicals manufacturers and suppliers; advertisers' index.

Genetic Engineering News Guide to Biotechnology Companies

Mary Ann Liebert, 1651 Third Avenue, New York, NY-10128, USA
Telephone: 212 289 2300

(English)

A directory of companies in the biotechnology/bioprocess industry. The guide is available as a desk-top book or diskette and contains information about each company such as company officials, date founded, financial sources and number of employees.

A Guide to the British Food Manufacturing Industry - 3rd Edition

Nova Press, available from P R Sheard, University of Bristol, Dept of Meat Animal Science, Langford, Bristol, BS18 7DY, UK
Telephone: 934 852661

ISBN: 0 948786 03 5
(English)

Sections on: industry statistics including marketing data, top 100 food companies, acquisitions and mergers, companies taken into receivership; company information for large, medium and small companies including secretary and registered office, directors, activities, products, divisions, acquisitions, UK and overseas operations, recent developments, five year record of turnover, profit, number of employees, operating revenue and net earnings; company addresses for manufacturers and ingredient suppliers, trade and research associations, consultants and places of education.

The International Biotechnology Directory

The Macmillan Press Ltd, 4 Little Essex Street, London, WC2R 3LF, UK
Telephone: 71 836 6633
Fax: 71 379 4980

Annual *ISBN: 0 333 56553 3*
ISSN: 0265-3877

The Directory is divided into three parts. Part I - International Organizations gives an alphabetical listing with full contact details. As many of the organizations are listed under acronyms, a classification index at the end of the section lists the organizations under subject headings. Part II - Companies, Universities and Government Departments, Institutes and Societies constitutes the core of the Directory and is used most effectively in conjunction with the Buyer's Guide. Entries are listed alphabetically country by country and are described in terms of their products and/or research activities. Part III - Buyer's Guide to Products and Services consists of over 2,000 categories of product or service. Relevant entries are listed under each category, along with their location, so entries can be easily cross-referenced.

International Crop Protection Directory

PJB Publications Ltd, 18-20 Hill Rise, Richmond, Surrey, TW10 6UA, UK
Telephone: 81 948 3262
Fax: 81 948 6866

(English)

Provides a reference of potential business contacts worldwide. Includes agrochemicals, crop biotechnology and biotechnology firms as well as government regulatory agencies, research institutes, trade and other relevant organizations. Three volumes available separately: Europe, The Americas, Rest of the World.

Linscott's Directory of Immunological and Biological Reagents

Linscott's Directory, 4877 Grange Road, Santa Rosa, CA-95404, USA
Telephone: 707 544 9555

(English)

Includes monoclonal antibodies and hybridomas, polyclonal antibodies, tissues, blood products, biochemicals, cytokines, enzymes, lectins, venoms, living organisms and their derivatives and cell cultures, immunoassay kits, other kits and systems, contract and custom service and assay and testing services, as well as supplier names in alphabetical order with their addresses and phone numbers.

Macmillan Directory of Business Information Sources

Macmillan Publishers Ltd, Distributed by Globe Book Services Ltd, Brunel Road, Houndmills, Basingstoke, Hants, RG21 2XS, UK
Telephone: 256 292 42
Fax: 256 479 476

ISBN: 0 333 48982 9
ISSN: 0267-498X
(English)

Provides a guide to publications and organizations relevant to the needs of the non-professional business researcher.

The Merck Index

Merck & Co Inc, Merck Sharp & Dohme Research Labs, PO Box 2000, Rahway, NJ-07065, USA
Telephone: 201 574 4000
Fax: 201 574 4720

New Editions published irregularly. An Encyclopedia of Chemicals, Drugs and Biologicals. Each entry includes: common name; chemical abstracts name; alternative names and/or trademarks; drug code number; molecular formula; molecular weight; percentage composition; literature references; patent and chemical information; biological, pharmacological, etc information; review articles, structure, physical data, derivative, physical data for derivative, trademarks and/or generic names of derivatives, toxicity data, therapeutic category (in humans). Also contains periodic chart, organic name reactions, miscellaneous tables (selected biochemical and immunological abbreviations, code letters used by companies for experimental substances, company register, abbreviated terms, non-proprietary names for radicals and groups proposed by WHO, radioactive isotopes, standard tables, etc), chemical abstracts names and registry numbers for listed common names, formula index, cross index of names and atomic weights.

Northern Ireland Trade Directory, Marketing Development Division

HMSO, HMSO Publications Centre, PO Box 276, London, SW8 5DT, UK
Telephone: 71 873 0011
Fax: 71 873 8200

(English)

IDB Northern Ireland, Industrial Development Board for Northern Ireland, 64 Chichester Street, Belfast BT1 4JX, UK
Telephone: 0232 233233
Fax: 0232 231328

Consists of three volumes: Manufacturing, Services, Directory Index. The Manufacturing volume covers textiles, clothing, food and drink, engineering and general manufacturing sectors. Each entry in the Manufacturing and Services volumes lists company, address, contact name and numbers, name of managing director or proprietor, number of employees, products, brand names and services. The Directory Index volume has three master indexes of company names, brand names and products or services. It also contains addresses for IDB offices worldwide, other support in business, and general institutes and associations, as well as for clothing, engineering, textiles, timber and furniture, and agriculture, food and drink products trade associations, guilds and institutes.

Polmark Directory

Frost and Sullivan Ltd, Sullivan House, 4 Grosvenor Gardens, London, SW1W 0DH, UK

Annual
(English)

Frost and Sullivan Inc, 106 Fulton Street, New York, NY-10038, USA

The European Pollution Control and Waste Management Industry Directory, produced by ECOTEC Research and Consulting Ltd. The Market Analysis discusses industry characteristics, main players, marketing and competition, research and development, finance and performance, market opportunities and industry trends for the European pollution control industry. Organizations are listed by country (Austria, Belgium, Denmark, Finland, France, Germany, Greece, Ireland, Italy, Luxembourg, The Netherlands, Norway, Portugal, Sweden, Switzerland, UK) and directory entries give organization name, address, contact name and numbers, key words, activities, turnover and number of employees. A products and services thesaurus and a technology index ares also given.

The Sitefile Digest 1991 Edition

Aspinwall and Company, Walford Manor, Baschurch, Shropshire, SY4 2HH, UK
Telephone: 939 261144

ISBN: 0 951287 81 8
(English)

A digest of authorised waste treatment and disposal sites in Great Britain. Includes entries for waste management, land and property development, water and effluent management, environmental audits and monitoring, planning, corporate environmental affairs, environmental information and data services. Fuller information on sites is contained in the SITEFILE database and a SITEFILE SEARCH service is available.

State-by-State Biotechnology Directory, Centres, Companies and Contacts

The Bureau of National Affairs Inc, 1231 25th Street NW, Washington, DC-20037, USA

ISBN: 1 55871 173 2
(English)

Prepared by The Biotechnology Information Division of The North Carolina Biotechnology Center. Entries are listed by US state. The Contacts section identifies the person in state government responsible for six areas related to biotechnology: agriculture - plants, agriculture - animals, environment, education/public instruction, science/technology, commerce/business development. In addition it determines who is responsible in each state for overseeing and/or coordinating compliance with certain federal regulations involving biotechnology including, where available, contact name, institution and telephone number. The Regulation section gives details of legislation relevant to biotechnology and presents a synopsis of state regulatory activity, where available. The section on Biotechnology Centres lists non-profit organizations with activities that promote the regional development of biotechnology. Entries include centre name, address, telephone and fax numbers, director, contact, aims, numbers of personnel, funding, programmes, main areas of activity. The Companies section gives names, locations and telephone numbers of more than 750 small and privately financed companies. Large corporations are not shown.

The UK Biotechnology Handbook

BioCommerce Data Ltd, Prudential Buildings, 95 High Street, Slough, Berks, SL1 1DH, UK
Telephone: 753 511777
Fax: 753 512239

(English)

Produced in collaboration with The BioIndustry Association. A useful reference work giving up-to-date, detailed information on over 675 organizations involved in British biotechnology. Divided into five sections (primary producers, resource base, financial sector, academic sector, government framework) covering companies, services, research laboratories and government in the fields of pharmaceuticals, diagnostics, equipment, genetic engineering, agriculture, cell culture, reagents and waste treatment. A subject index lists organizations under 57 topics.

UK Green Growers Guide

CPL Press, CPL Scientific Ltd, Science House, Winchcombe Road, Newbury, Berks, RG14 5QX, UK
Telephone: 635 524064
Fax: 635 529322

ISBN: 1 872691 10 2
(English)

A complete and independent compilation of non-chemical products for plants and animals in the UK aimed at gardeners and farmers wishing to reduce their dependence on synthetic chemicals. The 107 page Introduction includes chapters on Need, Safety, Organisations and Laws as well as How Chemicals and Agrobiologicals Work. The Products Section has a page on each of 300 products - a detailed description including use, suppliers etc. The products in the Guide are for use in homes and gardens, agriculture, glasshouses and for protected crops, horticulture, orchards, amenity woodlands, animal nutrition and forestry in conjunction with systems of organic, low-input and conservation grade farming and integrated pest management.

World Directory of Collections of Cultures of Microorganisms

World Data Center for Microorganisms, University of Queensland, Brisbane, Queensland, Australia

(Mainly English)

Mainly English with contents and introduction also in French, German, Italian, Chinese, Russian, Japanese and Arabic. Produced with the support of UNEP, UNESCO, Australian Development Assistance Bureau and the University of Queensland and endorsed by the World Federation for Culture Collections (WFCC), International Union of Biological Societies (IUBS), International Union of Microbiology Societies (IUMS), Committee on Data for Science and Technology (CODATA) and the International Cell Research Organisation (ICRO). Attempts to include all collections, regardless of size, from all parts of the world. The index lists species and a reference to the collections they are in under the following categories: algae, bacteria (valid and invalid names), fungi, yeasts, lichens, tissue cultures and viruses (animal and bacterial).

World Directory of Collections of Microorganisms

World Data Centre on Microorganisms,
Contact: Dr K Komatagata,
RIKEN Institute, 2-1 Hirosawa Wako,
Saitama, 351-01, Japan
Telephone: 484 62 1111
Fax: 484 64 5651

Also available on-line.

The Worldwide Directory of Agrobiologicals 1993/94 3rd Edition

CPL Press, CPL Scientific Ltd,
Science House, Winchcombe Road,
Newbury, Berks, RG14 5QX, UK
Telephone: 635 524064
Fax: 635 529322

ISBN: 1 872691 80 3
(English)

A comprehensive Directory of biological and other non-synthetic chemical products for agriculture, horticulture, forestry, home and garden, public health and other uses. The Introduction features a Market Summary, a Current Legislative Review and an Overview of Corporate Activities, where there has been vigorous development recently. The Product Directory has been reorganized and numbered to improve product referencing and indexing, making this edition even more 'user-friendly' than before. The Directory includes products for control of insects, weeds and diseases, as silage inoculants, rhizobium inoculants, probiotics for animals, compost, soil improvers and fertilizers. The Appendices contain an index of companies and their products as well as an Active Ingredient Index with the names of all products containing each active ingredient. Companies are also listed alphabetically with addresses, contact names and numbers. With over 1,000 products, 300 companies, and 270 active ingredients, this Directory is essential for every company in the crop protection or crop production business.

The Yearbook of Renewable Energies

EUROSOLAR Germany, Plittersdorfer Strasse 103, Bonn 2, D-5300, Germany
Telephone: 228 36 23 73
Fax: 228 36 12 79

Annual
(English)

Published by the European Solar Energy Association. A source of information for all people interested in current actions, conceptions, events and initiatives concerning renewable energies. Publishes new strategies for the implementation of renewable energies in order to replace fossil and nuclear energies.

DATABASES AND ABSTRACTS

AANSPII

Amino Acid and Nucleotide Sequences of Proteins of Immunological Interest, NIH Bethesda, Building 8/126, MD-20892, USA
Telephone: 301 496 0316

A database of the functional features of amino acid sequences, nucleotide sequences and sequence alignments.

Abstracts in BioCommerce - ABC

BioCommerce Data Ltd, 95 High Street, Slough, Berkshire, SL1 1DH, UK
Telephone: 753 511777
Fax: 753 512239

24 issues per year
(English)

Rapid news indexing and abstracting bulletin published twice monthly with quarterly cumulated editions. Biotechnology newsletters are covered as well as newspaper articles, business publications, trade journals and popular science magazines. Available as an online database through Data-Star as well as through Dialog as BioCommerce Data.

Abstracts of Mycology

BioScience Information Service (BIOSIS), 2100 Arch Street, Philadelphia, PA-19103-1399, USA
Telephone: 215 587 4800

12 issues per year
(English)

Subjects covered include agronomy, air pollution, allergy, biological control, chemotherapy, cytology, food, forestry, genetics, immunology, medicine, neoplasms, pharmaceutical botany, plant biochemistry, biophysics and physiology, plant diseases, public health radiation biology, soil microbiology and toxicology.

AGDEX

Edinburgh School of Agriculture, The Library, West Mains Road, Edinburgh, EH9 3JG, UK
Telephone: 31 668 3471
Fax: 31 667 2601

Covers aspects of practical agriculture and horticulture, including land use, rural development and economics through abstracts from the farming press, trade and business literature and semi-scientific journals. Includes some coverage of Europe and North America. Host-CABI. Updated weekly.

AGREP - Agricultural Research Projects

Commission of the European Communities DG-VI, 200 rue de la Loi, Brussels, B-1049, Belgium

(English)

Available through ECHO. Holds publicly financed research projects in the EC in agriculture, forestry, fisheries, foodstuffs, land use and development, veterinary medicine, food science and technology. The files are updated quarterly. The database is current only since terminated projects are deleted. Host, DIMDI.

Agricultural Engineering Abstracts

See CAB Abstracts

(English)

Agroforestry Abstracts

See CAB Abstracts

(English)

Agrow OnScreen

PJB Online Services, 18/20 Hill Rise, Richmond, Surrey, TW10 6UA, UK
Telephone: 81 948 0751
Fax: 81 332 8991

(English)

PJB Online Services, c/o Mary Dalia, PharmaBooks Ltd, 1775 Broadway, Suite 511, New York, NY-10019, USA
Telephone: 212 262 8230
Fax: 212 262 8234

An electronic news service allowing user to see Agrow articles before they are published and to find any relevant article published since October 1985. Updated daily, Agrow OnScreen brings users Agrow's coverage of the world agrochemical industry including competitor news, key markets, regulations and legislation, and new products and research. Also gives full text of articles from other PJB Newsletters: Scrip (World Pharmaceutical News), Clinica (World Medical Device and Diagnostic News) and Animal Pharm (World Animal Health News). Can be accessed through Data-Star, Dialog or BRS.

American Type Culture Collection - ATCC

12301 Parklawn Drive, Rockville, MD-20852-1776, USA
Telephone: 301 881 2600/800 638 6597
Fax: 301 231 5826

Has detailed catalogues on bacteria, yeasts and fungi as well as cell lines etc for its collections. Offers on-line searching of its hybridoma databank and ATCC on-line.

Analytical Abstracts

The Royal Society of Chemistry, University of Nottingham, Nottingham, NG7 2RD, UK
Telephone: 602 507411

Monthly
(English)

Covers biochemistry, pharmaceutical chemistry, food chemistry, agrochemistry, environmental chemistry, apparatus and techniques, general analytical chemistry and inorganic chemistry.

Animal Breeding Abstracts

See CAB Abstracts

(English)

The Applied Environmetrics Meteorological Tables

Lewis Publishers Inc, 2000 Corporate Blvd NW, Boca Raton, FL-33431, USA

(English)

Software package supplied on one 5.25 inch MS-DOS diskette. The tables include pressure, density, temperature, potential temperature with height, geodetic tables and the reference atmosphere, solar position, times of sunrise and sunset, incident radiation on slopes, blackbody radiation, density, viscosity, fall velocity of particles, saturation pressure and latent heats for pure water and ice, psychometric tables of moisture parameters given dry and wet or ice bulb readings, and thermodynamic temperatures for moist air, including equivalent potential and wet bulb potential temperatures.

BDT/FTPT

Rua Latino Coelho 1301, CP 1889, Campinas, SP-13085, Brazil

Tropical database developed in Brazil covering national information networks on culture collection holdings, research programmes and projects as well as a directory of experts in microbiology.

Beilstein

The Beilstein Institute,
Springer-Verlag, Heidelberger Platz 3,
Berlin, D-1000 33, Germany
Telephone: 30 8207 1

Produced by the Beilstein Institute and Springer-Verlag Inc. The on-line version of the Beilstein Handbook on organic chemistry. Contains information on around 5 million organic compounds.

BEST - British Expertise in Science & Technology

Longman Cartermill Ltd,
Technology Centre, St Andrews, Fife,
KY16 9EA, UK
Telephone: 334 77660

Bibliography of Agriculture

Oryx Press,
2214 North Central at Ecanto, Phoenix,
Arizona, AZ-85004, USA

12 issues per year
(English)

Covers journal articles, pamphlets, government documents, special reports and proceedings of conferences on all aspects of agriculture worldwide.

BIKE - Biotechnology Information Knot for Europe GBF mbH

Gesellschaft fur Biotechnologishe
Forschung, Mascheroder Weg 1,
Braunschweig, D-3300, Germany
Telephone: 531 6181
Fax: 531 6181

(Mainly German)

A biotechnology information service using the BIKE database developed and maintained by GBF, covering biotechnology information from Germany, Switzerland and Austria.

Biochemistry Abstracts

Cambridge Scientific Abstracts,
5161 River Road, Bethesda,
MD-20816, USA

12 issues per year
(English)

Comprises three parts: Part 1, Biological Membranes; Part 2, Nucleic Acids; Part 3, Amino Acids, Peptides and Proteins.

BioCommerce Abstracts and Directory Online

BioCommerce Data Ltd, 95 High
Street, Slough, Berks, SL1 1DH, UK
Telephone: 753 511777
Fax: 753 512239

(English)

Features: product, patent and market news; multiple citations; twice monthly updates; abstracts and profiles; detailed company information; contact names and addresses. Accessible via the BioCommerce Data Bureau Service, DATA-STAR and DIALOG.

Biodeterioration Abstracts

See CAB Abstracts

(English)

Provides a regular survey of the world literature on biodeterioration of materials, artifacts of facilities of economic importance, including foodstuffs, timber, fuels, construction materials etc. Covers effects of all living organisms including microorganisms, insects, rodents and higher plants. Also deals with biodegradation of waste materials to less objectionable, more easily disposable or higher value products by living organisms.

Biological Abstracts

BioScience Information Service (BIOSIS), 2100 Arch Street, Philadelphia, PA-19103-1399, USA
Telephone: 215 587 4800

24 issues per year
(English)

Covers all biological subjects. Contains a Biosystematic Index, a Generic Index, a Concept Index and a Subject Index which uses a keyword system.

BIOREP

Library of the Royal Netherlands Academy

(English)

Database describing biotechnological projects carried out in the twelve member states of the European Community. Covers approximately 70% of the publicly funded research projects in the EC. Available through the ECHO-HOST service of the Commission of the European Communities.

BIOSIS Previews

BioSciences Information Service, 2100 Arch Street, Philadelphia, PA-19103-1399, USA
Telephone: 215 587 4800
Fax: 215 587 2016

On-line database covering bibliographic material printed in Biological Abstracts and BioResearch Index including material on all life science topics such as agriculture, biochemistry bioengineering, biophysics, cell biology, clinical medicine, genetics immunology, microbiology, pathology, physiology, pharmacology and toxicology. Available through BRS, CAN/OLE, Data-Star, DIMDI, Dialog, ESA-IRS, Mead and SDC. Also produces BioBusiness which contains news on economic trends and forecasts, US patents etc.

BIOSIS/CAS Selects

Chemical Abstracts Service, 2540 Olentangy River Road, PO Box 3012, Columbus, OH-43210, USA
Telephone: 614 421 3600
Fax: 810 482 1608

(English)

A series of 29 biweekly current awareness publications covering the literature in biological chemistry, pharmacology and medicine. Topics include Allergy, Antiviral agents, Bacterial and Viral Genetics, DNA Replication, Food and Drug Replication, Genetic Manipulation in Plants, Hormones and Gene Expression, Interferon, Leukotrienes, Lymphokines, Monoclonal Antibodies, Peptide and Protein Sequences, Plant Genetics and Vitamins.

Biotech Knowledge Sources - BKS

BioCommerce Data Ltd, 95 High St, Slough, Berks, SL1 1DH, UK
Telephone: 753 511777
Fax: 753 512239

12 issues per year
(English)

A monthly bulletin alerting subscribers to over 600 books, 150 market surveys and 1,000 conferences on biotechnology each year. Also includes feature articles on key information resources such as databases and CD-ROM services, as well as details of periodicals, software and videos, government reports and training courses. Available on-line from MSDN.

Biotechnology Abstracts

Derwent Publications Ltd, Rochdale House, 128 Theobalds Road, London, WC1X 8RP, UK
Telephone: 71 242 5823

24 issues per year
(English)

Prepares and provides abstracts of literature in all areas of biotechnology, including patents.

Biotechnology Citation Index

Institute for Scientific Information,
3501 Market Street, Philadelphia,
PA-19104, USA
Telephone: 215 386 0100
Fax: 215 386 6362

Bimonthly
(English)

A CD-ROM database. Recently published research from all areas of biotechnology, including basic research in genetics, molecular biology and microbiology; and applications in medicine, agriculture, industry and the environment.

Biotechnology Japanese Patent Alert

A Fujiwara Scientific Liaison Office,
1059-7 Rueta, Kamakura-shi,
Kanagawa-ken 248, Japan
Telephone: 467 31 7833
Fax: 467 31 5491

32 issues per year
(English)

Contains a concise summary of the Japanese gazette 'Published Unexamined Patent Applications' on biotechnology.

Biotechnology Patent Fast-Alert

Current Patents Ltd, 34-42 Cleveland Street, London, W1P 5FB, UK
Telephone: 71 323 0323
Fax: 71 580 1938

52 issues per year
(English)

Weekly bulletin launched September 1990 providing summaries of relevant patents and patent applications from the USA, British, World (PCT) and European collections within 10 days of patent availability as well as a 'same-day' delivery service for every patent listed.

Biotechnology Research Abstracts

Cambridge Scientific Abstracts,
5161 River Road, Bethesda,
MD-20816, USA

7 issues per year
(English)

Covers all aspects of biotechnology from basic genetic engineering to medical, chemical, pharmaceutical, agricultural and aquacultural applications as well as process engineering and production methods. 7 issues per year plus cumulative index.

The British Library Biotechnology Information Service

The British Library, Science Reference Library, 25 Southampton Bldgs,
Chancery Lane, London,
WC2A 1AW, UK
Telephone: 71 405 8721

Provides a search service for documents in biotechnology and related fields.

The British Library Science Reference and Information Service - SRIS

Holborn Reading Rooms,
25 Southampton Buildings, London,
WC2A 1AW, UK
Telephone: 71 323 7494/7496
Fax: 71 323 7495

British and EPO patent equiries
Telephone: 71 323 7919
Business enquiries
Telephone: 71 323 7454

Offer business information on companies, markets and products; physical sciences and technologies; engineering; British, European and Patent Co-operation Treaty patents and trade marks. Also offer access to related CD-ROM databases.

The British Library Science Reference and Information Service - SRIS

Chancery House Reading Room,
Chancery House, London,
WC2A 1AW, UK
Telephone: 71 323 7902

Information available on foreign patents and trade marks.

The British Library Science Reference and Information Service - SRIS

Aldwych Reading Room, 9 Kean Street, London, WC2B 4AT, UK
Telephone: 71 323 7288
Fax: 71 323 7217

Provide information on the life sciences and technologies, including biotechnology, medicine and agriculture.

CA Search

Chemical Abstracts Service,
2540 Olentangy River Road,
PO Box 3012, Columbus,
OH-43210, USA
Telephone: 614 421 3600
Fax: 810 482 1608

On-line database covering material printed in chemical abstracts since 1967 including journals, patents, technical reports, conference proceedings and dissertations from all areas of chemistry. Available on :BRS, CAN/OLE, CAS online, STN Data-Star, Dialog, CNIC, VCH and JAIC.

CA Selects

Chemical Abstracts Service,
2540 Olentangy River Road,
PO Box 3012, Columbus,
OH-43210, USA
Telephone: 614 421 3600
Fax: 810 482 1608

(English)

A series of 164 current awareness bulletins published once every two weeks selected by computer from the CAS database according to a precise special interest profile. Areas of life sciences covered include topics such as: Amino Acids, Peptides and Proteins; Beta-Lactam Antibiotics; Electrophoresis; Flavours and Fragrances; Prostaglandins; Waste Recycling; Ultrafiltration etc.

CAB Abstracts

CAB International Information Services, Wallingford, Oxfordshire, OX10 8DE, UK
Telephone: 491 32 111
Fax: 491 33 508

Contains worldwide information on agriculture, covering aspects of food science, crop science, animal science, biotechnology and economics. Over 12,000 serials and many other publications are scanned, and about 150,000 bibliographic records are compiled each year. From this computerized database is derived a wide range of information services and products in printed and electronic form, designed to meet a variety of customer needs. Copies of most articles cited in the CAB Abstracts journals and database are available from the CAB International Library Services Centre (see International Organizations). Documents can also be ordered via DIALOG's DIALORDER or ESA/IRS's PRIMORDIAL. Subscribers can also receive abstract journals in computerized format. Electronic journals are available in comma-delimited, ISO2709 or Pro-Cite compatible format. The diskettes (either 3.5 inch or 5.25 inch) are supplied without retrieval software but in standard format so they can be easily used with most commercial database management software. Sample diskettes are available on request. The database is also available on CD-ROM (CAB Abstracts on Compact Disk - CABCD) suitable for use with IBM-compatible PCs or Macintosh computers. Volume One contains bibliographic citations and abstracts for 1984-86; the second volume contains the 1987-89 data; the third volume contains the 1990-91 data. CABCD will be updated annually. Organizations with access to mini or mainframe computers may lease all or part of CAB Abstracts on magnetic tape. Update tapes are issued monthly, and retrospective tapes as required. The CAB SDI (Selective Dissemination of Information) Service is a current awareness service based on the most recent input to the database. Subscribers decide their own search profile and receive search results at intervals of their choice, either as computer printout or on diskettes suitable for most microcomputers. CAB Alerts is a monthly SDI service providing subscribers with the search results from standard profiles chosen by CAB International. Numerous titles are available.

CABS - Current Awareness in Biological Sciences

Pergamon Journals Ltd,
Headington Hill Hall, Oxford, Oxon, OX3 0BW, UK
Telephone: 865 794141
Fax: 865 60285

12 issues per year
(English)

Provides monthly abstracts service on pure and applied science in a series of publications under the general title of Current Advances. Topics are listed under the entry for the on-line services offered by CABS. Based on scanning several thousand source journals. Also offers a CABS-Text service which makes available photocopies of the full text of any article listed through 'Information on Demand', an information retrieval company associated with the Pergamon Group.

CABS Database Online

Pergamon Infoline Inc,
1340 Old Chain Bridge Road, McLean, VA-22101, USA

Offers direct access to the following sections of the CABS database published as hardcopies as Current Advances in: Biochemistry; Cell and Development Biology; Clinical Chemistry; Ecological Sciences; Endocrinology; Genetics and Molecular Biology; Immunology; Microbiology; Neurosciences; Pharmacology and Toxicology; Physiology; Plant Sciences.

CARBBANK

Contact: Dr S Doubet, Complex Carbohydrate Research Centre, Russell Lab, PO Box 5677, Athens, GA-30613, USA
Telephone: 404 546 3555

Covers index terms, literature citations and sources of carbohydrate sequences, especially those associated with structures with biological activity of proteins, cell surface recognitions, immunology and site-specific binding.

CAS Biotech Updates

Chemical Abstracts Service,
2540 Olentangy River Road,
PO Box 3012, Columbus,
OH-43210, USA
Telephone: 614 421 3600
Fax: 810 482 1608

26 issues per year
(English)

A series of current awareness bulletins covering a wide range of biotechnology applications in agriculture, chemical manufacture, energy, pharmaceuticals and environmental protection. Sections cover agriculture, biopolymer sequences, biosensors, environmental biotechnology, genetic engineering and pharmaceutical applications. Also include 'Biotech Industry Notes' which cover news of production, pricing, processing, legislation, government activities and people. Also available from RSC (UK), CNIC (France), VCH (Germany) and JAICI (Japan).

CAS ONLINE

Chemical Abstracts Service,
2540 Olentangy River Road,
PO Box 3012, Columbus,
OH-43210, USA
Telephone: 614 421 3600
Fax: 810 482 1608

Provide access to two patent information files, AgPat, for chemical and biochemical patents related to agriculture and pest control, and PharmPat, for patents in pharmaceuticals, medicinal chemistry, biotechnology and drug-related fields as well as abstracts from Chemical Abstracts, substance and use indexes, summaries of substance claims etc, available only through STN International.

CASORF

Chemical Abstracts Service, Customer Service, PO Box 3012, Columbus, OH-43210, USA
Telephone: 800 848 6538

Provides details of chemical substances, index terms, molecular formulae, molecular struictures and literature citations.

CBA Database

Royal Society of Chemistry,
University of Nottingham, Nottingham,
NG7 2RD, UK
Telephone: 602 507411

On-line equivalent of Current Biotechnology Abstracts, available through Data-Star, Pergamon Infoline or ESA-IRS or by direct interaction with the RSC help desk.

CCD - Cambridge Structural Database

Contact: Dr O Kennard, Cambridge Crystallographic Data Centre,
University Chemical Laboratory,
Lensfield Rd, CB2 1EW, UK
Telephone: 223 336 409

CEA Database

Royal Society of Chemistry,
University of Nottingham, Nottingham,
NG7 2RD, UK
Telephone: 602 507411

On-line equivalent of Chemical Engineering Abstracts, available through Data-Star, Pergamon Infoline or ESA-IRS or by direct interaction with the RSC help desk.

CFISM

Institut National Agronomique, 16 rue Claude Bernard, Paris Cedex 05,
F-75321, France
Telephone: 1 43 37 15 50

French database of microbial strains.

Chemical Abstracts

Chemical Abstracts Service,
2540 Olentangy River Road,
PO Box 3012, Columbus,
OH-43210, USA
Telephone: 614 421 3600
Fax: 810 482 1608

52 issues per year
(English)

Covers papers of chemical, biochemical and chemical engineering interest as well as patents in these fields. Includes Keyword, Patent and Author Indexes. 52 issues per year in 2 vol.

Chemical Engineering Abstracts

Royal Society of Chemistry, Thomas Graham House, Science Park, Milton Road, Cambridge, CB4 4WF, UK
Telephone: 602 507411

12 issues per year
(English)

Chemical Engineering and Biotechnology Abstracts

Dechema, Postfach 970146, Theodor Hause Allee 25, Frankfurt am Main, D-6000, Germany
Telephone: 69 75 64 235

Contains theoretical, applied and business material on chemical engineering and biotechnology.

CHEMTRACTS - Biochemistry and Molecular Biology

Data Trace Chemistry Publishers Inc, PO Box 1239, Brooklandville, MD-21022, USA
Telephone: 800 342 0454

6 issues per year
(English)

Designed to give scientists rapid access to key research articles in a unique format intended to initiate and stimulate new research activity. Current research areas covered include membranes, biochemistry, protein structure, RNA, DNA, enzymology, genome sequencing, immunology and cell biology. Each research study is accompanied by a full range of analysis, including purpose of study, what the researchers accomplished, background information on methods, researcher's approach, commentary on research implications, references. Includes an annual cumulative index.

CHEMTRACTS - Inorganic Chemistry

Data Trace Chemistry Publishers Inc, PO Box 1239, Brooklandville, MD-21022, USA
Telephone: 800 342 0454

6 issues per year
(English)

This new journal gives scientists rapid access to the most exciting recent research articles, along with the benefit of indepth commentary on the significance of the article from the world's leading authorities in the field of inorganic chemistry. Among other topics covered is the important area of bioinorganic chemistry.

CHEMTRACTS - Organic Chemistry

Data Trace Chemistry Publishers Inc, PO Box 1239, Brooklandville, MD-21022, USA
Telephone: 800 342 0454

6 issues per year
(English)

Brings to light the types of molecules that are biologically active and the mechanisms of how they operate. The journal is an all-important seeding ground between new academic research and research at pharmaceutical houses. Among the topics covered are catalytic antibodies, enzyme mechanisms and medicinal chemistry.

Chromatography Abstracts

Elsevier Applied Science, Crown House, Linton Road, Barking, Essex, IG11 8JU, UK
Telephone: 81 594 7272

10 issues per year *ISSN: 0268-6287*
(English)

Covers all major world journals for papers reporting advances in chromatographic techniques and their application to specific problems.

CNIC - Centre National de l'Information Chimique

La maison de la Chimie, 28 ter rue St Dominique, Paris, F-75007, France
Telephone: 1 45 51 37 40

Serves France, Monaco and many African nationals for all CAS services.

COMPASS - Computer Assisted Yeast Identification System

National Collection of Yeast Cultures (NCYC), AFRC/IFR, Colney Lane, Norwich, NR4 7UA, UK
Telephone: 603 56122
Fax: 603 507723

(English)

Consists of a probabilistic system for yeast identification that allows the user to enter the results of up to 46 morphological and physiological tests and receive the probability of match with any of 460 species. Available through NCYC (UK).

COMPENDEX

Engineering Information

Bibliographic database on all fields of engineering, including information on biosensors, biology, electrochemistry and biophysics. Corresponds to the Engineering Index Monthly.

CORDIS

ECHO Customer Service, BP 2373, L-1023, Luxembourg

European Community R&D Information Service. Available through the ECHO host system, the CORDIS databases include: information on programmes under preparation; descriptions of programmes, key references, internal contacts etc; descriptions of projects and the participating organizations; abstracts of documents and technical reports resulting from projects; information contacts within member states of the EC; expressions of partnership interests; descriptions of results and R&D prototypes; news and announcements (call for proposals, etc).

Crop Physiology Abstracts

See CAB Abstracts

(English)

CSA Biochemistry Abstracts

Cambridge Scientific Abstracts, 5161 River Road, Bethesda, MD-20816, USA

12 issues per year
(English)

Abstracts are divided into the following parts: Part 1, Biological Membranes; Part 2, Nucleic Acids; Part 3, Amino Acids, Peptides & Proteins. 12 issues per year plus cumulative index.

CSRS - Compilation of Small RNA Sequences

Contact: Dr R Reedy, Dept of Pharmacology, Baylor College of Medicine, Houston, TX-77030, USA
Telephone: 713 799 4458

Current Advances in Genetics & Molecular Biology

Pergamon Press Inc,
395 Saw Mill River Road, Elmsford, NY-10523, USA
Telephone: 914 592 7700
Fax: 914 592 3625

12 issues per year
(English)

Abstracting service produced from the CABS Database, covering all aspects of genetics and molecular biology, listing titles, authors and addresses.

Current Advances in Microbiology

Pergamon Press Inc,
395 Saw Mill River Road, Elmsford,
NY-10523, USA
Telephone: 914 592 7700
Fax: 914 592 3625

12 issues per year
(English)

Abstracting service produced from the CABS Database, covering all aspects of microbiology, listing titles, authors, bibliographic details and addresses.

Current Awareness in Biological Sciences

Current Awareness in Biological Sciences, Pergamon Press plc, Oxford, OX3 OBW, UK
Telephone: 865 64881

Bibliographic database of information on the biological sciences including biochemistry, microbiology, physiology and plant science.

Current Biotechnology Abstracts

Royal Society of Chemistry,
University of Nottingham, Nottingham,
NG7 2RD, UK
Telephone: 602 507411

12 issues per year
(English)

Searches from over 200 journals and newsletters covering biotechnology, microbiology, genetics, medicine, biochemistry, immunology, pharmaceuticals, fermentation etc, supplemented by computer searches of Chemical Abstracts and Biological Abstracts monitoring over 20,000 journals on a worldwide basis.

Current Clinical Chemistry

Pergamon Press Inc,
395 Saw Mill River Road, Elmsford,
NY-10523, USA
Telephone: 914 592 7700
Fax: 914 592 3625

12 issues per year
(English)

Includes congress commentaries, short reviews and comments on instrumentation advances.

Current Contents

Institute for Scientific Information -
ISI, 3501 Market Street, Philadelphia,
PA-19104, USA
Telephone: 215 386 0100
Fax: 215 386 6362

52 issues per year
(English)

ISI European Branch, 132 High Street,
Uxbridge, UB8 1DP, UK
Telephone: 895 270016
Fax: 895 256710

Reproduces the contents pages of over 850 journals worldwide as well as multi-author books on: Agricultural, Biological and Environmental Sciences; Chemical and Earth Sciences; Clinical Practice; Engineering, Technology and Applied Sciences. Topics can be identified using the Title World Index. There is also an Authors Index, and a Publishers' Address Index. Now also available on diskette for IBM PC or Apple Macintosh.

Current Contents on Diskette - Life Sciences

Institute for Scientific Information -
ISI, 3501 Market Street, Philadelphia,
PA-19104, USA
Telephone: 215 386 0100\800 336 4474
Fax: 215 386 6362

Weekly
(English)

ISI European Branch, 132 High Street, Uxbridge, UB8 1DP, UK
Telephone: 895 2700116
Fax: 895 256710

Database of bibliographic information from the world's leading journals. Seven multidisciplinary editions are available, including: life sciences (available in editions covering either 600 or 1200 different journals); agriculture, biology and environmental sciences; clinical medicine. Includes comprehensive search facilities.

CUTG

Contact: Dr T Ikemura, National Institute of Genetics, Mishima, Shizuoka, 411, Japan
Telephone: 559 75 0771

Dairy Science Abstracts

See CAB Abstracts

(English)

DCT - *Drosophila* Codon Tables

Contact: Dr M Ashburner, Department of Genetics, University of Cambridge, Cambridge, UK
Telephone: 223 333 969

Covers codon usage, nucleotide sequences (including cross-references), base composition and base comparisons.

DECHEMA - Dechema Chemical Engineering and Biotechnology Abstracts Database

Dechema, Postfach 970146, Theodor Hause Allee 25, Frankfurt am Main, D-6000, Germany
Telephone: 75 64 235

(German with English titles in most cases)

On-line database containing references to the German and International scientific and technical literature in areas of chemical technology, biotechnology, chemical equipment, chemical plant, pollution control and safety control. May be accessed through STN.

DEQUIP

DECHEMA, Theodor Hause Allee 44, Postfach 97 01 46, Frankfurt a/Main, D-6000, Germany
Telephone: 69 75 64 46

On-line product database listing manufacturers and products of around 2800 companies from all over the world which supply apparatus and chemical equipment for use in chemical engineering and biotechnology. Product areas include chemical plant, pumps and compressors, process and control instrumentation, automation equipment, materials testing equipment, laboratory equipment, analytical chemicals, computer aided engineering software and hardware as well as industrial safety equipment. May be accessed through STN.

Diagnostics Business

Cowper House, 67 High Street, Tring, Herts, HP23 4AB, UK
Telephone: 282 8504
Fax: 282 7272

12 issues per year
(English)

International digest of the latest activity in *in vitro* diagnostics.

DIANEGUIDE

Information Market Development Group

(English)

Information on databases and databanks, database producers and host organization available on DIANE (Direct Information Access Network for Europe). Only on-line databank that gives the users immediate access, to a wide range of information on on-line services offered in Europe. Covers more than 850 databases in all subject areas. Available through ECHO-HOST

Dnastar Inc

1228 South Park Street, Madison, WI-53715, USA
Telephone: 608 258 7420
Fax: 608 258 7439

Supply a comprehensive sequence analysis software package available for IBM PC compatible and Mackintosh computers. Quarterly updates keep investigators abreast of nine databases (GenBank, NBRF-PIR, SwissProt, EMBL, AIDS(HIV), Berlin RNS, Genetic Medline, Roberts Enzyme and Brookhaven Protein DataBank).

Dnastar Ltd

St James House,
105-113 The Broadway, West Ealing, London, W13 9BL, UK
Telephone: 81 566 1200
Fax: 81 566 3377

Design, develop, sell and support molecular biology/genetic sequence analysis software and databases on CD-ROM. Currently developing database resources, CD-ROM resources for molecular biology, DNA sequence analysis, sequencing products and protein sequence analysis and database resources. Subsidiary of Dnastar Inc, US.

DOMIS

Commission of the EC - DG-XIII, Boite Postal 777, Luxembourg

(English)

Directory of Materials Data Information Sources. On-line directory of information sources, materials and services in Europe (eg databanks and bases, technical centres, scientific and technical laboratories, experts, etc.). Areas include: metals, alloys and steel; ceramics and glass; composite materials, coating and joints, plastics and rubber, etc. Covers more than 250 information sources and services. Available through ECHO-HOST.

DRHPL - Database for the Repository of Human Probes and Libraries

ATCC, 12301 Parklawn Drive, Rockville, MD-20852-1776, USA
Telephone: 301 231 5586

Cotains details of relevant clones and RFLPs (restriction fragment length polymorphisms), including literature citations and (gene) library characterizations.

EABS

Commission of the European Communities DG-XIII,
Boite Postal 777, Luxembourg

(English, French, German)

On-line version of the monthly journal Euro-Abstracts. References to the published results of scientific and technical research programmes wholly or partly sponsored by the Commission of the European Communities. Wide range of subject areas (eg nuclear research, new sources of energy, environmental research, etc). Covers more the 50000 citations since 1966. Available through ECHO-HOST.

ENREP

Commission of the European Communities DG-XIII,
Boite Postal 777, Luxembourg

(English)

On-line directory of environmental research projects in the member states of the European Community. Covers 50000 research projects from over 5000 organisations from 1980 onwards. Available through ECHO-HOST.

ENZIDEX

Biocatalysts Ltd, Main Avenue, Treforest Industrial Estate, Pontypridd, CF37 5UT, UK

An enzyme databank.

EPA's Sampling and Analysis Methods Database

Lewis Publishers Inc,
2000 Corporate Blvd NW, Boca Raton, FL-33431, USA

(English)

Compiled by EPA chemists to permit rapid searches of sampling and analytical method summaries. Organized into a three-volume set. Features a menu-driven program with free text searches of many organic and inorganic analytes. The three volumes are Industrial Chemicals; Pesticides, Herbicides, Dioxins and PCPS; Elements and Water Quality Parameters.

EUREKA

EUREKA Secretariat

(English and French)

On-line directory of more than 10000 theses and studies, both current and completed, which have been conducted since the early 1950s. The theses and studies deal with Community policies (competition law, external relations, European institutions, etc). Information on over 5000 professors and university researchers who are studying the construction of Europe. Since 1952. Available through ECHO-HOST.

European Bank of Computer Programs in Biotechnology

Delft University of Technology,
Julianlaan 67, BC Delft, NL-2628,
The Netherlands
Telephone: 15 782353
Fax: 15 782355

European Molecular Biology Laboratory - EMBL

Postfach 102240, Heidelberg 1,
D-6900, Germany

Has established a European nucleotide sequence data library.

Excerpta Medica

Excerpta Medica, Postbus 1126,
BC Amsterdam, NL-2628,
The Netherlands
Telephone: 20 26 44 38

(English)

Produces 47 separate abstracting journals covering all fields of medical interest including: Immunology, Microbiology, Bacteriology, Mycology and Parasitology, and Toxicology. 10-20 issues per year for each section.

The Farmstat System

(Farmstat International)

Internationally compatible information database about the size and composition of the market for a number of important crop inputs: seeds, agrochemicals and fertilizer. Also generates agronomic evidence about the relationship between inputs and yields as a result of a field-based recording system.

Field Crop Abstracts

See CAB Abstracts

(English)

Focus On: Global Change

Institute for Scientific Information,
3501 Market Street, Philadelphia,
PA-19104, USA
Telephone: 215 386 0100
Fax: 215 386 6362

Biweekly
(English)

Bibliographic diskette database with abstracts regarding human interaction with the environment. Brings together published information from disciplines including the environmental and earth sciences, meteorology and atmospheric science, agriculture, and ecology.

Food Science and Technology Abstracts

International Food Information Service,
Lane End House, Shinfield, Reading,
RG2 9BB, UK
Telephone: 734 883895

12 issues per year
(English)

Covers all aspects of food science and technology and includes information on patents, standards and legislation. FSTA is also available on-line (CISTI, DATA-STAR, DIALOG, DIMDI, ESA-IRS, ORBIT, STN) and on a CD-ROM, which holds all the records ever published in FSTA for use with IBM PC compatibles. Other IFIS addresses are listed under International Organizations.

Forest Products Abstracts

See CAB Abstracts

(English)

Forestry Abstracts

See CAB Abstracts

(English)

FROSTI - Food RA Online Scientific and Technical Information Database

Leatherhead Food RA, Randells Road,
Leatherhead, Surrey, KT22 7RY, UK
Telephone: 372 376761

Produced by the Food RA, which also acts as host. Provides reference to literature on scientific and technical aspects of food manufacturing. Covers meat, confectionary, edible oils, fruits, vegetables, etc from the points of view of microbiology, quality control, health and nutrition. Updated weekly.

GBSOFT - GenBank Software Clearing House

Intelligenetics Inc, 700 El Camino Real
East, Mountain View, CA-94040, USA
Telephone: 415 926 7364

GC - Gene Communications

Gene Communications,
Contact: Dr J Schmidtke
Box 3426, Gottingen, D-3400,
Germany
Telephone: 551 397 599

GENBANK

Bolt Beranek & Newman Inc,
Computer Systems Division, 10
Moulton Street, Cambridge,
MA-02238, USA
Telephone: 415 962 7364

Repository for all published nucleic acid sequences greater than 50 base pairs, catalogued and annotated for sites of biological interest. Available for a nominal fee as printed compendium or computer magnetic tape; limited on-line access. Sponsored by the US Government and Los Alamos National Laboratory. Includes functional features, index terms, literature citations and taxonomic identification.

GeneSeq

Derwent Publications Ltd,
Rochdale House, 128 Theobalds Road,
London, WC1X 8RP, UK
Telephone: 71 242 5823

Database comprising information on nucleic acid and protein sequences derived from patent applications and granted patents. Includes all nucleotide sequences longer than nine bases, all protein sequences longer than three amino acids and probes of any length. Covers all patents issued since 1989, with sequences from earlier patents entered in batches, starting with 1988 and going back at least to 1981.

Genetics Abstracts

Cambridge Scientific Abstracts,
5161 River Road, Bethesda,
MD-20816, USA

12 issues per year
(English)

Covers all aspects of molecular, viral, bacterial, fungal, algal, plant, animal and human genetics. 12 issues per year plus cumulative index.

Hazardous Chemical Regulations Database

Lewis Publishers Inc,
2000 Corporate Blvd NW, Boca Raton,
FL-33431, USA

(English)

Contains regulatory information for over 500 chemicals. Seven major regulatory groups of standards, including the Resource Conservation & Recovery Act (RCRA), Comprehensive Environmental Response, Compensation & Liability Act (CERCLA) and Superfund Amendment & Reauthorization Act (SARA), Clean Water Act (CWA), Safe Drinking Water Act (SDWA), 1990 Clean Air Act (CAA), Worker Exposer Information, and other information. Revised annually.

Helminthological Abstracts

See CAB Abstracts

(English)

Herbage Abstracts

See CAB Abstracts

(English)

HGIR - Human Genome Information Resource

Contact: Dr T Marr, MS K710,
Los Alamos National Laboratory,
Los Alamos, NM-875, USA
Telephone: 505 665 0838

Contains information concerning human genetic sequences including details of probes, restriction fragments and physical maps.

HGMCR - Human Genetic Mutant Cell Repository

Catalogue of Cell Lines, Coriell Institute for Medical Research,
Copewood and Davis Streets, Camden,
NJ-08103, USA
Telephone: 609 966 7377

Details cells lines together with index terms.

HGML - Human Gene Mapping Library

Howard Hughes Medical Institute,
25 Science Park, New Haven,
CT-06511, USA
Telephone: 203 786 5515

Covers DNA probes, restriction length fragment polymorphisms and genetic maps.

Horticultural Abstracts

See CAB Abstracts

(English)

Hybridoma Data Bank

National Research Council, Piazzale Aldo Moro 7, Roma, I-00100, Italy
Telephone: 6 49 93

Provides detailed information on hybridomas and their properties.

Hybridoma Data Bank

ATCC, 12301 Parklawn Drive, Rockville, MD-20852, USA
Telephone: 301 231 5585
Fax: 301 231 5826

Provides detailed information on hybridomas and their properties. An international databank on cloned cell lines and their immunoreactive products.

Hybridoma Data Bank

CERDIC, 2 C A I Ave des Muarettes, Velleneuve-Loubet, F-06270, France
Telephone: 93 20 01 80
Fax: 93 20 01 81

Provides detailed information on hybridomas and their properties.

Hybridoma Data Bank

RIKEN, 2-1 Hirosawa, Wako-shi, Saitama 351-01, Japan
Telephone: 48 462 1111
Fax: 48 464 5651

Provides detailed information on hybridomas and their properties.

Hybridoma Data Bank

PHLS Centre for Applied Microbiology, ECACC Biologics Division, Porton Down, Salisbury, SP4 0JG, UK
Telephone: 980 610 391
Fax: 980 611 315

Provides detailed information on hybridomas and their properties.

Hybridoma Database

University of Nice/CODATA, Parc Valrose, F-06034, France
Telephone: 93 52 98 98

ICDB - Immunoclone Database

CERDIC, 2 C A I Ave des Muarettes, Velleneuve-Loubet, F-06270, France
Telephone: 93 20 01 80
Fax: 93 20 01 81

A database of immunoclones and their related products.

ICGEBnet

Padriciano 99, Trieste, I-34012, Italy
Telephone: 40 226 556
Fax: 40 226 556

IES-DC

Commission of the European Communities DG-XIII,
Boite Postal 777, Luxembourg

(English)

Information Exchange System - Data Collections. Directory and reference service to the European IT (Information Technology) Community. EIS-DC currently covers three separate domains:
1. Publicly funded research and development programmes carried out in Europe by institutions active in the field of information technology.
2. Resources and facilities for network services to be used by institutions active in this domain.
3. Addresses in the various electronic mail systems of individuals involved in research and management, mainly within the ESPRIT programme.
Covers more than 2000 projects, 1000 resources and facilities and 4200 addresses. Available through ECHO-HOST.

Immunology Abstracts

Cambridge Scientific Abstracts,
5161 River Road, Bethesda,
MD-20816, USA

12 issues per year
(English)

Covers immune systems of man and animals, including methodology, response to infections, tumours and transplantation, genetic characteristics and disorders of the immune system. Each issue contains an author and a subject index with one cumulative index per year. Also available on magnetic tape.

In Vitro Conservation Database

Internat Board for Plant Genetic Resources, Via Delle Terme di Caracalla, Rome, I-00100, Italy

Covers plant cell culture techniques.

Index Medicus

US Department of Health & Human Services, National Library of Medicine, 8600 Rockville Pike, Bethesda, MD-20209, USA

12 issues per year
(English)

Covers about 2600 of the world's biomedical journals. Includes Bibliography of Medical Reviews, Subject Section and Author Section.

INIST Diffusion

Institut de l'Information Scientifique et Technique, 2 alle du parc de Brabois, Vandoeuvre l Nancy, F-54514, France
Telephone: 83 50 46 64
Fax: 85 50 46 66

International Pharmaceutical Abstracts

American Society of Hospital Pharmacists, 4630 Montgomery Avenue, Bethesda, MD-20814, USA

24 issues per year
(English)

Includes journal list, subject index and index guide.

Irrigation and Drainage Abstracts

See CAB Abstracts

(English)

ISI-Europe

132 High Street, Uxbridge,
UB8 1DP, UK
Telephone: 895 270016
Fax: 895 256710

Japan Technology

Scan C2C

Monthly

Bibliographic database with abstracts in English from Japan's scientific, technical and business journals.

JICST-E

(English citation and abstracts)

Japan Info Centre of Science and Technology. Contains scientific, technical and medical information on publications in Japanese scientific literature.

Lewis Publishers

2000 Corporate Blvd NW, Boca Raton, FL-33431, USA

Produce database software packages including: Chemical protective clothing permeation/degradation database;

EPA's clean air act - air toxics database; EPA's pesticide fact sheet database; The National Toxicology Program's chemical database; National toxicology program's chemical solubility database; Emergency response to chemical spills; Hazardous chemical regulations database; Environmental fate databases.

LIFO-DISK - LIposome inFOrmation Electronic Liposome Library

MM Developments, 60 MacLaren St, Suite 404, Ottawa, K2P 0K7, Canada
Telephone: 613 234 7709
Fax: 613 563 8241

(English)

Contains details of books, conferences, proceedings, information on the manufacturers of liposomal products, drugs, cosmetics, equipment, etc published since 1965. Includes keyword searching.

LiMB Database

Contact: Dr C Burks, Los Alamos National Laboratory, Theoretical Biology and Biophysics, Los Alamos National Laboratory, NM-87545, USA
Telephone: 505 667 6683

(English)

Listing of Molecular Biology Databases. Intended to facilitate locating and accessing databases relevant to molecular biology. Each entry in LiMB currently consists of 54 fields, including: database staff names and addresses; database maintenance hardware and software; scope of coverage and database goals; details about submission and access to data sets, database size; and types of data covered by the database.
Available through electronic network mail, MS-DOS based floppy disk, or printed form.

Maize Abstracts

See CAB Abstracts

(English)

MEDLINE

National Library of Medicine, MEDLARS Management Section, National Library of Medicine, Bldg 38A, NIH, Bethesda, MD-20894, USA
Telephone: 800 638 8480

(English)

Bibliographic database providing abstracts, standardized index terms and key words which may be used to retrieve citations which include references to sequence database entries.

Merck Index Online

Merck & Co Inc, Merck Sharp & Dohme Research Labs, PO Box 2000, Rahway, NJ-07065, USA
Telephone: 201 574 4000
Fax: 201 574 4720

Descriptions of chemicals. Searchable by CAS Registry Number, CAS name, alternative names, trademarks and physical properties.

MiCIS - Microbial Culture Information Service

Laboratory of the Government Chemist, Biotechnology Unit, Queens Road, Teddington, Middlesex, TW11 OLY, UK
Telephone: 81 943 7000
Fax: 81 943 2767

(English)

A Bibliographic Life-Science reference database source (Textual-numeric). Contains strain data on approximately 30000 microorganisms held in national and private culture collections in the UK. Includes name and synonyms, source, hazards, morphology, enzymes, culture conditions, metabolites, sensitivities, maintenance requirements, industrial applications and properties, alternative strain designations, patent status, and literature references. Contributing collections are: National Collection of Type Cultures (NCTC); National Collection of Industrial and Marine Bacterial (NCIMB); National Collection of Yeast Cultures (NCYC); CAB International Mycological Institute (Incorporating the National Collection of Fungal Cultures); National Collection of Feed Bacterial (NCPF); National Collection of Pathogenic Fungi (NCPF); and National Collection of Plant Pathogenic Bacteria (NCPPB). Available through the MSDN.

MiCIS - Information Centre for European Culture Collections

Collections, Mascheroder Weg 1b,
Braunschweig, D-3300, Germany
Telephone: 531 618 715
Fax: 531 618 718

Microbial Strain Data Network - MSDN

MSDN Secretariate, Biotechnology Centre, Cambridge University,
307 Huntingdon Road, Girton,
Cambridge, CB3 0JX, UK
Telephone: 301 231 5586

(English)

Microbial Strain Data Network Central Directory. An on-line referral database to centres or laboratories having information on the properties of microorganisms and cell lines. The MSDN Secretariat will make searches of the Directory for interested persons without access to computer facilities. The MSDN provides an integrated 'one-stop-shopping' service through a system of gateways to other databases and information centres. Includes descriptions and references to sources of bacteria, yeasts, actinomycetes, filamentous fungi, algae, vectors, hybridomas and animal cell lines.

Microbial Strain Data Network - MSDN

Bioinformatics Department, ATCC,
12301 Parkland Drive, Rockville,
MD-20852, USA
Telephone: 301 231 5586

See previous entry.

Microbiological Resource Databank

Excerpta Medica

Database on available cell lines and viruses. Set up as a joint venture of the European Federation of Cell and Virus Collections, World Foundation for Cell Cultures and Excerpta Medica. Sections include: animal cells (general, genetic, mutants, hybrid cells, hybridomas and genetically engineered); plant cells (general and genetically engineered); plant and animal viruses.

Microbiology Abstracts

Cambridge Scientific Abstracts,
5161 River Road, Bethesda,
MD-20816, USA

12 issues per year
(English)

Abstracts are divided into the following sections: Section A, Industrial and Applied Microbiology; Section B, Bacteriology; Section C, Algology, Mycology and Protozoology. 12 issues per year plus cumulative index in each section.

MIM - Mendelian Inheritance in Man

Contact: Dr V McKusick, Moore Clinic, John Hopkins Hospital,
Baltimore, MD-21205, USA
Telephone: 301 955 6641

Covers human inherited disease.

MINE - Microbial Information Network in Europe

International Mycological Institute,
Ferry Lane, Kew, Richmond, Surrey,
TW9 3AF, UK
Telephone: 81 940 4086
Fax: 81 332 1171

A developing database network set up to improve awareness of microbial strains available in European to facilitate ordering. All participating national nodes and participating culture collections will be able to access MINE. Individual collections in all EC countries are invited to participate and contribute data. Sponsored by the Biotechnology Action Programme of the European Communities.

MINE - Microbial Information Network in Europe, French Office

Bureau des Ressources Genetiques,
57 rue Cuvier, Paris Cedex 05,
F-75231, France
Telephone: 1 47 07 15 75
Fax: 1 45 35 70 15

See previous entry.

MIRDAB - Microbiological Resource Databank

Excerpta Medica, Postbus 1126,
BC Amsterdam, NL-1000,
The Netherlands
Telephone: 20 26 44 38

NADL - Nucleic Acid Data Library

European Molecular Biology
Laboratory, Meyerhoffstrasse 1,
Heidelberg, D-6900, Germany.

The National Toxicology Program's Chemical Database

Lewis Publishers Inc,
2000 Corporate Blvd NW, Boca Raton,
FL-33431, USA

(English)

Consists of eight volumes containing 50 fields detailing information on 2,270 potentially toxic chemicals. Each volume consists of one 3.5 inch diskette or two 5.25 inch diskettes. May be purchased complete or as individual volumes. Volumes are chemical names and synonyms, chemical and physical properties, standards and regulations, medical hazards and symptoms of exposure, medical first aid, personal protective equipment, hazardous properties and uses, shipping classifications and regulations.

NCFB Catalogue

(English)

Contains references to approximately 2000 bacterial cultures available from the National Collection of Food Bacteria, UK. Available through NCYC.

Nematological Abstracts

See CAB Abstracts

(English)

NEWAT

University of California, Chemistry
Department D006, San Diego, La Jolla,
CA-92093, USA
Telephone: 619 534 4417

Contains amino acid sequences, including literature citations and sources.

NTIS Bibliographic Database

US National Technical Information Service, Department of Commerce, 5285 Port, Royal Road, Springfield, VA-22161, USA
Telephone: 804 487 4600

Details of government-sponsored research reports and studies in the physical sciences, technology, engineering, biological sciences, medicine and agriculture (mainly in the USA).

Nutrition Abstracts and Reviews, Series A - Human and Experimental

See CAB Abstracts

(English)

Nutrition Abstracts and Reviews, Series B - Livestock Feeds and Feeding

See CAB Abstracts

(English)

Oncology Abstracts

Cambridge Scientific Abstracts, 5161 River Road, Bethesda, MD-20816, USA

12 issues per year
(English)

Covers all aspects of physical, chemical and biological carcinogenesis, tumour studies, diagnosis and therapy, and clinical studies. 12 issues per year plus cumulative index.

Packaging Science and Technology Abstracts

IFIS Publishing, Lane End House, Shinfield, Reading, RG2 9BB, UK
Telephone: 734 883895
Fax: 734 885065

Compiled by the Fraunhofer-Institut fuer Lebensmitteltechnologie und Verpackung (ILV) in Germany in collaboration with IFIS. IFIS has specific responsibility for marketing PSTA on-line. Covers literature on all aspects of packaging, food and non-food. Subjects covered include materials, machinery, packages, packaging aids, specialized packs, testing, transport and storage. Other IFIS offices are listed under International Organizations.

Pascal Biotechnology

CNRS

12 issues per year

Covers 13,000 journals relating to science and technology, conference proceedings, theses and reports.

PatentImages

MicroPatent

Weekly
(English)

Fully searchable database of complete facsimile US patents on CD-ROM. Contains complete information on up to 1,000 patents three weeks after their publication by the US Patent Office.

PC-ABC

BioCommerce Data Ltd, Prudential Buildings, 95 High St, Slough, UK, Sl1 1DH
Telephone: 753 511777
Fax: 753 512239

24 issues per year

Floppy disk version of Abstracts in BioCommerce (ABC) which provides over 350 abstracts in each update. Operates on IBM PC compatibles.

PESTDOC - Pest Control Literature Documentation

Derwent Publications, Rochdale House, 128 Theobalds Road, London, WC1X 8RP, UK
Telephone: 71 242 5823
Fax: 71 405 3630

A series of databases covering pesticides, herbicides and plant protection including biochemistry, toxicology etc but excluding fertilizers. Various sub-files cover journal literature, conference reports, etc. Updated quarterly. Available on ORBIT.

PESTMAN - Pesticide Databank

CAB International, Wallingford, Oxon, OX10 8DE, UK
Telephone: 491 32111
Fax: 491 33508

Covers chemical compounds used in pest control with lists of names, properties, toxicology, uses, formulations and analysis.

PG-TRANS

Computer Science Unit, Institut Pasteur, France

A protein sequence databank.

Pharmaceased Discontinued Drugs File

PJB Online Services, 18/20 Hill Rise, Richmond, Surrey, TW10 6UA, UK
Telephone: 81 948 0751
Fax: 81 332 8991

(English)

PJB Online Services, c/o Mary Dalia, PharmaBooks Ltd, 1775 Broadway, Suite 511, New York, NY-10019, USA
Telephone: 212 262 8230
Fax: 212 262 8234

Contains details of over 7,000 discontinued drugs. Can be accessed through Data-Star, Dialog or BRS.

Pharmacontacts Directories File

PJB Online Services, 18/20 Hill Rise, Richmond, Surrey, TW10 6UA, UK
Telephone: 81 948 0751
Fax: 81 332 8991

(English)

PJB Online Services, c/o Mary Dalia, PharmaBooks Ltd, 1775 Broadway, Suite 511, New York, NY-10019, USA
Telephone: 212 262 8230
Fax: 212 262 8234

An onscreen directory of over 20,000 agrochemical, pharmaceutical and animal health companies. Can be accessed through Data-Star, Dialog or BRS.

Pharmacontacts OnScreen

PJB Online Services, 18/20 Hill Rise, Richmond, Surrey, TW10 6UA, UK
Telephone: 81 948 0751
Fax: 81 332 8991

(English)

PJB Online Services, c/o Mary Dalia, PharmaBooks Ltd, 1775 Broadway, Suite 511, New York, NY-10019, USA
Telephone: 212 262 8230
Fax: 212 262 8234

Based on the printed publication Pharmaprojects, and updated on a monthly basis, Pharmaprojects OnScreen gives details of the current progress of drugs in active development throughout the world. Gives an overview of the R&D pipeline worldwide and allows the activities of a particular manufacturer can be followed and licensing opportunities in specific therapeutic fields can be identified. Can be accessed through Data-Star, Dialog or BRS.

PHIND - Pharmaceutical and Healthcare Industries News Database

Online Services, PJB Publications Ltd,
18/20 Hill Rise, Richmond, Surrey,
TX10 6UA, UK
Telephone: 81 948 3262
Fax: 81 948 6866

Provides a news service via Data-Star with three files available; PHID, recent clinical news; PHIC, news covering the last three weeks; PHIN, archival data back to October 1980.

PHYTOMED

Quarterly

Produced by Biologische Bundesanstalt fur Land-und Fortwirtschaft, FRG. Bibliographic database on plant science information. Areas covered include plant diseases, plant and stored products protection, air pollution damage and toxicology.

Plant Biotechnology

University of Sheffield, Biomedical
Information Service, Yorkshire,
S10 2TN, UK

(English)

A bibliography, produced by the university's Library/Information Science Department.

Plant Breeding Abstracts

See CAB Abstracts

12 issues per year
(English)

Produced by the CAB International Bureau of Plant Breeding and Genetics.

Plant Genetic Resources Abstracts

See CAB Abstracts

4 issues per year
(English)

A quarterly journal from the CAB ABSTRACTS database. Produced by the CAB International Division of Crop Protection and Genetics and published in association with the International Board for Plant Genetic Resources (IBPGR). Topics covered include: genetic resources; taxonomy, evolution and origin of crops; genetic diversity; seed conservation; *in vitro* conservation; biotechnology; gene banks and their management; germplasm documentation; networks, information transfer and training; germplasm enhancement and use; plant health, quarantine and safe movement of germplasm; reports; conferences; books. The journal also includes an author and subject index. Each entry gives the author, title in English, original title if in a Roman script language, name of journal or other document, publication details (year of publication, volume, part number and pagination), language of text and summaries, bibliographic details, address of first author, abstract.

Plant Growth Regulator Abstracts

See CAB Abstracts

(English)

Plant Protection and Quarantine Import Data

United States Department of Agriculture,
1301 New York Avenue NW,
Washington, DC-20005-4788, USA

(English)

US imports of fresh fruit and vegetables under the Plant Quarantine and Inspection Program (PQ280) of APHIS. Includes monthly and annual data by commodity and by country of origin and port of entry for 1983-88.

Potato Abstracts

See CAB Abstracts

(English)

Poultry Abstracts

See CAB Abstracts

(English)

PPR - Plasmid Prefix Registry

Stanford University School of Medicine, Department of Medical Microbiology, Sherman Fairchild Building 5402, Stanford, CA-94305-2499, USA
Telephone: 415 723 1772

A database covering the nomenclature of plasmids.

PRF/LITDB

Protein Research Foundation, 4-1-2 Ina, Minoh-shi, Osaka, 562, Japan

Covers literature abstracts and citations.

PRF/SEQDB

Protein Research Foundation, 4-1-2 Ina, Minoh-shi, Osaka, 562, Japan

Covers amino acid sequences.

Protozoological Abstracts

See CAB Abstracts

(English)

PSEQIP

Institut Pasteur,
25 rue du Docteur Roux,
Paris Cedex 15, F-75724, France
Telephone: 1 45 68 85 09

An amino acid sequence databank.

PTG - Protein Translation

Los Alamos National Laboratory, Theoretical Biology and Biophysics, T-10, MS K710, Los Alamos, NM-87545, USA
Telephone: 505 667 6683

Contains amino acid sequences including functional features, literature citations, index terms and taxonomic identification.

QTDGPD-2D Gel Protein Database

Cold Spring Harbor Laboratory, PO Box 100, Cold Spring Harbor, NY-11724, USA
Telephone: 516 367 8356

Contains data from gel electrophoretic separations of proteins as well as details of molecular properties.

RED - Restriction Enzyme Database

Cold Spring Harbor Laboratory, Cold Spring Harbor, NY-11724, USA
Telephone: 516 367 8388

Covers details of restriction enzymes as well as citations, abstracts and relevant methodologies.

Review of Medical and Veterinary Mycology

CAB International Mycological Institute, Ferry Lane, Kew, Surrey, TW9 3AF, UK
Telephone: 81 940 4086
Fax: 81 581 1676

4 issues per year
(English)

An abstract journal compiled from the world literature on medical and veterinary mycology, first issued in

1943. Covers mycoses of man as well as farm, domestic and wild animals including fish and large invertebrates but excluding insects. Also covers allergic disorders of man and animals associated with fungi, poisoning of man and animals by fungi or mould-contaminated feed, systematic mycology, and use of fungicides and antibiotics in relation to mycotic disorders.

Review of Plant Pathology

CAB International Mycological Institute, Ferry Lane, Kew, Surrey, TW9 3AF, UK
Telephone: 81 940 4086
Fax: 81 581 1676

12 issues per year
(English)

An abstract journal compiled from the world literature on plant pathology since 1922. Subjects covered include diseases of crops, ornamentals and forest trees caused by fungi, bacteria, mycoplasma and viruses as well as application of fungicides and antibiotics in disease control. Also covers general and systematic mycology as it relates to fungi of agricultural importance.

Rice Abstracts

See CAB Abstracts

(English)

RFLPD-RLFP Database

Centre d'Etude du Polymorphisme Humain, 3 rue d'Ulm, Paris, F-75005, France
Telephone: 1 43 29 12 11

Covers details of restriction fragment length polymorphisms.

Seed Abstracts

See CAB Abstracts

(English)

SEQANALREF

Deparment de Biochimie Medicale CMU, 1 Rue Michel Servet, Geneva 4, CH-1211, Switzerland
Telephone: 22 61 84 92

Contains literature citations and reference terms for protein and nucleic acid sequences.

SESAME

Commission of the EC - DG-XVII

(English)

On-line source of information on energy projects supported by the EC. Contains two areas: Demonstration projects and hydrocarbon technology. Contains up-to-date view of current projects financed by the EC, and detailed information on all completed projects carried out under the 'Hydrocarbon Technology and Energy Demonstration' programmes since 1975 and 1978. Available through ECHO-HOST.

SIGPEP

Royal Institute of Technology, Theoretical Physics, Stockholm, S-10044, Sweden
Telephone: 8 787 7172

A database of amino acid sequences based on literature citations.

Sorghum and Millets Abstracts

See CAB Abstracts

(English)

Soyabean Abstracts

See CAB Abstracts

(English)

SPF - Standard Pesticides File

Derwent Publications Ltd,
Rochdale House, 128 Theobalds Road,
London, WC1X 8RP, UK
Telephone: 71 242 5823

Covers the properties of chemical compounds used in pest control. Available on-line or as hardcopy.

SRRSD - 16S Ribosomal RNA Sequence Database

University of Houston,
Department of Biochemical Science,
SR-1, 4800 Calhoun, Houston,
TX-77004, USA
Telephone: 713 749 3980

Covers nucleotide sequences determined from large ribosomal RNA.

SRSRSC - Small Ribosomal Subunit RNA Sequence Compilation

Contact: Dr R de Wachter,
Universiteit Antwerpen, Department
Biochimie, Universiteitsplein 1,
Antwerpen, B-2610, Belgium
Telephone: 3 828 2528

Covers the functional features and nucleotide sequences of small ribosomal RNA.

Sugar Industry Abstracts

See CAB Abstracts

6 issues per year
(English)

A bimonthly journal produced by the CAB International Division of Crop Production and published in association with Tate & Lyle plc. Topics covered include: cane and beet sugar manufacture; sugar refining; liquid sugar and nutritive sweeteners; analysis and control; environmental issues; by-products; properties and uses of sugar; nutrition and health aspects; economics; terminology; plant sources of sugar other than cane or beet. Also includes reports, conferences and books, as well as an author, subject and annual indexes. Each entry includes author; English translation of title; title; journal or other document; year, volume, part and pagination; language of text and other language of summary; address of first author; abstract.

Sugar Statistical Compendium

United States Dept of Agriculture,
1301 New York Avenue NW,
Washington, DC-20005-4788, USA

(English)

A database of US beet and cane sugar statistics.

SWISS-PROT

Universite de Geneve, Dept de
Biochimie Medicale CMU,
1 rue Michel Servet, Geneva 4,
CH-1211, Switzerland
Telephone: 22 61 84 92

A protein sequence databank.

TED - Tenders Electronic Daily

Office for Official Publications of the European Communities

(All official languages of EC except Greek)

On-line version of the supplement (series S) to the Official Journal of the European Communities. Contains public calls for tender offered by more than 80 countries, including calls for proposals for EC research projects. The documents are available on the morning of their publication. Available through ECHO-HOST.

TELEGEN

(English)

A bibliographic database for biotechnology. Contains over 23,000 citations, with abstracts, to the worldwide literature on all aspects of biotechnology and genetic engineering. Covers business and economics; social impacts and constraints; patents and legal issues; policy and regulatory issues; general and historical aspects; international discussions and general policy;industrial microbiology; applications to chemistry, energy and mining, the environment, food processing and production, crops, livestock, pharmacy and medicine; and general research and development by specific type of gene (ie human, animal, plants, yeast and fungus, bacterial, or viral).

TOXLINE

US National Library of Medicine,
8600 Rockville Pike, Bethesda,
MD-20894, USA

Bibliographic files from the National Library of Medicine covering pharmacology, toxicology and physiology.

TOXNET

National Library of Medicine,
8600 Rockville Pike, Bethesda,
MD-20894, USA
Telephone: 301 496 6531

Covers relevant chemicals, their toxic properties and literature citations.

TRF - Taxonomic Reference File

BIOSIS, 2100 Arch Street,
Philadelphia, PA-19103-1399, USA
Telephone: 215 587 4917

A database of organisms including taxonomic classification and index terms.

TRNAC - tRNA Compilation

University of Bayreuth, Department of Biochemistry, Universitatsstrasse 30,
Bayreuth, D-8580, Germany
Telephone: 921 552 668

A database of nucleotide sequences and sequence alignments based on literature citations, secondary structures and sources.

Tropical Data Base

(English)

Developed in Brazil, a national information network maintaining data on culture collection holdings, research programmes and projects, and a directory of experts in microbiology.

VECTOR - Cloning Vector Sequence Database

Whitehead Institute, Room 211, Nine
Cambridge Center, Cambridge,
MA-02142, USA
Telephone: 617 258 5139

Contains literature citations and nucleotide sequence data of cloning vectors including base compositions, cross references and index terms.

Virology and AIDS

Cambridge Scientific Abstracts,
5161 River Road, Bethesda,
MD-20816, USA

12 issues per year
(English)

Covers all aspects of viruses, including their effects on humans, animals and plants. 12 issues per year plus cumulative index.

Weed Abstracts

See CAB Abstracts

(English)

Wheat, Barley and Triticale Abstracts

See CAB Abstracts

(English)

World Agricultural Economics and Rural Sociology Abstracts

See CAB Abstracts

(English)

World Data Centre on Microorganisms

Contact: Dr K Komagata, RIKEN Institute, 2-1 Hirosawa Wako, Saitama, 351-01, Japan
Telephone: 484 62 1111
Fax: 484 64 5651

Produces a World Directory of Collections of Microorganisms available on-line. Holds Japanese Collection of Microorganisms.

PART IV

SERVICES

COMPUTER SOFTWARE

Abel Electronics

Viale C Casella 48, Roma, I-00122, Italy
Telephone: 6 844 2338
Fax: 6 844 2338

Supply an integrated DNA, RNA and protein analysis software package and complete range of compatible hardware including advanced gel assembly program. Offer annotated databases for genetics and molecular biology.

American Synthesis

1177-C Quarry Lane, Pleasanton, CA-94566, USA
Telephone: 415 462 2232
Fax: 415 462 6247

Produce sequencing software.

Analytical Imaging Concepts Inc

12 Woodfall, Irvine, CA-92714, USA
Telephone: 714 857 1269/800 535 2425

Manufacture computerized image analyzers.

Anville Instruments

Watchmoor Trade Centre, Watchmoor Road, Camberley, Surrey, GU15 3AJ, UK
Telephone: 276 684613
Fax: 276 29141

Supply systems and software for data logging and acquisition as well as temperature sensors and calibration systems.

Asahi Kasei Information Systems Co Ltd

Imperial Tower 17F 1-1, Uchisaiwai-cho 1-chome, Chiyoda-ku, Tokyo 100, Japan
Telephone: 3 507 2316
Fax: 3 507 2464

Market computer software packages for graphics and molecular design.

Astra Scientific Instruments Inc

PO Box 611088, 1961 Concourse Drive, San Jose, CA-95161-1088, USA

Manufactures a cell-culture bioreactor and other fermentation equipment as well as data-collection, control and analysis software for use with all analog signals from the bioreactors and associated accessories.

Autoscribe

PO Box 5, Riseley, Reading, Berkshire, RG7 1YL, UK
Telephone: 734 883125
Fax: 734 885604

Produce molecular modelling software.

Baird Europe BV

Produktieweg 30, PO Box 81, AB Zoeterwonde, NL-2380, The Netherlands
Telephone: 71 413151
Fax: 71 414899

Market a multichannel analyzer and a range of accessories and software.

Balzers High Vacuum Ltd

Bradbourne Drive, Tilbrook, Milton Keynes, MK7 8AZ, UK
Telephone: 908 373333
Fax: 908 377776

Supply gas monitoring equipment for fermentation processes with computer control and data recording software.

Balzers Inc

8 Sagamore Park Road, Hudson,
NH-03051-4914, USA
Telephone: 603 889 6888
Fax: 603 889 8573

Supply gas monitoring equipment for fermentation processes with computer control and data recording software.

BBN Software Products Corp

10 Fawcett St, Cambridge,
MA-02238, USA
Telephone: 617 864 1780

Supply computers for use in experimental design and data analysis including facilities for selection and application of statistical methods and the interpretation of statistical results.

Beckman Instruments

2500 Harbor Blvd, Fullerton,
CA-92634, USA
Telephone: 714 8714848
Fax: 714 8708083

Products include a VAX-based chromatography system designed to acquire, analyse and report data.

Becton Dickinson

383 Hillen Rd, Towson,
MD-21228, USA
Telephone: 301 337 8700/800 638 8656

Supply instrumentation and reagents for clinical assay including automated analyzers and microbiology instruments and associated computer software, prepared culture media as well as standards and reagents for EIA and RIA and clinical kits.

Betagen Corp

100 Beaver St, Waltham,
MA-02154, USA
Telephone: 617 899 3400/800 421 4213
Fax: 617 899 7063

Supply autoradiographic equipment, blot analyzers, image analysis computers and software, and automated electrophoresis systems.

Biocom

Immeuble Le Nath, 24 Avenue de la Baltique, ZA Courtaboeuf,
F-91940, France
Telephone: 1 64 46 27 60
Fax: 1 69 82 95 18

Supply systems for image analysis for biological and medical applications; suitable for use in pathology, autoradiography, electrophoresis, cytology, etc as well as in process control.

Bioconcept Sarl

28 avenue de la Californie, Nice,
F-06200, France
Telephone: 93 96 44 00

Supply systems for image analysis for use in biological, food technology and medical applications. Products include a combined fluorescent microscope and image analysis system for use in the rapid counting and recognition of microorganisms.

BioImage Corp

1460 Eisenhower Place, Ann Arbor,
MI-48108, USA
Telephone: 313 971 7500/800 624 5894
Fax: 313 971 1061

Manufacture electrophoresis apparatus and analytical imaging instruments, densitometers, scanners and associated computer software. Subsidiary of Eastman Kodak.

Bio-Ion Nordic AB

PO Box 1504, Seminariegatan 30,
Uppsala, S-75015, Sweden
Telephone: 18 55 23 15
Fax: 18 55 11 14

Supply laboratory equipment including mass spectrometry system with software.

Biometallics Inc

PO Box 2251, Princeton,
NJ-08543, USA
Telephone: 609 275 0133
Fax: 609 275 9485

Produce test kits for progesterone and ovulation detection. Develop software for data acquisition of ELISA analysis.

Bio-Rad Laboratories

1000 Alfred Nobel Drive, Hercules,
CA-94547, USA
Telephone: 510 724 7000
Fax: 510 724 5024

Produce sequencing software.

Bioscan

170 West Dayton, Suite 204, Edmonds,
WA-98020, USA
Telephone: 206 7758000
Fax: 206 7753640

Products include image processing, measurement and analysis software.

Biosoft

22 Hills Rd, Cambridge, CB2 1JP, UK
Telephone: 223 68622
Fax: 223 312873

Supply a wide range of software with applications in biotechnology including programs for recombinant DNA, radioimmunoassay and ELISA techniques, presentation graphics etc.

Biostructure

Les Algorythmes, parc Innovation,
Bat Euclide, Illkirch Graffenst,
F-67400, France
Telephone: 88 67 98 00
Fax: 88 67 98 01

Provide software for graphic workstations and computers for protein engineering and determination of the structure of macromolecules.

Biosym Technologies Inc

10065 Barnes Canyon Rd, San Diego,
CA-92121, USA
Telephone: 619 458 9990/619 546 5507
Fax: 619 458 0136

Market a 3-dimensional, real-time molecular modelling program which can selectively display and manipulate molecules and their components. Offer contract research in computer-aided molecular design.

Biosystems

9 rue des Mardors, Couternon,
F-21560, France
Telephone: 80 47 57 62
Fax: 80 47 50 71

Provide electronics and information systems for data acquisition and process control for the fermentation industry.

Biotage Europe

Harforde Court, Foxholes Business
Park, Hertford, SG13 7NW, UK
Telephone: 992 501535
Fax: 992 501547

Develop and manufacture controlled protein purification systems. The systems have built-in safety features combined with high-level software to give maximum flexibility. Products include automated systems for protein chromatography at the laboratory and pilot scale. Developing monitoring and control instrumentation and expert systems for chromatographic methods development.

Biotechnology Computer Systems Ltd

South Bank Technopark,
90 London Rd, London, SE1 6LN, UK
Telephone: 71 922 8875
Fax: 71 922 8875

Develop and market computer software for the bioindustry.

Bio-Tek Instruments Inc

Highland Park, PO Box 998,
Winooski, VT-05404-0998, USA
Telephone: 802 655 4040/800 451 5172
Fax: 802 655 7941

Develop and manufacture ELISA and microplate readers accommodating all colorimetric assays and featuring computerized functions.

BMDP Statistical Software

1440 Sepulveda Blvd, Suite 316,
Los Angeles, CA-90025, USA
Telephone: 213 479 7799
Fax: 213 312 0161

Supply statistical software and computer equipment for statistics and statistical research, analytical research and biometrics, comprising an extensive range of data analysis programs.

BMDP Statistical Software

Cork Technology Park,
Model Farm Rd, Cork, Ireland
Telephone: 21 542722
Fax: 21 542822

Supply statistical software and computer equipment for statistics and statistical research, analytical research and biometrics, comprising an extensive range of data analysis programs.

Bruker Analytische Messtechnik GmbH

Wikingerstrasse 13, Karlsruhe 21,
D-7500, Germany
Telephone: 721 5967 0
Fax: 721 5967 173

Manufacture high-resolution automated nuclear magnetic resonance spectrophotometers, mass spectrometers and high performance liquid chromatographs as well as associated data processing systems.

Bruker-Franzen Analytik GmbH

Fahrenheitstrasse 4, PO Box 330126,
Bremen 33, D-2800, Germany
Telephone: 421 2205 0
Fax: 421 2205 100

Manufacture mass spectrometers and high performance liquid chromatography as well as data processing systems.

Chemical Design Ltd

7 West Way Unit 12, Oxford,
OX2 OJB, UK
Telephone: 865 251483
Fax: 865 250270

Supply software and databases for computer-aided molecular modelling and protein engineering.

Chemical Design US Ltd

200 Rte 17 S, Suite 120, Mahwah,
NJ-07430, USA
Telephone: 201 529 3323
Fax: 201 529 2443

Produce computer systems and software for modelling the structure and properties of both small molecules and proteins, and studying their interactions.

Clontech Laboratories

4030 Fabian Way, Palo Alto,
CA-94303, USA
Telephone: 415 424 8222
Fax: 415 424 1352

Produce sequencing software.

CompuGene

223 Renaldo Drive, Chesterfield,
MO-63017-2210, USA
Telephone: 314 434 7250/800 231 4545

Produce computer software packages for use in molecular biology and analysis of DNA suitable for IBM PC or VAX systems.

Computer General

13 Campbell Court, Bramley,
Basingstoke, Hants, RG26 5EG, UK
Telephone: 256 881377
Fax: 256 882905

Market image processing equipment for laboratory data analysis and presentation.

The Core Store Ltd

The Studio, Hawthorn Cottage,
Marbury Road, Comberbach, Cheshire,
CW9 6AU, UK
Telephone: 606 891980
Fax: 606 891980

Supply science and engineering computer software for data acquisition processing, plotting, statistical analysis and image analysis.

Coulter Corp

PO Box 2145, Hialeah,
FL-33012-0145, USA
Telephone: 305 885 0131/800633 7427
Fax: 305 883 6841

Design and manufacture flow cytometry instrumentation, cell sorters, computer data systems and analytical software.

Coulter Electronics of Canada Ltd

905 Century Drive, Burlington,
Ontario, L7L 5J8, Canada
Telephone: 416 639 4701
Fax: 416 333 3787

Supply blood cell counters, flow cytometry equipment, clinical chemistry instruments, scientific instruments and related diagnostic reagents and consumables.

Coultronics France SA

29 avenue Georges Pompidou,
Margency, F-95580, France
Telephone: 1 34 16 10 30

Supply equipment for haematology, flow cytometry, particle analysis and surface area analysis with computer control and software systems, designed for use in medical research and biotechnical applications. Agents for Micrometrics, US and Chrono-Log Corporation, US.

Cybertech

Bergmannstrasse 102, Berlin 61,
D-1000, Germany
Telephone: 30 693 57 97
Fax: 30 322 89 73

Manufacture an electronic camera system for the documentation of electrophoresis gels. Develop software for image analysis of gels.

Cyborg Corp

97 Bridge St, Newton,
MA-02158, USA
Telephone: 617 964 9020
Fax: 710 335 7706

Produce combined hardware and software systems with recorder, data logger alarm monitoring, automatic data collection and data reduction facilities designed for use in the monitoring of processes, waste and effluent production, energy use, etc, in food, pharmaceutical and chemical industries. Also suitable for accumulation and analysis of data from repeated laboratory procedures with scans of up to 64 parameters and frequencies varying from minutes to days. Calculations include first and second derivatives, areas under curves and statistical treatments.

Daikin Industries Ltd

Shinjuku-Sumitomo Bldg,
6-1 Nishi-Shinjuku 2-Chome,
Shinjuku-ku, Tokyo 163, Japan
Telephone: 3 344 8148
Fax: 3 344 8113

Market computer software packages for molecule simulation.

Difco Laboratories Inc

PO Box 331058, Detroit,
MI-48232-7058, USA
Telephone: 313 961 0800/800 521 0851
Fax: 313 591 3530

Products include computer software for clinical testing systems.

Digital Equipment Corp

3 Results Way, Marlboro,
MA-01752, USA
Telephone: 508 467 5111
Fax: 508 467 5640

Produce a range of computers, microcomputers and interfaces for data acquisition and management designed for locally controlled real time input/output functions without the need to access the host computer, suited to automated laboratory uses.

Digital Instruments Corp

PO Box 9901, Newark,
DE-19714-9901, USA
Telephone: 215 255 4149

Supply densitometers, image analyzers and associated software.

Dnastar Inc

1228 South Park Street, Madison,
WI-53715, USA
Telephone: 608 258 7420
Fax: 608 258 7439

Supply a comprehensive sequence analysis software package available for IBM PC compatible and Mackintosh computers. In addition to database searching the software can accomplish sequence comparison, circular and linear restriction maps, sequence analysis and presentation. Also offers sequence gel assembly software - currently being used in one of the *E coli* genome sequencing projects.

Dnastar Ltd

St James House,
105-113 The Broadway, West Ealing,
London, W13 9BL, UK
Telephone: 81 566 1200
Fax: 81 566 3377

Design, develop, sell and support molecular biology/genetic sequence analysis software and databases on CDROM. Currently developing database resources, CDROM resources for molecular biology, DNA sequence analysis, sequencing products and protein sequence analysis and database resources. Subsidiary of Dnastar Inc, US.

Dorand Electronics Ltd

8-9 Allens Lane, Hamworthy, Poole,
Dorset, BH16 5DA, UK
Telephone: 202 622006

Supply gas chromatographs as well as complete GC mass spectrometer systems, pyrolysis systems, computing integrators and GC/MS data systems.

Du Pont Medical Products

Barley Mill Plaza (16-2120),
Wilmington, DE-19898, USA
Telephone: 302 992 4783/800 527 2601
Fax: 302 992 4442

Supply computer software for clinical testing systems.

Dynamic Data Links Ltd

PO Box 31, Elsworth, Cambridge,
Cambridgeshire, CB3 8LG, UK
Telephone: 95 47 309
Fax: 95 47 89703

Software consultants offering services in development of custom windows based on Windows 3-based software for the scientific and multimedia markets. Configure custom image analysis and image processing systems.

Dynatech Laboratories Inc

14340 Sullyfield Circle, Chantilly,
VA-22021-1617, USA
Telephone: 703 631 7800/800 336 4543
Fax: 703 631 7816

Products include computer image analysis systems and software.

Elsevier Scientific Software

Postbus 330, AH Amsterdam,
NL-1000, The Netherlands
Telephone: 20 5803 911

Distribute scientific software packages for use with IBM PC and Apple computer systems.

Enprotech Corp/BEA

1 Westinghouse Plaza, Hyde Park,
MA-02136, USA
Telephone: 617 361 5700/800 433 6433
Fax: 617 361 5708

Supply a wide range of fermentation equipment including computerized controllers and computer software as well as protein separation equipment for pilot plant or production facilities and associated software.

ESA Inc

45 Wiggins Ave, Bedford,
MA-01730, USA
Telephone: 617 275 0100
Fax: 617 275 5529

Supply analytical chromatography equipment including HPLC systems, detectors etc, as well as chemical and element analyzers and related computer software.

Etronics

Postfach 23 Sonnenberg,
Unterengstringen, CH-8103,
Switzerland
Telephone: 1 750 3511

Produce scientific computer programs and systems for biometrics, organic chemistry, statistics, pharmacology, drug design and project management

Eurogenetics Italia Srl

52 Corso Vittorio Emanuele II, Torino,
I-10123, Italy
Telephone: 11 561 20 60
Fax: 11 54 12 48

Develop and market diagnostic products and computer software packages. Products include microreaders and microtitre plates as well as programs for analysis of ELISA tests which may be networked through RS232C using IBM systems.

Eurogenetics NV

Luikersteenweg 232, Hasselt,
B-3500, Belgium
Telephone: 11 2744 30

See previous entry.

Evans & Sutherland

540 Arapeen Drive, PO Box 8700,
Salt Lake City, UT-84108, USA
Telephone: 801 582 5847

Supply computer graphics systems designed for the creation, manipulation, modification and dynamic viewing of complex three-dimensional molecular structures, as well as a variety of molecular graphics software. Systems are used by researchers in crystallography, drug design, protein research engineering and polymer research.

Evans & Sutherland

Stahlgruberring 32, Munich 82, D-8000, Germany
Telephone: 89 429041

Supply computer graphics systems designed for the creation, manipulation, modification and dynamic viewing of complex three-dimensional molecular structures, as well as a variety of molecular science and molecular graphics software. Systems are used by researchers in crystallography, drug design, protein research, genetic engineering and polymer research.

FPS Computing Canada Ltd

Suite 285, 5945 Airport Rd, Mississauga, Ontario, L4V 1R9, Canada
Telephone: 416 673 0911
Fax: 416 673 2714

Produce computer control systems and analytical software for laboratory workstations.

Genesearch Pty

14 Technology Dr, Gold Coast Technology Park, Labrador, Queensland, 4215, Australia
Telephone: 75 940019
Fax: 75 940562

Produce sequencing software.

GeneSystems Computer Software

871 Partridge Ave, Menlo Park, CA-94025, USA
Telephone: 415 325 4943
Fax: 415 325 4943

Produce software to create publication-quality restriction map drawings with Macintosh interface.

Genofit

C Postale 6GS, Grand-Lancy 1, CH-1212, Switzerland
Telephone: 22 794 84 00
Fax: 22 794 72 23

Produce sequencing software.

Genomic

441 Route de la Combe, PO Box 43, Collonges Sous Saleve, F-74160, France
Telephone: 5043 6765
Fax: 5043 6870

Produce molecular modelling and sequencing software.

HCS Industrial Automation BV

Postbus 88, AB Hattem, NL-8050, The Netherlands

Produce hardware and software systems for process control and data handling.

Helena Laboratories

1530 Lindbergh Drive, PO Box 752, Beaumont, TX-77704-0752, USA
Telephone: 409 842 3714/800 231 5663
Fax: 409 842 6241

Supply materials and accessories for use in electrophoresis, spectrophotometry, densitometry, chemical analysis and blood testing as well as computer accessories which provide a concise presentation of densitometrically generated data from electrophoretic and chromatographic separations, antibodies, reagents and test kits.

Heyden Datasystems

Spectrum House, Hillview Gardens, London, NW4 2JQ, UK
Telephone: 81 203 5171

Supply software for processing analytical data from spectrophotometers as well as associated hardware.

Hitachi Software Engineering Co Ltd

6-81 Onoe-cho, Naka-ku
Yokohama-shi, Kanagawa 231, Japan
Telephone: 45 681 2111
Fax: 45 681 5847

Produce a computer package (DNASIS) designed for molecular biologists which includes computer-generated graphics and software programs for use in collection, analysis and presentation of nucleic acid and protein sequences. Adaptable for use with IBM (US) or compatible machines.

Hoefer Scientific Instruments

Unit 12 Croft Road Workshops, Newcastle, Staffordshire, 5T5 0TW, UK
Telephone: 782 617317
Fax: 782 617346

Produce sequencing software.

Hunter Associates Laboratory Inc

11495 Sunset Hills Rd, Reston, VA-22090, USA
Telephone: 703 471 6870

Manufacture a computerized colour analysis system which simulates visual observations.

Hypercube Inc

Suite 7, 419 Phillip St, Waterloo, Ontario, N2L 3X2, Canada
Telephone: 519 725 4040
Fax: 519 884 0115

Develop hardware and software for molecular modelling. Projects include simulation of proteins and nucleic acids for biotechnical and pharmaceutical industries.

IBM Corp

Orchard Road, Armonk, New York, NY-10504, USA

Have developed bench-top computers for laboratory use, with real-time operating system software to both control instruments and acquire, as well as analyse, data from multiple instruments. The system, which offers printing, plotting and data storage facilities, includes a chromatography applications program that can be used with both liquid and gas chromatographs.

IBM Instruments Inc

PO Box 332, Danbury, CT-06810, USA
Telephone: 203 796 2500

Manufacture laboratory computer systems, spectrophotometers, liquid chromatographs and columns for molecular weight analysis and other gel permeation equipment.

IMA GmbH

Admiral-Rosendahl-Strasse 16, Zeppelinheim, D-6078, Germany
Telephone: 69 6934 61

Develop process control systems for use in laboratory and industrial fermentation and protein separation including computerized control systems and data loggers for use with bioreactors as well as controllers for electrophoretic separations. Products include a bench-top airlift fermenter, nucleosides for synthesis of oligonucleotides, enzymes, coenzymes, soil and groundwater analytical and in situ decontamination products.

Image Technology Ltd

Green Meadows, Pulborough,
West Sussex, UK
Telephone: 798 872703
Fax: 798 873185

Supply computer-assisted imaging systems for blood-grouping, bacterial agglutination tests and particle counting as well as monitoring systems for animal behaviour activity.

Imaging Research Inc

c/o Brock University, 600 Glenridge Ave, St Catharines, Ontario, LS2 3A1, Canada
Telephone: 416 688 2040
Fax: 416 685 5961

Products include computers and software for image analysis and densitometry and autoradiographic equipment.

Imstar

60 rue Notre-Dame-des-Champs, Paris, F-75006, France
Telephone: 1 46 34 53 52
Fax: 1 46 34 51 57

Supply workstations for image analysis in relation to biomedical and clinical tests.

IntelliGenetics Inc

700 E El Camino Real, Mountain View, CA-94040, USA
Telephone: 415 962 7300
Fax: 415 962 7302

Market software for analysis and management of DNA and protein sequence information. Typical applications include structural predictions, motif determination, sequence alignments, DNA sequencing fragment assembly, database searching, site-directed mutagenesis, cloning simulation and probe design. Products are offered for use on IBM PC, Macintosh, Sun and VAX computer systems. A semi-automatic DNA sequence film reading and analysis device is also available. Offer their own database: Keybank, a database of structural and functional patterns for protein and nucleic acid sequences and vectorBank, a cloning vector data bank, in addition to distribution of other public databases. Have just released GeneSeq, a proprietary database of DNA and protein sequences appearing in patents.

IntelliGenetics Inc

Amocolaan 2, Geel, B-2440, Belgium
Telephone: 14 58 99

Market software for molecular biology and biochemistry including cloning simulation and data management as well as protein and nucleic acid sequence analysis, selection and assembly.

Intracel Ltd

Unit 4, Station Rd, Shepreth, Royston, Herts, SG8 6PZ, UK
Telephone: 763 262 680
Fax: 763 262 676

Supply computer operating systems and software for scientific instrumentation and data analysis including spectrophotometry equipment and control systems.

Jandel Scientific

65 Koch Rd, Corte Madera, CA-94925, USA
Telephone: 415 924 8640
Fax: 415 924 2850

Manufacture and market microcomputer software products for research. Products include programs for: creating quality charts and graphs for scientific journal publication; performing one-step automated curve fitting; separating and analyzing mutiple peak or functional form data; making two-dimensional measurements from strip charts, X-rays, photomicrographs etc on a digitizing tablet; performing video measurement and analysis including densitometry, automatic object counting, edge and line digitizing and morphometric measurements; creating three-dimensional computer models from serial sections.

Jandel Scientific

Schimmelhuschstrasse 25, Erkrath,
D-4006, Germany
Telephone: 2104 36098
Fax: 2104 33110

See Jandel Scientific, USA.

Jeol Ltd

3-1-2 Musashino, Akishima 196, Japan
Telephone: 425 431 111
Fax: 425 46 3353

Supply automated spectrometers offering high-speed data processing.

Keithley MetraByte/Asyst/DAC

440 Myles Standish Blvd, Taunton,
MA-02780, USA
Telephone: 734 575666
Fax: 734 596469

Products include data acquisition and control software. Software specifically for the 80386 powered PCs allows data acquisition, analysis and graphics in a single multitasking system. Multitasking allows such activities as the setting up of new experiments while others are running.

Kratos Analytical

Barton Dock Road, Urmston,
Manchester, M31 2LD, UK
Telephone: 61 865 4466
Fax: 61 864 3424

Manufacture mass spectrometers for GC-MS, LC-MS and electrospray-MS as well as acquisition and processing software for mass spectrometry.

Kubota Computer Inc

8-8 Shinjuku 2-chome, Shinjuku-ku,
Tokyo 160, Japan
Telephone: 3 225 0741
Fax: 3 225 0740

Market computer graphics packages with applications in polymer research.

Laboratory Software Associates Pty Ltd

Suite 6, 19-35 Gertrude St, Fitzroy,
Victoria 3056, Australia
Telephone: 3 419 7400

Supply data acquisition and analysis software for laboratory instrumentation.

Laboratory Technologies Inc

1517 S Wright Blvd, Schaumburg,
IL-60193-4509, USA
Telephone: 312 529 3112/800 542 1123
Fax: 312 529 3141

Supply process monitoring software for acquisition and manipulation of laboratory data, equipment for EIA, standards and controls for RIA, radioactivity counters and radiation safety equipment.

Large Scale Biology Corp

9620 Medical Center Drive, Rockville,
MD-20850, USA
Telephone: 301 424 5989
Fax: 301 762 4892

R & D company providing services and systems for large scale work in two-dimensional gel electrophoresis. These include control research, startup support, consulting, workstation-based analytical software, computer-controlled instrumentation, and databases (under ongoing development) of information on gene expression and xenobiotic effects in mammalian and plant systems.

Lewis Publishers

2000 Corporate Blvd NW, Boca Raton, FL-33431, USA

Produce software packages including: Atmospheric oxidation rate program; Biodegradation probability program; Henry's law constant program; Hydrolysis rate program; Soil sediment adsorption constant program; Analytical groundwater modelling; Drinking water and groundwater remediation cost evaluation.

Maihak AG

Semperstrasse 38, Postfach 601709, Hamburg, D-2000, Germany
Telephone: 40 27 160
Fax: 40 27 16242

Supply a wide range of analytical equipment including total carbon, oxygen and gas analyzer systems designed for measurement of air pollution an emission sources. Also supply data acquisition and evaluation systems as well as laboratory and process plant analyzers.

Milton Roy Co

PO Box 10235, Riviera Beach, FL-33404, USA
Telephone: 407 844 5241

Manufacture liquid chromatography equipment including high performance liquid chromatography pumps, detector systems, sample injectors and data recording and integration devices and software.

Mitsui & Co Ltd

PO Box 822, 2-1 Otemachi-1-chome, Chiyoda-ku, Tokyo 100, Japan
Telephone: 3 285 7887
Fax: 3 285 9809

Products include software packages for computer-aided molecular design.

Molecular Design Ltd

2132 Farallon Drive, San Leandro, CA-94577, USA
Telephone: 415 895 1313
Fax: 415 352 2870

Supply software for molecular structure comparisons allowing the user to manipulate two 3-dimensional molecular models on the screen of a computer graphics terminal in order to compare their geometric and topographic features. Also offer a chemical database management software system that can store, search, retrieve and display molecules and chemical reactions.

Molecular Devices Ltd

110 Strand (1st Floor), London, WC2R OAA, UK
Telephone: 71 836 8918
Fax: 71 836 4286

Supply software for interface with a Macintosh computer to provide a user-friendly interface for complete microplate reader control and data analysis.

Molecular Dynamics

880 East Argues Ave, Sunnyvale, CA-94086, USA
Telephone: 408 773 1222
Fax: 408 733 8343

Manufacture densitometers, image analyzers and associated software and autoradiography equipment. Products include: a computing densitometer using a high speed laser for fast, high resolution scanning of samples up to 35cm x 42cm with quantitative data systems for correction of film non-linearity: a phosphorimager for radioisotopes which eliminates the need for film in autoradiography except for tritium; a confocal scanning laser microscope designed for high resolution one-dimensional and three-dimensional reconstruction.

Molecule Co Ltd

1-8-20-601 Akebono, Kashiwa 277, Japan
Telephone: 471 46 8172
Fax: 471 46 8172

Develop and market computer software for the molecular design of drugs.

Dr I Molnar

Blucherstr 22, Berlin 61, D-1000, Germany
Telephone: 30 6943068
Fax: 30 6931061

Products include software or method development in HPLC and GC. Two experimental runs can be used to simulate a large number of further runs. These are used to determine whether separation of the sample is feasible with this column. Many parameters such as retention times of every peak in the sample for any condition, influence flow rate, column dimension, particle size and expected separation time and system pressure are provided.

Nachet Vision

125 Boulevard Davout BP 128, Paris Cedex 20, F-75963, France

Supply computerized image processing equipment.

NAG Ltd

Wilkinson House, Jordan Hill Rd, Oxford, OX2 8DR, UK
Telephone: 865 511245
Fax: 865 310139

Supply and market a wide range of products and services worldwide to a variety of industries, including sophisticated mathematical tools for biotechnology for data retrieval and processing as well as monitoring of processing.

Nelson Analytical Inc

10040 Bubb Road, Cupertino, CA-95014, USA
Telephone: 408 725 1107
Fax: 408 725 0825

Supply chromatographic data systems and computer software packages developed for use with gas, liquid, ion and ion-exchange chromatography.

Omega

One Omega Drive, Box 4047, Stamford, CT-06907, USA
Fax: 203 359 7700

Packages include icon-driven data acquisition software for IBM compatible PCs.

Oxford Molecular

Oxford Science Park, Sandford-on-Thames, Oxford, OX4 4GA, UK
Telephone: 865 784600
Fax: 865 784601

Produce a range of software packages for modelling and analysis of small molecules and larger molecules.

Oxford University Press

Walton Street, Oxford, OX2 6DP, UK
Telephone: 865 56767

Publish a range of software packages across the disciplines. The most recent include a desktop molecular modeller and accompanying structures libraries as well as a molecular structures database on compact disc containing information derived from the protein data bank compiled at Brookhaven National Laboratory.

Paedia Corp

499 Carolina St, San Francisco, CA-94107, USA
Telephone: 415 861 8097
Fax: 415 861 6683

Supply densitometry software for analysis of data from electrophoretic separations from stained gels, blots and films.

Perkin-Elmer

Rotenhofgasse 17, Vienna, A-1014, Austria
Telephone: 1 534 24 0
Fax: 1 534 24 520

Supply products of Perkin Elmer, USA.

Perkin-Elmer Corp

Analytical Instruments, 761 Main Ave, Norwalk, CT-06859-0012, USA
Telephone: 203 762 1000/800 762 4000
Fax: 203 762 6000

Manufacture laboratory instruments for spectroscopy, chromatography and thermal and elemental analysis. Products include: complete systems for gas, high performance liquid and liquid chromatography; liquid chromatography-mass spectrometry; dispersive and Fourier Transform infrared spectroscopy; atomic absorption, inductively coupled plasma and inductively coupled plasma-mass spectrometry; gas chromatography-mass spectrometry; and UV/Visible and fluorescence spectroscopy, including microprocessor systems, video display, data-processing capabilities as well as a complete line of accessories and supplies for these systems.

Perkin-Elmer Corp

2771 North Garey Avenue, Pomona, CA-91767, USA
Telephone: 714 593 3581

Produces equipment for monitoring and analysis of fermentation reactions capable of handling inputs from up to 30 fermenters on a sequential basis determining levels of oxygen and carbon dioxide.

Perkin-Elmer GmbH

Postfach 101164, Askaniaweg 1, Ueberlingen, D-7770, Germany
Telephone: 7551 81 0
Fax: 7551 1612

Offer computerized laboratory data handling systems.

Pharmacia LKB Biotechnology

9238 Gaither Road, Gaithersburg, MD-20877, USA
Telephone: 301 963 3200
Fax: 301 963 7780

Produce sequencing software.

Phase Separations Sales

Deeside Industrial Park, Deeside, Clwyd, CH5 2NU, UK
Telephone: 244 816444
Fax: 244 821232

Products include a molecular modelling package comprising software and plug-in transputer boards for IBM compatible PCs.

Pias Co Ltd

Takahashi Bldg, 6-3 Sotokanda 5-chome, Chiyoda-ku, Tokyo 101, Japan
Telephone: 3 834 3751
Fax: 3 834 3754

Market image analyzing systems.

Porton Instruments

18740 Oxnard Street, Suite 303,
Tarzana, CA-91356, USA
Telephone: 818 996 0220
Fax: 818 342 9738

Produce sequencing software.

Protein Databases Inc

405 Oakwood Rd, Huntington Station,
New York, NY-11746, USA
Telephone: 516 673 3939/800 777 6834
Fax: 516 673 4502

Develop software for analysis of biological data including analysis and identification of two-dimensional gel protein separations. Offer custom software services.

Proteus Molecular Design Ltd

Proteus House, 48 Stockport Rd,
Marple, Cheshire, SK6 6AB, UK
Telephone: 61 426 0191
Fax: 61 426 0192

Products include integrated highly advanced computer software for molecular modelling with special features for protein modelling and drug design. Undertake contract research in design of new drugs, conformational studies and protein sequence analysis, epitope prediction and protein redesign. An integrated service is offered whereby consequent synthesis, physiological testing, and pharmacological and toxicological testing may be performed. Complete drug/diagnostic/vaccine development projects are undertaken. Several potential new drugs, synthetic vaccines or AIDS therapeutics developed in house are the subject of patent applications.

Rainin Instrument Co Ltd

Mack Road, Woburn, MA-01801, USA
Telephone: 617 935 3050
Fax: 617 938 8157

Products include a chromatographic software tool that stores and reprocesses chromatographic data while running other applications on an Apple Macintosh.

Research Information Systems Inc

Camino Corporate Center,
2355 Camino Vida Roble, Carlsbad,
CA-92009, USA
Telephone: 619 438 5526/800 722 1227
Fax: 619 438 5573

Develop and market scientific information management software.

Rikei Corp

Shinjuku Nomora Bldg,
1-26-2 Nishi-Shinkuku, Shinjuku-ku,
Tokyo 163, Japan
Telephone: 3 345 1411

Market computer programs for simulation of chemical and enzyme catalyzed reactions capable of real-time manipulation.

Scanalytics/CSPI

40 linnell Circle, Billerica,
MA-01821, USA
Telephone: 617 272 6020
Fax: 508 663 0150

Produce an image analysis system for interpretation of gel electrophoretograms including a comprehensive software package.

Scientific Device Laboratory Inc

PO Box 88, 508 Zenith Drive,
Glenview, IL-60025, USA
Telephone: 312 998 6860

Supply computer software for clinical testing systems as well as diagnostics, reagents, standards and controls.

SES GmbH

Goldbergstrasse 15, Postfach 45,
Nieder-Olm, D-6501, Germany
Telephone: 61 36 4994
Fax: 61 36 42917

Supply advanced analytical and data management systems including chromatography data handling systems, laboratory information management systems, continuous flow analyzers, flow injection analyzers, TLC/FID analyzers etc.

Sierra Instruments

5 Harris Court, Building L, Monterey,
CA-93940, USA
Telephone: 408 373 0200
Fax: 408 373 4402

Produce molecular modelling software.

Sigma Chemical Company

3050 Spruce Street, St Louis,
MO-63103, USA
Telephone: 314 771 5750

Produce molecular modelling software.

SLT Labinstruments Austria

Untersbergstrasse 1A, Groedig,
Salzburg, A-5082, Austria
Telephone: 6246 3438-0
Fax: 6246 2770

Products include supply software for data analysis and laboratory operations.

Societe Bioinformatique et Biotechnologie - SBB

45 rue Emile Zola, Tours, F-37000,
France
Telephone: 47 05 27 04

Offer services in image analysis, modelling, artificial intelligence, expert systems, process simulation, etc with applications to biotechnology.

Soft Gene Berlin

Offenbacher Strasse 5, Berlin 33,
D-1000, Germany
Telephone: 30 821 1407
Fax: 30 821 9764

Market software for molecular biology and gene technology, including DNA and protein sequence interpretation programs, in collaboration with the Free University Berlin.

Spectra Physics GmbH

Siemenstrasse 20,
Darmstadt-Kranichstein, D-6100,
Germany
Telephone: 6151 708 0
Fax: 6151 75000

Supply equipment for liquid chromatography: complete systems, monitoring and detection, data logging and processing, columns and accessories.

Spectra Physics Inc

Autolab Division, 3333 N 1st St,
San Jose, CA-95134, USA
Telephone: 408432 3333/800 424 7666
Fax: 408 0203

Laser Analytics Division,
25 Wiggins Ave, Bedford,
MA-01730, USA
Telephone: 617 275 2650

Provide chromatographic data systems and electronic control systems for combined gas chromatography/mass spectrometry. Also supply complete range of equipment and accessories for chromatography, detection and quantitation of products including complete systems and fluorescent and spectrophotometric column monitors.

Spectrex Corp

3580 Haven Ave, Redwood City,
CA-94063, USA
Telephone: 415 365 6567
Fax: 415 365 5845

Manufacture a laser-based particle counter and sizer based on the principle of near angle light scattering compatible with Apple or IBM PC to provide complete data from *in situ* counting of sealed bottled liquids or flow through systems.

Spezielle Elektronische Systeme GmbH

Goldbergstrasse 5-7, Nieder-Olm, Mainz, D-6501, Germany
Telephone: 61 31 49 94

Supply the Trivector Trinet Laboratory Information Management System including both hardware and software modules. The system allows networking of many different types of inputs, from integrators or balances to dedicated analyzers. Software available includes sample tracking and report generation; or the user can modify the packages using Basic, Pascal or C. Systems can be tailored to individual needs.

SPI Synthetic Peptides Inc

12285 Yonge St, PO Box 417, Richmond Hill, Ontario, L4C 4Y6, Canada
Telephone: 416 773 2820
Fax: 416 773 2489

Activities include the development of specialized computer software for peptide and protein structure prediction and simulation of HPLC.

Spiral System Instruments Inc

7830 Old Georgetown Rd, Bethesda, MD-20814, USA
Telephone: 301 657 1620
Fax: 301 652 5036

Products include computer software for clinical assays.

Tecan UK Ltd

18 The High St, Goring-on-Thames, Reading, RG8 9AR, UK
Telephone: 491 875087
Fax: 491 875432

Supply a wide range of robotic sample processors, single or dual arm, for fast and accurate performance of all laboratory liquid handling tasks. IBM-based software enables user-defined control over work surface usage, pipetting protocols, liquid handling and data management. Hardware includes liquid level sensing sampling probes, positive sample identification and a complete range of microplate readers and washers with 'on board' or PC-based software.

Tietz Video and Image Processing System GmbH

Herbstrasse 7, Gautin, D-8035, Germany
Telephone: 89 8506567
Fax: 89 8509488

Manufacture, sell and support a range of TV-based fast digital image processing systems for varied applications in science, medicine and industry. Products include: complete image processing systems for microscopy; real-time, floating point and array processors; supporting software and instrument feedback control; large-area slow-scan CCD-cameras.

TosoHaas GmbH

Zettachring 6, Stuttgart 80, D-7000, Germany
Telephone: 711 715 6031
Fax: 711 715 6034

Supply HPLC, ion and liquid chromatography equipment, accessories and associated computer software as well as filtration equipment.

Tripos Associates

3 allee des Garays, ZI, Palaiseau,
F-91124, France
Telephone: 1 69 20 16 47

Supply molecular modelling and analysis systems for proteins and nucleic acids providing biopolymers and 'small' molecules software in a single package (see Tripos Associates, US).

Tripos Associates

Gallop House, Hasler's Lane,
Great Dunmow, Essex, CM6 1XS, UK
Telephone: 371 6191

Supply molecular modelling and analysis systems for proteins and nucleic acids providing biopolymer and 'small' molecules software in a single package.

Tripos Associates Inc

1699 South Hanley Road, Suite 303,
St Louis, MO-63144, USA
Telephone: 314 647 1099/800 323 2960
Fax: 314 647 9241

Develop molecular modelling and analysis software and provide related scientific support services for the pharmaceutical, biotechnology and chemical research industries. Provide biopolymer and 'small' molecules software in a single package, and rapid and flexible construction and manipulation of macromolecules, including insertions, deletions and point mutations. Subsidiary of Evans & Sutherland.

U-Microcomputers Ltd

Winstanley Industrial Estate,
Long Lane, Warrington, Cheshire,
WA2 8PR, UK
Telephone: 925 54117

Supply a computerized chemical modelling workstation which allows steriochemical modelling designed for synthetic chemists.

Varian Sample Preparation Products

St John's Innovation Centre,
Cowley Rd, Cambridge,
CB4 4WS, UK
Telephone: 223 421140
Fax: 223 421182

Supply instruments and accessories for chromatography and spectroscopy including atomic absorption spectrophotometers, UV/VIS spectrophotometers, electrothermal atomizers, gas chromatographs, liquid chromatographs, amino acid analyzers, diode array detectors, GC-mass spectrometers, NMR spectrometers and imaging systems and data systems for chromatography.

VG Instruments (Canada) Inc

115 Place Frontenac, Pointe-Clarie,
Quebec, H9R 4Z7, Canada
Telephone: 514 695 6257
Fax: 514 695 8077

Supply laboratory and industrial data systems.

VG Laboratory Systems Ltd

Tribune Avenue, Altrincham, Cheshire,
WA14 5TP, UK
Telephone: 61 941 6159

Supply automated, integrated laboratory management systems.

Vision Systems Ltd

Second Avenue, Technology Park,
The Levels, Adelaide, South Australia
Telephone: 618 349 5988
Fax: 618 349 5989

Products include an image analysis system for electrophoresis and other patterns. Comprises sophisticated image processing and computer hardware together with a suite of software modules and powerful relational database. Software modules include forensic identification, paternity testing, track and pattern matching, population statistics, sequencing and densitometry.

VWR Scientific

1400 N Providence Road, Suite 100, Media, PA-19063, USA
Telephone: 215 891 2770

Produce sequencing software.

CONSULTANTS

ABE Trading Co Ltd

PO Box 262, Central Osaka, Osaka, Japan
Telephone: 443 8801 8

Supply apparatus and reagents for the synthesis of oligonucleotides and have an engineering group which offers consultancy in instrument use as well as the supply of custom-synthesis and sequencing for DNA and peptide.

Adams Business Associates

Kelvin House, Totteridge Avenue, High Wycombe, Bucks, HP13 6XG, UK
Telephone: 494 465244
Fax: 494 446788

Offer services in biotechnology including international market analysis, management and implementation of projects and business development and marketing programmes.

Advanced Breeding Technology Pty Ltd

PO Box 4, c/o Dept of Biochemistry, Australian National University, Canberra 2601, Australia
Telephone: 62 476 596
Fax: 62 485 641

Consultants offering services in research and development in the application of recombinant DNA technology to animal breeding and other biotechnological areas including DNA probes and sequencing.

AEM Consultants

30 Federal Street, Salem, MA-01970, USA
Telephone: 617 744 5923

Management consultancy firm serving high technology industries on an international basis. Provides information, process evaluation, custom studies, marketing and operational reports. Experience includes large-scale fermentation and other biotechnology facilities, market evaluation of filtration systems, systems analysis of microprocessor control requirements, product assessment of control instruments, evaluation of heat transfer system, market planning for process industries etc.

Ag-West Biotech Inc

105-15 Innovation Blvd, Saskatoon, Saskatchewan, S7N 2X8, Canada
Telephone: 306 975 1939
Fax: 306 975 1966

Non-profit company concerned with promoting technology transfer of agricultural biotechnology being developed in Saskatchewan, in particular in the areas of cell culture, crop breeding, vaccines, biological pesticides and plant growth regulators.

Agri Contact

Torupvejen 97, Hundested, DK-3390, Denmark
Telephone: 47 92 38 00
Fax: 47 92 48 48

Act as consultants to the agricultural, food and bioindustries serving as a link between research and development advances. Have both personal and postal contacts with a wide circle of research institutes at home and abroad.

Agrisearch Services Pty Ltd

50 Leewood Drive, PO Box 972, Orange, NSW 2800, Australia
Telephone: 63 624 539
Fax: 63 627 844

Offer services in agricultural research, development and product testing in biological control, agrochemicals and plant propagation.

Allied Colloids Manufacturing GmbH

Tarpenring 23, Hamburg 62, D-2000, Germany
Telephone: 40 527208 0
Fax: 40 5270915

Supply selected bacteria and equipment for environmental biotechnology. Offer environmental services.

Amdel

PO Box 114, Eastwood, SA 5063, Australia
Telephone: 8 372 2700

Offer analytical services and research and development to the agricultural, food and environment industries.

AMEC Design and Construction Ltd

Frobisher House, Nelson Gate, Commercial Rd, Southampton, SO1 OGX, UK
Telephone: 703 634400
Fax: 703 636116

Engineering company, undertake project management, engineering design, feasibility studies and design and build projects within the chemical, pharmaceutical, food, biotechnology and energy industries.

Analytical Technologies Inc

PO Box 36385, Tucson, AZ-85740, USA
Telephone: 602 744 0012

Offer analytical services ranging from toxicological testing to trace organic and inorganic analyses for a wide range of materials including drinking water and the profiling of hazardous wastes through a network of environmental chemistry laboratories providing QA testing.

Aplibio SA

4 bis boulevard Stalingrad, Choisy le Roi, F-94600, France
Telephone: 1 48 52 36 29

Carry out studies on the tolerance to and efficiency of new medicines. Carry out clinical studies (phase I to IV) as well as undertaking immunological and haematology investigations.

Applied Bioscience International

Mettlers Rd, PO Box 2360, East Milstone, NJ-08875-2360, USA
Telephone: 908 873 2550

Provide a broad range of services including biological safety testing, clinical research and development, management of agrochemical research and development, and the assessment and management of chemical and environmental risk.

APV Baker Pty Ltd

10 Blissington St, Springvale, Melbourne, Victoria, 3171, Australia
Telephone: 3 549 2222
Fax: 3 548 1985

Supply process technology to the food, dairy and brewing industries including custom-built plant for all stages of production from ingredients to packaging.

Aquaculture Nederland BV

Gloudriaankade 1, GM s'-Gravenhage, NL-2516, The Netherlands
Telephone: 70 388 3743
Fax: 1749 44440

Offer consultancy to companies involved in aquaculture.

Arthur D Little Inc

25 Acorn Park, Cambridge,
MA-02140, USA
Telephone: 617 864 5770

Contract research and consulting organization offering a wide range of services including R&D and consulting in biotechnology such as immobilized proteins, biosensors, downstream processing and biotechnology-based diagnostics.

Ashact Ltd

11 Wycombe Rd, Prestwood, Great Missenden, Bucks, HP16 ONG, UK
Telephone: 494 891100
Fax: 494 890320

Consultants in environmental management and effluent treatment.

The Austin Co UK Ltd

35 Ballards Lane, London,
N3 1XW, UK
Telephone: 81 346 9351
Fax: 81 349 9116

Offer regulatory compliance and validation services of facilities and processes in both the pharmaceutical and biotechnology industries as well as design, engineering and construction of 'state of the art' facilities and remodelling of existing facilities through 47 offices worldwide.

Bachmann Favereau Glover & Associates

Hekstraat 30, Heme, B-1540, Belgium

Carry out multi-client studies on areas which include immunoassay methods, diagnostic products, biological reagents etc.

BASE International Biotechnology Ltd

Midsummer House, 443 Midsummer Blvd, Central Milton Keynes, Bucks, MK9 3BN, UK
Telephone: 908 664315
Fax: 908 665312

Biotechnology consultants.

Battelle Institute Ltd

15 Hanover Square, London,
W1R 9AJ, UK
Telephone: 71 493 0184
Fax: 71 629 9705

Contract research and consultancy group offering single and multi-client studies in most areas of science and technology, including assessments of prospects of biotechnology in the following areas: chemicals, fine chemicals, food, animal feed, pharmaceuticals and fuel industries.

Beak Consultants Ltd

14 Abacus Rd, Brampton, Ontario,
L6T 5B7, Canada
Telephone: 416 458 4044
Fax: 416 458 7303

Offer consultancy in environmental biotechnology in: regulation; bioremediation assessment; toxicity and genotoxicity testing and tracing of specific microorganisms and genes in the environment. Develop and distribute diagnostic kits of chemical and biological environment contaminants.

BellBiotech Consulting

6/40 Lillimur Rd, Ormond,
Victoria 3163, Australia
Telephone: 3 578 9179

Offer consulting services in biotechnology including engineering design and construction, equipment supply at laboratory and pilot scale, R&D and product testing.

Benezech-Simpson & Co

25 Paget Rd, Langley, Slough,
SL3 7QP, UK
Telephone: 753 40995
Fax: 753 40995

Biotechnology consultants.

Bennett & Company (Europe) Ltd

Field House, Newton Tony, Salisbury,
Wilts, SP4 OHF, UK
Telephone: 980 64488
Fax: 980 64327

Supply a very wide range of laboratory equipment and instrumentation for analysis, chromatography, cell and tissue culture, microscopy, filtration, spectrophotometry, monitoring and control. Consultants and contractors in the provision of complete, confidential, turnkey laboratory projects. The associate company, Biopak Ltd, undertake taylor-made packaging for scientific projects, from reagent kits to instrumentation.

Bernard Wolnak & Associates

180 North Michigan Avenue,
Suite 1016, Chicago, IL-60601, USA
Telephone: 312 782 4926

Management consultancy firm specializing in the economics and technology of fermentation and biotechnology products, eg antibiotics, enzymes, vitamins, acids, polysaccharides and proteins, based on rDNA technology.

Bethyl Laboratories Inc

PO Box 850, Montgomery,
TX-77356, USA
Telephone: 409 597 6111

Provide a custom antiserums production service to supplied antigens. Producer of commercial antiserum to human, rodent and domestic animal serum proteins. Carry out large-scale drug studies and research projects at their animal facilities and supply radial immunodiffusion kits to the bovine, equine, canine and feline immunoglobulins.

Bideco AG

Im Schoeri 3, Dubendorf, CH-8600,
Switzerland
Telephone: 1 820 21 30
Fax: 1 821 14 44

Offer consulting, engineering and construction of biotech facilities; risk analyses, environmental impact reports and validation services.

Bio Techno Systems Inc

Shyuwa 2nd Toranomon Building,
21-19-1 chome Toranomon, Minato-ku,
Tokyo, Japan
Telephone: 3 595 0969

Offer consultancy in biotechnology in relation to use of biological raw materials for chemical and fuels.

Biocatalysts Ltd

Main Avenue, Treforest Industrial
Estate, Pontypridd, Mid Glamorgan,
CF37 5UT, UK
Telephone: 443 843712
Fax: 443 841214

Specialize in industrial enzymes for the chemical, food and pharmaceutical industry and offer a full enzyme consultancy; including sourcing, cost-effectiveness studies, process development and modification.

Biochemische Synthesetechnik GmbH & Co

Stresemannstrasse 268, Hamburg 50,
D-2000, Germany
Telephone: 40 850 00 51

Provide a customer service to support genetic engineering programmes. Undertake custom synthesis of primers and probes including incorporation of modified bases, synthesis of linker molecules and total or partial synthesis of structural genes and regulatory DNA sequences. Also design and develop genetically engineered microorganisms for production of proteins useful for medicine, pharmacy and agriculture. Offer long-term contract research opportunities.

BIOCLEAR Environmental Biotechnology BV

Zernike Park 2, Postbus 2262,
CG Groningen, NL-9704,
The Netherlands
Telephone: 50 718455
Fax: 50 643556

Offer consultancy and research services in the field including biodegradation testing and information, feasibility studies, selection, isolation and application of microorganisms for specific biodegradations and biological monitoring.

BioCommerce Data Ltd

95 High Street, Slough, Berks,
SL1 1DH, UK
Telephone: 753 511777
Fax: 753 512239

Main products and services include abstracts, directories, online databases, advertising and mailing. However, offer strategic advice and research on any aspect of the biotechnology industry and marketplace. Specialize in confidential profiles of biotechnology companies, reports on specific technologies and advice on managing information services. Undertake projects of varying sizes and geographic spread.

Bio/dynamics Inc

PO Box 2360, Mettlers Road,
East Milstone, NJ-08875-2360, USA
Telephone: 908 873 2550
Fax: 908 873 3992

Offer services in toxicology testing including acute, subchromic and chronic studies in all species, pathology, metabolic, analytical residue and environmental fate chemistry as well as in quality assurance auditing.

Biointelligence Service

34 rue Perier, Montrouge, F-92120,
France
Telephone: 1 40 84 07 40
Fax: 1 40 84 07 48

Provide information and consultancy in technology and marketing covering biotechnology, the agrofood industry and the environment.

BioLabs Inc

15 Sheffield Court, Lincolnshire,
IL-60069-3161, USA
Telephone: 312 945 2767

Offer consulting services in biotechnology including company start-ups and turnarounds.

Biolink Developments Ltd

6th Floor, NZ House, 58 Queens
Drive, PO Box 30584, Lower Hunt,
New Zealand
Telephone: 4 662 181
Fax: 4 699 155

Consultants in research, development and marketing of speciality chemicals produced by cell culture techniques.

Biological Monitoring Inc

PO Box 184, Blacksburg,
VA-24063, USA
Telephone: 703 953 2821
Fax: 703 951 1481

Carry out environmental monitoring of water quality using computer monitoring equipment and fish as indicators.

Bioquest Resources Corp

98 Heritage Rd, Suite 100, Montreal, Quebec, H9W 3V3, Canada
Telephone: 514 630 9840
Fax: 514 630 9840

Consultants offering business development, market research and analysis, technology transfers and strategic alliance searches.

BIOS (Consultancy and Contract Research) Ltd

Pinewood, College Ride, Bagshot, Surrey, GU19 5ER, UK
Telephone: 276 73363
Fax: 276 73928

Contract research and consultancy organization offering services for health care, biotechnological and related products. Undertake: clinical research Phase I bioavailability, pharmacokinetics and dermatological tests; clinical trials Phases II-IV to UK GP, European and FDA standards; data management and statistics support to trials; regulatory affairs, specializing in assessment of data; laboratory services both analytical and bioanalytical including method development, stability studies and microbiological testing; project management and consultancy including development planning, safety evaluation and aspects of commercialization.

Biosurvey SA

15 rue d'Egmont, Brussels, B-1050, Belgium
Telephone: 2 517 23 68

Offer the following services in the field of biotechnology: studies, surveys and monographs and organization of seminars and symposia; assessment of investment projects including scientific, technical and financial as well as marketing and management aspects; recruitment of technicians and managers.

Biosystech Consulting Inc

2383 Edenhurst Drive, Mississauga, Ontario, L5A 2L1, Canada
Telephone: 416 279 5337/416 978 7096
Fax: 416 978 8548

Biotechnology consultants.

Biotechnica

Commite d'Expansion Aquitaine, 2 place de la Bourse, Bordeaux Cedex, F-33076, France
Telephone: 56 52 65 47

Supply information as well as technical and development assistance on a regional basis. Promote research and industrial activities.

Biotechnology Consultants Ltd

1 Northfield Rd, Reading, Berkshire, RG1 8AH, UK
Telephone: 734 566066
Fax: 734 596859

International management consultancy specializing in agriculture and biotechnology worldwide.

Biotechnology South West

Reed Hall, Streatham Water, Exeter, EX4 4QR, UK
Telephone: 392 214085

Regional biotechnology centre established in 1988 to promote investment and activity in biotechnology in the South of England. Offer companies and organizations, whether or not they are located in the region, consultancy, contract R&D, quality control, in service training and education in microbial fermentation and biotransformation, recombinant DNA technology, diagnostics and therapeutics.

Biotics Ltd

27 Abbotswood, Guildford, Surrey, GU1 1UY, UK
Telephone: 483 505252/60331

Consultants on screening, evaluation and possible exploitation of biological resources of developing countries, particularly plant products.

Biotransfer

Lab de Biochimie et Pathologie Vegetale, Tour 53, 4 Place Jussieu, Paris, F-75005, France
Telephone: 1 44 27 40 02

Offer analytical services and advice concerning protection against fungal diseases to agricultural organizations.

Biotropic

2214 Blvd de la Lironde, parc Agropolis, Montferrier s Lez, F-34980, France
Telephone: 67 52 65 66
Fax: 67 52 40 76

Carry out research and offer consultancy on processing of products of tropical agriculture. Offer feasibility studies, realization and commercialization of machines and processes.

BIRD Engineering BV

De Brauwweg 13, Postbus 149, AC Schiedam, NL-3100, The Netherlands
Telephone: 10 4157822
Fax: 10 4379648

Offer consultancy in bioprocess engineering including design and engineering of processes and installations, microbiology, equipment and instrumentation. Company is associated with research at Delft University of Technology and Applikon Instruments.

Bogart Delafield Ferrier (Europe) Ltd

7A Howe Street, Edinburgh, EH3 6TE, UK
Telephone: 31 557 0700
Fax: 31 557 2851

International management consultancy company specializing in the fields of healthcare and biotechnology. Services include technology assessment, market research and corporate partnering.

Boston Biomedical Consultants

100 Fifth Avenue, Waltham, MA-02154, USA
Telephone: 617 890 5060

Provide consultancy services to the health care industry.

Braunschweiger Biotechnologie

Mascheroder Weg 3, Braunschweig, D-3300, Germany
Telephone: 531 61684

An association offering the service of several international companies. Carry out research, process and product development in screening; microorganism, yeast and mammalian cell fermentation; purification of intra- and extra-cellular products; and environmental problems. Manufacture and sell products to pharmaceutical, chemical, food and environmental markets. Also carry out contract manufacturing of microorganisms, enzymes, proteins and peptides. Organized in four divisions: peptide synthesis, pharmaceutical production, environmental biotechnology and biocatalysis based on Bactem Inc, SIBA and Phillips Biosciences, all US.

Bromont Economic Development Corporation

2 Compton St, Bromont, Quebec,
JOE 1LO, Canada
Telephone: 514 534 2974
Fax: 514 534 0360

Support agency for emergent biotechnological R&D companies, industries already in operation and research laboratories. Offer financial and scientific analyses of R&D projects and access to sources of risk capital. Also provide laboratories for lease, associated with various networked service links.

Bureau European de Recherches SA

173 Rue Stevin, Brussels, B-1040,
Belgium
Telephone: 2 230 3449/2481

Consultancy dealing with policy issues related to food and agriculture in Europe.

Business Communications Co Inc

25 Van Zant St, Norwalk,
CT-06855, USA
Telephone: 203 853 4266
Fax: 203 853 0348

Independent market research and analysis firm with cross-discipline capabilities in biotechnology, health, agriculture, food, chemicals and many other fields emphasizing the future business implications of technical and market changes. Publish in-depth, multi-client technical-economic studies, host commercial biotechnology conferences, publish Applied Genetics News and other newsletters, and offer confidential, custom consulting to start-up as well as established firms.

CAB International - CABI

Wallingford, Oxfordshire,
OX10 8DE, UK
Telephone: 491 32111

Offer pest and disease identification services, consultancy to agriculture and biocontrol services. Also collect, collate and disseminate information through abstracts, publications and operation of databases.

Cambio

34 Millington Rd, Cambridge,
Cambridgeshire, CB3 9HP, UK
Telephone: 223 66500
Fax: 223 350069

Offer services as brokers in molecular biology, consultants in biotechnology and technology transfer and the organization of contract R&D as well as supplying biotechnology kits and reagents.

Cambridge Biomedical Consultants

Bolerostraat 3, RG, NL-6544,
The Netherlands
Telephone: 80 77 47 16
Fax: 80 77 47 16

Undertake commissioned consultancy research and conference organization in the biomedical sciences and biotechnology particularly in relation to manpower supply, training, recruitment, regulatory affairs and public information.

Cambridge Consultants Ltd

Science Park, Milton Road,
Cambridge, CB4 4DW, UK
Telephone: 223 358855
Fax: 223 861 373

Applying expertise in applied physics, engineering and computing to commercialization of biotechnology discovers in areas of biosensors, toxic effect monitoring, automation of biochemical analysis, fermentation process control and immobilized biocatalyst systems. Member of the Arthur D Little consultancy group.

Cardiff University Industry Centre

University College, 57 Park Place, PO Box 78, Cardiff, CF1 1XL, UK
Telephone: 222 874000

Supply consultancy to industry. Undertake contract research analysis and testing as well as industrial short courses on topics such as applied microbiology, microbiological aspects of effluent treatment and waste utilization.

CBR International Biotechnologies Corp

101-9865 West Saanich Rd, PO Box 2010, Sidney, British Columbia, V8L 353, Canada
Telephone: 604 655 1644
Fax: 604 655 7131

Consultants offering environmental research services including audit, assessment and monitoring. Developing DNA, antibody and chemical probes and methods of biodegrading and mineral binding. Developed an in-field detection kit for the mining industry. Formerly Genprobe Technologies Ltd.

Center for Science Information

4252-20th St, San Francisco, CA-94114-2815, USA
Telephone: 415 553 8772

Non-profit organization offering information on biotechnology advances and their role in industry and the environment.

CETEC

Av Jose Candido da Silveira 2000, Cidade Nova, Belo Horizonte, MG 30 000, Brazil
Telephone: 31 461 5619/461 7933

Carry out biochemical research and offer consulting services to the food, waste treatment, biometallurgy, and fermentation industries.

Chandler-Lorenz International

Im Buehl 1, Bad Krozingen, D-7812, Germany
Telephone: 76 33 1811
Fax: 76 33 14895

Biotechnology and healthcare consultants on market research, marketing and investment as well as management of clinical trials with special expertise in immunodiagnostics and immunomodulation.

Chapel Laboratories

Cliff Road, Hythe, Kent, UK
Telephone: 303 69511

Contract research organization offering services in the areas of health care. Developing monoclonal antibodies for use in diagnostics of antibodies and similar compounds.

Charles River Biotechnical Services Inc

251 Ballardvale Street, Wilmington, MA-01887, USA
Telephone: 508 657 6500/800 356 2544
Fax: 508 658 7132

A full service in bioprocessing for large-scale production of monoclonal antibodies and other biomolecules from mammalian cell cultures using *in vitro* and *in vivo* procedures. Provide biotechnology testing in accordance with good laboratory practices including MAP and RAP testing murine viruses. A Bausch & Lomb, US company.

Chathamborough Research Group Inc

PO Drawer 929, Pittsboro,
NC-27312, USA
Telephone: 919 542 5600

Offer R&D clinical services to medical and pharmaceutical sectors in relation to the gaining of FDA approval for products.

Chesnut Hill Consultants Inc

55 Hallwood Road, Chesnut Hill,
MA-02167, USA

Consultants in the area of immunology.

Civi Consultancy

Vlietweg 17, Postbus 185,
Leidschendam, NL-2260 ADZ,
The Netherlands
Telephone: 70 209393
Fax: 70 203671

Offer consultancy in biotechnology.

Clinical Science Research International Ltd

Alexandra House, Hinchingbrooke Hospital, Hinchingbrooke Park,
Cambs, PE18 8NT, UK
Telephone: 480 432133
Fax: 480 411646

Carry out clinical development studies from phase I through to phase IV and conduct phase I clinical trials as well as phases II, III and IV clinical trials on a national and international basis. Offer services in consumer product testing in human volunteers for the cosmetic and toiletries industry. Member of the Applied Bioscience International Group. Formed by the merger of Cantab Ltd and CQTC International Ltd.

Coast Biologicals Ltd

PO Box 58103, East Tamaki
Auckland, New Zealand, Australasia
Telephone: 9 274 5622
Fax: 9 274 8043

Manufacture agar and seaweed extracts. Offer consultancy in aquaculture, and research and development of marine products for agriculture and the food industry.

Consulting Resources Corporation

Six Northbrook Park, Lexington,
MA-02173, USA
Telephone: 617 863 1222
Fax: 617 863 1441

International management consultants providing professional services to executives involved in the biotechnology and chemical industry. Specialize in product and technologies assessment, business/market analysis and strategic planning. Knowledgeable in science and engineering as well as management and economics.

Corporation BDG

Biotechnology Research Institute,
6100 avenue Royalmount, Montreal,
Quebec, H4P 2R2, Canada

Consultants in the development and testing of medical diagnostic products.

CPL Scientific Ltd

Science House, Winchcombe Road,
Newbury, Berkshire, RG14 5QX, UK
Telephone: 635 524064
Fax: 635 529322

Provide information services, special studies, books, directories and consultancy in agriculture and food biotechnology including use of wastes and residues for fuel and chemicals. Also offer permanent and temporary employment services in biotechnology, pharmaceuticals, food and other life sciences.

Cranfield Biotechnology Ltd

Cranfield, Bedford, MK43 OAL, UK
Telephone: 234 750665
Fax: 234 750907

Biotechnology consultancy providing a full design and development service for all types of biosensors and other novel analytical devices and reagents. Supply expert business services to the biotechnology industry including technology assessment, market research and marketing trial management.

Croon Business Support BV

Edisonbaan 10, MN Nieuwegwein, NL-3439, The Netherlands
Telephone: 3402 45154
Fax: 3401 43554

Consultants in biotechnology, marketing and strategy as well as in development of business ventures.

CSIRO Division of Biotechnology

343 Royal Parade, Melbourne, Victoria 3052, Australia
Telephone: 3 342 4200
Fax: 3 347 5481

Offer consulting services in molecular and cell biology including DNA probes, diagnostics, enzymes, fermentation and cell culture, immunoassay, recombinant DNA technology etc. Work includes contract research and collaboration with industry.

De Ruiter Environment Technology BV

Haarlemmerstraatweg 79, Postbus 14, AA Zwanenburg, NL-1160, The Netherlands
Telephone: 2907 8011
Fax: 2907 2142

Offer consultancy and contracting in aspects of environmental biotechnology including microbiological soil, waste water and air treatment.

Delft Geotechnics

Stieltjesweg 2, Postbus 69, AB Delft, NL-2600, The Netherlands
Telephone: 15 693500
Fax: 15 610821

Specialize in consultancy and research in soil pollution, particularly industrial, including development and evaluation of soil protection techniques, environmentally safe waste storage systems, soil cleaning procedures, soil pollution monitoring, storage systems for low and medium level radioactive waste and the design of biological on-site remedial pollution treatment with reference to assessment and modelling of the rate of natural biodegradation.

DELPHI Associates SPRL

44 Avenue de Roodebeek, Brussels, B-1040, Belgium
Telephone: 2 732 57 73
Fax: 2 732 63 91

Produce and supply biotechnology-related market studies in the following fields: human and veterinary diagnostics, nucleic acid probes, biopharmaceuticals, cytokines, blood substitutes, thrombolytics, oral health care pharmaceuticals, angiogenic factors, drug delivery systems, biodiagnostic products, monoclonal antibodies and transgenic animals.

Dextra Laboratories Ltd

The Philip Lyle Building, The University, PO Box 68, Reading, Berkshire, RG6 2BX, UK
Telephone: 734 861361
Fax: 734 861894

Consultancy and specialist analytical services for carbohydrates in the food, pharmaceutical and chemical industries including oligosaccharides, monosaccharides, holo sugars, amino sugars, sugar esters, glycosides, deoxy sugars, polysaccharides and glycoproteins.

Drug Development (Scotland) Ltd

Ninewells Hospital & Medical School, Dundee, UK
Telephone: 382 646317
Fax: 382 645606

Offer contract research services in clinical testing for drug development from preclinical investigations, through phase I studies to the identification of therapeutic efficacy in phase II/III. Advise on drug development strategy, protocol design and drug regulatory affairs.

Duskin Institute for Biomedical Research

Naniwa Building 4-12 Banzai-cho, Kita-ku, Osaka 530, Japan
Telephone: 6 314 0528

Offer custom service for testing biological material for mutagenicity, carcinogenicity and toxicity.

Dvorkovitz and Associates

PO Box 1748, Ormond Beach, FL-32074, USA
Telephone: 904 677 7033
Fax: 904 677 7113

Act as an exchange for license of technology on a subscription basis in areas of science and technology which include biotechnology and genetic engineering.

Euro BioMed

Postbus 210, AE Maastricht, NL-6200, The Netherlands
Telephone: 43 887122
Fax: 43 258183

International network of consultants set up to facilitate technology transfer and business cooperation in biotechnology and medical technology.

Euro Laboratories Ltd

Cranfield Biotechnology, Cranfield, Bedford, MK43 OAL, UK
Telephone: 234 750665
Fax: 234 750907

Provide services in biodegradability, eco-toxicity and environment fate assessments for new and existing chemicals as well as advice on regulatory and registration requirements.

EuroForum

14 Bowden Street, London, SE11 4DS, UK
Telephone: 71 793 1230
Fax: 71 793 1250

Monitor developments in law, finance, business, human resources, the environment and public policy. Produce, together with its affiliates in the United States and Canada, in excess of 300 conferences per year which are attended by more than 20,000 lawyers, other professionals and business executives.

Frank Maine Consulting Ltd

71 Sherwood Drive, Guelph, Ontario, N1E 6E6, Canada
Telephone: 519 823 1465

Offer consultancy on mushroom technology and production of chemicals from peat.

Frost & Sullivan Inc

106 Fulton Street, New York, NY-10038, USA
Telephone: 212 233 1080
Fax: 212 619 0831

Carry out surveys analyzing and forecasting market developments; these include a study of the existing and potential fermentation products markets within the European Community in relation to antibiotics, enzymes, organic acids, amino acids, vitamins, organic solvents, single-cell protein, oils, fats and polysaccharides. Offices in UK, Germany, France, Italy, Spain, Portugal, Hungary, Japan and Brazil.

GB Biotechnology Ltd

4 Beaconsfield Court, Sketty, Swansea, West Glamorgan, SA2 9JU, UK
Telephone: 792 208190
Fax: 792 297402

Offer research and development consultancy in enzymes, biotechnology products and equipment; the assessment of biotechnology ventures; national conference, exhibition and industrial seminars; magazine development and publication; industry/education liaison.

GEM

58 rue du Dessous des Berges, Paris, F-75013, France
Telephone: 1 45 84 01 94

Carry out studies on technoeconomics in areas of biomass energy, fermentation, the environment, water and the agro-food industries. Consult on financing of biotechnology processes.

Griffith Hack & Co

71 York St, Sydney, NSW 2000, Australia
Telephone: 2 261 3311
Fax: 2 290 1578

Biotechnology consultants on patents, licensing and all legal services. Also supply information services.

Grontmij

Huize Houdringe, Postbus 203, AE De Bilt, NL-3730, The Netherlands
Telephone: 30 767911

Offer consultancy and process management in the area of environmental biotechnology.

H G Engineering International (1980) Ltd

34 Yitzhaq Sadeh St, Tel Aviv, Israel
Telephone: 3 330132

Consultant engineers to the pharmaceutical and biotechnology industries.

H O M Consultancy BV

Scheveningseweg 9, KS 's-Gravenhage, NL-2517, The Netherlands
Telephone: 70 3469643
Fax: 70 3614695

Offer consultancy in technology, marketing and finance in areas of medical and biotechnology innovation.

Hazleton Europe

Otley Road, Harrogate, North Yorks, HG3 1PY, UK
Telephone: 432 500011
Fax: 423 525620

Specialize in contract testing of biopharmaceuticals/monoclonal antibodies for *in vivo* use. Areas of expertise include virological validation of cell lines and the purification process; cell line characterization; microbial and mycoplasma contamination; toxicology and genetic toxicology studies on the final product. Work is carried out to meet regulatory requirements including worldwide 'points to consider' in the manufacture of products from rDNA technology and from rodent or human hybridomas.

Hazleton SA

Les Oncins - BP, l'Arbresle, F-69210, France
Telephone: 74 01 10 10

Offer analytical and consultancy services in testing for toxic substances, carcinogens, mutagens etc.

Heineken NV

Postbus 500, BA Zoeterwoude,
NL-2380, The Netherlands
Telephone: 71 456111
Fax: 71 410013

Offer technical services in brewing, distilling and soft drink production.

Helix CV

St Pietersnieuwstraat 136, Gent,
B-9000, Belgium
Telephone: 91 330133

Biotechnology consultants.

HGF Hamburgische GmbH

Hamburger Strasse 11, Hamburg 76,
D-2000, Germany
Telephone: 40 2270190

Promote new businesses and serve as a consultant to international and national companies; establishing a biotechnology-based industrial park in Hamburg.

Hill & Knowlton (UK) Ltd

5-11 Theobalds Rd, London,
WC1X 8SH, UK
Telephone: 71 413 3000
Fax: 71 413 3111/2

International public relations company with a specialist biotechnology practice offering a full range of services to the bioscience industries.

Huntingdon Research Centre Ltd

Huntingdon, Cambs, PE18 6ES, UK
Telephone: 480 890431
Fax: 480 890693

Provide a complete service of biological testing to evaluate safety and efficiency of chemicals and compounds with reference to effects in humans, animals and the environment.

IBRD

2525 Campus Drive, Irvine,
CA-92715, USA
Telephone: 800 358 0700
Fax: 714 752 7297

Offer consultancy in clinical product development and project management, licensing, clinical data processing etc, for pharmaceutical and biotechnology industries.

IMC Netherlands BV

Achter Emer 17, Breda, NL-4824 ZA,
The Netherlands
Telephone: 76 411 964
Fax: 76 419216

Offer management consultancy in industrial chemicals, food, feed and environment industries. Services include marketing, licensing, take-overs, R&D project management and innovation, technology transfer and commercialization.

Innovation 128

24 rue du 4 Septembre, Paris,
F-75002, France
Telephone: 1 42 68 09 71
Fax: 1 42 65 47 76

Offer consultancy in areas of technical innovation, in particular in the biotechnology sector.

Institute of Clinical Pharmacology

PO Box 469, Dublin 8, Ireland
Telephone: 1 758932

Specialize in providing a service to the pharmaceutical industry offering a programme for the safety evaluation of drugs in humans.

Institute of Drug Technology Australia Ltd

45 Wadhurst Drive, Boronia,
Victoria, 3155, Australia
Telephone: 3 801 8888
Fax: 3 801 8773

Offer consultancy in regulatory procedures relating to drug production.

Integrated Explorations

189 McCurdy Rd, Box 1385, Guelph,
Ontario, N1H 6N8, Canada
Telephone: 519 822 2608
Fax: 519 822 0284

Environmental consultants specializing in aquatic habitat assessment, on-site biomonitoring and mapping of underwater contaminated sediments as well as offering aquatic laboratory services.

Intek Pty Ltd

PO Box 8, Cowandilla, Adelaide,
South Australia, 5033, Australia
Telephone: 8 234 0022
Fax: 8 234 0058

Offer consultancy and contract research services in brewing, fermentation and food technology including product testing and development.

Intron BV

Dr Nolenslaan 26, Postbus 5187,
PD Sittard, NL-6130, The Netherlands
Telephone: 4490 28888
Fax: 4490 29060

Consultants in environmental biotechnology.

Inveresk Research International Ltd

Tranent, EH33 2NE, UK
Telephone: 875 614545
Fax: 875 614555

Offer control services in biotechnology (including exploitation of microorganisms for industrial use, recombinant DNA work and ancillary genetic engineering services) as well as custom production of monoclonal antibodies for the food, pharmaceutical, chemical and oil industries.

IWACO BV

Schiekade 189, Postbus 183,
AD Rotterdam, NL-3000,
The Netherlands
Telephone: 10 4241641
Fax: 10 4330977

Carry out research and feasibility studies as well as design and supervision of environmental technology projects.

J F Coates Inc

3738 Kanawha Street, Washington,
DC-20015, USA
Telephone: 201 966 9307

Consultants on the implications of biotechnology and the consequences of regulation on the commercialization of biotechnology as well as scientific and technological forecasts.

J R Snyder Inc

PO Box 3638, Champaign,
IL-61826, USA
Telephone: 217 352 9880

Biotechnology consultants offering services in assessment, management and commercialization.

John Brown Engineers & Constructors Ltd

1 Buckingham Street, Portsmouth,
Hants, PO1 1HN, UK
Telephone: 705 822300
Fax: 705 733028/830452

Provide complete contracting services to the bioprocess, pharmaceutical, fine chemicals and polymer process industries. The Portsmouth office is also the source of this technology to other offices in the John Brown Group in London, the Hague, Paris, Madrid, Stanford and Houston. Part of the Trafalgar House Group. Provide a collaborative approach to clients, with innovative translation of R&D data into commercial scale plant. Many novel processes and equipment designs and engineering techniques have been developed in bioprocessing and large plant constructed.

Korda & Company Ltd

5th floor, 18-20 Farringdon Lane,
London, EC1R 3AU, UK
Telephone: 71 253 5882
Fax: 71 251 4837

Consultants in exploitation of life science industries and venture capital investments in start up enterprises.

Laboratoire Midac

40 boulevard de la Liberte, Lille,
F-59800, France
Telephone: 20 97 02 80

Offer consultancy on the use of microbiology systems in the food, pharmaceutical, cosmetic and hospital industries.

Land Restoration Systems (UK) Ltd

Galvin Rd, Slough, SL1 4DL, UK
Telephone: 753 71551
Fax: 753 694201

Offer consultancy services for site investigation and risk assessment for contaminated sites.

Landell Mills Commodities Studies - LMC

14-16 George Street, Oxford,
OX1 2AF, UK
Telephone: 865 791737
Fax: 865 791739

1841 Broadway, New York,
NY-10023, USA
Telephone: 212 586 2427
Fax: 212 397 4756

Provide a wide range of economic, marketing and planning services for enterprises involved in commodities and industrial raw materials including fermentable substrates and timber, pulp and paper; such services also cover by-products and manufactured products. LMC has a representative office in Japan and associates in major business centres.

Lansdowne Bodycombe and Partners

89/90 Gower Road, Sketty, Swansea,
SA2 9BZ, UK
Telephone: 792 204059

Consulting and structural engineers offering services to the chemical and biotechnology industries.

A Lassonde & Fils Inc

170 5th Avenue, Rougemont, Quebec,
J0L 1M0, Canada
Telephone: 514 878 1057
Fax: 514 583 2553

Offer technology and consulting on continuous processes developed for enzymatic clarification of fruit juices.

Leatherhead Food Research Association

Randalls Rd, Leatherhead, Surrey,
KT22 7RY, UK
Telephone: 372 376761
Fax: 372 386228

Offer information and research based services to manufacturers and retailers within the food market; data on all areas of food science and technology.

Life Sciences International

1818 Market St, Philadelphia,
PA-19103, USA
Telephone: 215 299 8506
Fax: 215 299 2273

Perform engineering, architectural, construction and validation/start-up services for the pharmaceutical and biotechnology industries. Expert in the design, construction and validation of facilities involving the application of biotechnology and pharmaceutical technology to large-scale fermentation and the manufacture of synthetic drugs, finished dosage pharmaceuticals, medical devices and related products. Experienced in cell culture, rDNA fermentation, monoclonal antibody engineering and purification design. Subsidiary of Day & Zimmermann Inc.

McKnight Laboratories GmbH

Osterstrasse 86, Hamburg 20, D-2000, Germany
Telephone: 40 499727

Offer consultancy in bio-availability studies, planning, non-clinical testing, biometrical evaluations and project reporting.

Modern Talking Picture Service Inc

5000 Park St North, St Petersburg,
FL-33709, USA
Telephone: 813 541 5736

Distributor of educational videos, films and documentaries including coverage of biotechnology in agriculture.

Monkman International Consulting

Breck's Mill on the Brandywine,
PO Box 3760, Wilmington,
DE-19807, USA

Provide consultancy in the following areas: patents, licensing and joint ventures, market and technology studies, business venture plans, strategy and long-range planning, for both US and international clients in fields which include biotechnology, chemicals, energy, electronics, health care pharmaceuticals and plastics.

NATEC

Behringstrasse 154, Postfach 501568,
Hamburg 50, D-2000, Germany
Telephone: 40 8 8309 0
Fax: 40 88 309-170

Offer services in problem solving in natural science and technology including studies and contract research. Operate internationally mainly within the fields of analytical chemistry in relation to toxicological safety, synthesis of radioactive and other chemicals and additives, as well as packaging materials particularly with respect to food legislation. Concerned with pollution control, environmental and consumer protection. Also offer assistance in establishing quality assurance systems especially for the food but also for the cosmetic and pharmaceutical industry.

NIZO Environmental Services

Kernhemseweg 2, Postbus 20, BA Ede,
NL-6710, The Netherlands
Telephone: 8380 59522
Fax: 8380 51200

Offer services in areas of process water supply and treatment, waste water reduction and treatment, soil monitoring, noise abatement and air pollution. Division of NIZO, The Netherlands.

North Carolina Biotechnology Centre

79 Alexander Drive, PO Box 13547,
Research Triangle Park,
NC-27709-3547, USA
Telephone: 919 541 9366
Fax: 919 549 9710

Private, non-profit corporation established to promote development and commercialization of biotechnology statewide by working to strengthen the biotechnological research capabilities of the state's universities, assist biotechnology business development, educate and inform the public about the science, issues and applications of biotechnology and encourage collaborations among the state's universities, industry and government.

Novabio

34 Grande-Rue, Commeny, F-95450, France
Telephone: 1 34 67 44 20
Fax: 1 34 67 41 20

Offer consultancy and information services to assist in strategic development of R&D policy as well as providing scientific and technical advice, help with patents and assistance in marketing of biological products.

Operations Analysis Centre A/S

Socfielundsvej 23, Glostrup, DK-2600, Denmark
Telephone: 42 96 80 00

Consulting company specializing in technology assessment, feasibility studies, implementation and management of new technologies.

Oranjewoud Engineers BV

Driehuizerkerweg 138, Postbus 439, Ijmuiden, NL-1970, The Netherlands
Telephone: 2550 34734
Fax: 2550 36128

Offer consultancy, research, design and engineering in aerobic and anaerobic treatment of municipal and industrial waste water as well as air, soil contamination and ground waste treatments.

Oxford Biomedical Ltd

The Spendlove Research Centre,
Enstone Rd, Charlbury, Oxfordshire,
OX7 3PQ, UK
Telephone: 608 811288
Fax: 608 811323

Offer consultancy to the independent medical sector and provide contract clinical research in diagnostics and phase I-IV testing facilities for the pharmaceutical industries.

P Lane & Associates Ltd

5439 Cogswell St, Halifax, Nova Scotia, B3J 1R1, Canada
Telephone: 902 423 8197
Fax: 902 429 8089

Environmental consultants on environmental risk/impact assessment, terrestrial ecology, aquaculture, soil pollution, heavy metals and organics and the application of microbiological and other solutions. Carrying out research on waste water treatment processes, oceanography, environmental modelling and laboratory and field studies.

PA Consultancy Group

279 Princeton Rd, Hightstown,
NJ-08520, USA
Telephone: 609 426 4700
Fax: 609 426 4046

International management and consulting services in development of advanced technology products, processes and computer systems.

PA Consulting Group

Avenue Louise 390/4, Brussels,
B-1050, Belgium
Telephone: 2 640 3100
Fax: 2 640 4963

Offer consultancy and implementation in biotechnology to the food, agriculture, health care, environmental and process industries.

PA Technology

Cambridge Laboratory, Melbourn,
Royston, Hants, SG8 6DP, UK
Telephone: 763 261222
Fax: 763 260023

Carry out contract research as well as all aspects of new product and process development in all areas of industry, including the biosciences.

Pambec Laboratories

36533 N Streamwood Drive, Gurnee,
IL-60031, USA
Telephone: 414 963 8594

Carry out custom and contract work in genetically engineered prokaryotes and eukaryotes, DNA probes, polypeptides, polynucleotides, conjugate synthesis and consulting.

Parexel International Corp

195 West St, Waltham,
MA-02154, USA
Telephone: 617 487 9900
Fax: 617 487 0525

Consultants in regulatory affairs and clinical trials management.

Phero Tech Inc

7572 Progress Way, RR 5
Delta, British Columbia, V4G 1E9,
Canada
Telephone: 604 940 9944
Fax: 604 940 9433

Offer pest management services to the forestry and agriculture sectors.

Carrying out research in areas of forest pest management including control of bark beetles, ambrosia beetles, flies and nematodes as well as the development of mammal repellants to protect saplings from rabbits and deer etc, and the synthesis and use of pheromones.

PJB Publications Ltd

18-20 Hill Rise, Richmond, Surrey,
TW10 6UA, UK
Telephone: 81 948 3262
Fax: 81 948 6866

PharmaBooks Ltd, 1775 Broadway,
Suite 511, New York, NY-10019, USA
Telephone: 212 262 8230
Fax: 212 262 8234

Shiryo Kenkyujo, 3-6-5-904 Sotokanda,
Chiyoda-ku, Tokyo 101, Japan
Telephone: 3 3258 9207
Fax: 3 3258 9209

Offer information services including books, directories, databases, market insights, competitor analysis, world market data, on-line systems, journals and multi-client and market studies in the following fields: biotechnology, animal health, veterinary R&D, agriculture, food, agroindustry, pesticides and biocontrol, drug delivery, EC regulatory guides (agrochemical registion, biotechnology, pesticide residues, good laboratory practice - GLP, veterinary products), diagnostics.

Plant Biotechnology (UCC) Ltd

University College, Cork, County
Cork, Ireland
Telephone: 21 26871

Offer services to agricultural, horticultural and related industries in areas of diagnostics, micropropagation and new variety production, including genetic engineering, mutation breeding, germplasm storage and disease indexing.

Pollutech Environmental Ltd

1149 Vanier Rd, Suite 4, Sarnia,
Ontario, N7S 3Y6, Canada
Telephone: 519 339 8787
Fax: 519 336 6965

Environmental consultants offering waste biotreatment services.

RauCon GmbH

PO Box 1069, Lieselottestrasse 4,
Dielheim, D-6912, Germany
Telephone: 6222 73562
Fax: 6222 74884

Biotechnology consultants offering market and information reports and newsletters.

Ravvan International BV

Postbus 1275, BG Aalsmeer, NL-1430,
The Netherlands

Advise biotechnology business and investors with regard to new genetic engineering advances in agriculture and horticulture.

Reading Scientific Services Ltd

Lord Zuckerman Research Centre,
PO Box 234, Whiteknights, Reading,
RG6 2LA, UK
Telephone: 734 868541
Fax: 734 868932

Offer biotechnology consultancy services in contract process research and analysis in relation to food technology, biochemistry, microbiology and fermentation, including advice on appropriate legislation on product safety testing and evaluation and an Emergency Response Service which supplies priority analytical resources in cases of deliberate or accidental contamination of products. Subsidiary of Cadbury Schweppes with access to additional facilities and expertise at the University of Reading.

Roan Ventures

2343 Halyard Drive, Merrick,
New York, NY-11566-5528, USA

Venture capital investment company with interests in biotechnology new business startup.

Saros Ltd

PO Box 805, Edinburgh,
EH14 7LS, UK
Telephone: 31 449 3814

Clinical research consultancy advising on product development and clinical research programmes.

Scansearch

31 Flemington St, Frewville, Adelaide,
South Australia, 5063, Australia
Telephone: 8 372 2735
Fax: 8 796 623

Consultants. Offer information services in biotechnology.

SEAI Technical Publications

PO Box 590-E, Madison,
GA-30650, USA
Telephone: 404 342 9638

Consultancy organization producing multiclient studies and technical reports on areas such as biosensors.

Sequal

University of Aberdeen, Marischal
College, Aberdeen, AB9 1AS, UK
Telephone: 224 273103
Fax: 224 273144

Offer a contract service in the sequencing of proteins and peptides.

Spie Batignolles

Tour Anjou, 33 quai de Dion Bouton,
Puteaux Cedex, F-92814, France
Telephone: 1 47 76 43 64

Major engineering consultant and contracting company. Associates include Speichim and Citrem, both in France.

SRI International

Pelikanstrasse 37, Zurich, CH-8001, Switzerland
Telephone: 1 211 06 36
Fax: 1 211 61 08

A contract research and consultancy organization offering single client and multi-client studies in the fields of biotechnology with an emphasis on applications in industry and business.

Stratagene GmbH

Postfach 105466, Im Weiher 12, Heidelburg, D-6900, Germany
Telephone: 6221 400634
Fax: 6221 400639

Offer custom services in preparation of gene libraries and related aspects as well as information services.

Strategic Technologies International Inc

1204 Allanson Rd, Mundelein, IL-60060, USA
Telephone: 708 362 5620
Fax: 708 362 5669

Management consulting firm providing professional services in market research and planning, strategic planning, technology and business opportunity assessments, product and process, licensing, acquisition evaluation and liaison services to biotechnology companies worldwide in the following industries: pharmaceutical, chemicals, petrochemicals and polymers, speciality chemicals, pesticides, veterinary products, enzymes, laboratory and medical instruments, clinical diagnostics and food additives.

Summit Biotechnology Associates

PO Box 58081, Salt Lake City, UT-84158, USA
Telephone: 801 581 1800

Offer consulting services to biotechnology and venture capital investment companies in R&D, technical evaluation, marketing and competitive analyses management consulting, strategic planning and financing.

Tauw Infra Consult BV

Handelskade 11, Postbus 479, AL Deventer, NL-7400, The Netherlands
Telephone: 5700 99911
Fax: 5700 99270

Offer consultancy in waste water treatment including detection of xenobiotic micropollutants in contaminated soil, groundwater and air. Also offer services in environmental analysis, biological treatment systems and supervision of soil remediation projects. Carrying out related research on both a laboratory and pilot plant scale.

TBZ Hamburg

Zippelhaus 5, Hamburg 11, D-2000, Germany
Telephone: 40 33 1186
Fax: 40 335449

Offer consultancy in technology transfer, finance and marketing as well as organizing seminars, workshops and information services.

Tebodin BV

Laan van Nieuw Oost Indie 25, Postbus 16092, BA 's-Gravenhage, NL-2500, The Netherlands
Telephone: 70 3480911
Fax: 70 3480645

Consultants in bio-engineering for the food and feed industry. Carry out feasibility studies and offer environmental engineering and contracting services including waste water treatment.

Technology Marketing Group

950 Lee Street, Des Plaines,
IL-60016, USA
Telephone: 708 297 1404
Fax: 708 297 1407

Consultants in areas of marketing research and development including phone survey, mail survey and focus group techniques.

Technoma GmbH

Im Neuenheimer Field 519,
Heidelberg, D-6900, Germany
Telephone: 6221 40 0624
Fax: 6221 41 0630

Offer consultancy and development advice in the application of mathematics and computer services to biotechnology. Services include supply of standard software for molecular biology and for use with laboratory equipment, as well as custom designed software. Also offer on-line search facilities of databanks, information on patents, 'state of the art', market reviews and executive summaries.

Theta Corp

Theta Building, Middlefield,
CT-06455, USA
Telephone: 203 349 1054
Fax: 203 349 1227

Carry out market research reports and consulting in those areas of biotechnology likely to affect medical diagnostic industries. Produce a monthly newsletter.

TNO

PO Box 297, BD Den Haag, NL-2501,
The Netherlands

Independent non-profit organization consisting of 35 institutes. Supporting research and development in biotechnology related to food and feed, nutritional, pharmaceutical, chemical, equipment and environmental industries. Also provide consultancy and develop databanks and information sources.

Triad Technologies Inc

101 Centrepoint Boulevard,
New Castle, DE-19720, USA
Telephone: 302 323 1550
Fax: 302 323 1557

Engineering/consulting firm specializing in cGMP process and facility design, qualification, validation and regulatory compliance in the biopharmaceutical and health care industries.

Trimach BV

Sophialaan 1, Postbus 851,
BR Leeuwarden, NL-8901,
The Netherlands
Telephone: 58 124916
Fax: 58 126449

Design and install equipment for the dairy and cheese industries. Offer consultancy in biotechnological processing to the food industry.

Twose Anthony & Associates

1 The Acreage, Goostrey, Cheshire,
CW4 8JY, UK
Telephone: 477 33741

Biotechnology consultants.

U-Gene Research BV

Heidelberglaan 8, CS Utrecht,
NL-3584, The Netherlands
Telephone: 30 533636
Fax: 30 515141

Offer contract services in development of recombinant DNA and RNA probes for rapid identification of bacterial infections in fields of industrial, environmental and food microbiology.

United Biotechnology Sc

25 Schemeringsweg, Sterrebeek,
B-1933, Belgium
Telephone: 2 731 90 15/733 21 23

Consultancy established in 1991 to undertake services to small biotechnology firms including market surveys, share funding, management and marketing.

Venture Information Services

6114 La Salle Ave, Suite 250,
Oakland, CA-94611, USA
Telephone: 415 530 5030
Fax: 415 531 4140

Supply information services including market research, competitive analysis, patent searching and industry studies, technical articles, patents and reports. Can produce custom information reports on biotechnology, pharmaceuticals and health care etc. Publish Directory of California Biotechnology.

Warren Spring Laboratory

Gunnels Wood Road, Stevenage, Herts,
SG1 2BX, UK
Telephone: 438 741122
Fax: 438 360858

Environmental Technology Executive Agency of the Department of Trade and Industry. Undertake contract research and consultancy in environmental technology for government and industry in areas of environmental technology, biosafety, workplace monitoring, bioseparations, biodesulphurization, biodegradation, treatment of wastes, solid/liquid separation, air pollution monitoring and modelling, pollution abatement, marine pollution and bulk materials handling, minerals and metals.

Webb Technical Group Inc

4320 Delta Lake Drive, Raleigh,
NC-27612, USA
Telephone: 919 787 9171

Carry out R&D and analysis related to microbiological processes and offer consultation in the applications of genetic engineering to the food and environmental sectors. Offer analytical services including chemical analysis of water and waste products and samples.

Western Chemical Research Corp

2300 North Highway 287,
PO Box 1255, Fort Collins,
CO-80522, USA
Telephone: 303 493 8474

Supply bulk immunodiagnostic reagents and antisera. Services include custom antiserum production, complete immunodiagnostic assay development, automated system development, competitive product evaluation, general consulting services and contract research and development.

CULTURE COLLECTIONS AND SUPPLIERS

ABC Research Corp

3437 SW 24th Avenue, Gainesville,
FL-32607, USA
Telephone: 904 372 0436
Fax: 904 378 6483

Manufacture custom cultures for food processing including lactic acid starter cultures, bacterial receptor proteins and biological products. Offer analytical services and undertake contract research.

Advanced Biotechnologies Inc

9108 Guilford Road, Columbia,
MD-21046, USA
Telephone: 901 792 9779/800 426 0764
Fax: 301 497 9773

Produce: purified viruses, viral proteins, viral and cellular nucleic acids, selected cell components purified for research purposes, in particular human retrovirus research reagents. Complete molecular biology services include: plasmid DNA purification, insert isolation, transformation, labelling and sequencing. Products for cell, tissue and bacteriological culture include: media, human and animal sera, buffered and balanced salt solutions, growth and attachment factors (human IL-2) and antibiotics/antimycotics for cell culture. Carrying out contract R&D and provide custom biological services.

AFRC Institute of Arable Crops Research

Rothamsted Experimental Station,
Harpenden, Hertfordshire,
AL5 2JQ, UK
Telephone: 582 763133
Fax: 582 760981

The institute maintains National Collections of Yeast Cultures and Food Bacteria.

AGA SA

Rua da Passagem 126 E 7 Andares CP
1823/1824, Rio de Janeiro 22 290,
Brazil
Telephone: 21 295 9432

Supply plant cell lines.

Agriculture Canada

200 Rene-Levesque Blvd W, East
Tower Suite 1002-D, Montreal,
Quebec, H2Z 1Y3, Canada
Telephone: 514 285 8888
Fax: 514 283 3143

Have a collection of plant viruses infecting fruit trees and grapevines. Maintain a large collection of virus-tested fruit tree and grapevine cultivars for international distribution. Also maintain a large collection of Kiwi and *Ipex* sp cultivars.

Agrifutur Srl

via Campagnole 8, Alfianello (BS),
I-25020, Italy
Telephone: 30 930 5697
Fax: 30 9305048

Manufacture and sell nitrogen-fixing bacterial inoculants (Rhizobium) for legume and gramineous roots, bacterial starter-cultures for salami-style meats, microorganisms for biological control of phytopathogens, bacterial inoculants for silage and microbial carriers. Also carry out R&D on monitoring of released microorganisms.

Agropur Cooperative Agro-Alimentaire

510 rue Principale, Granby, Quebec, J2G 7G2, Canada
Telephone: 514 375 1991
Fax: 514 375 2099

Manufacture cheese and other milk products. Carrying out research on lactic acid fermentations and proteases for cheese-making as well as waste water purification treatment. Supply lactic acid bacterial cultures.

Algensammling am Institut fur Botanik

Universitat Innsbruck, Sternwartestrasse 15, Innsbruck, A-6020, Austria

Hold culture collection of algae for general microbiology research and teaching with main interests in soil, airborne and lichen algae.

Alko Ltd

Salmisaarenranta 7 H, PO Box 350, Helsinki, SF-00101, Finland
Telephone: 0 13311
Fax: 0 1332386

Supply baker's yeast, fermented vinegar and process enzymes (cellulases, hemicellulases, pectinases and phytases).

American Type Culture Collection - ATCC

12301 Parklawn Drive, Rockville, MD-20852-1776, USA
Telephone: 301 881 2600/800 638 6597
Fax: 301 231 5826

One of the most diverse culture collections in the world with over 42,000 different strains of authentic cultures of bacteria, fungi, yeasts, algae, protozoa, viruses, cell lines, plant tissues, actinomycetes, recombinant DNA materials and antisera, with departments of bacteriology, cell culture, molecular biology, mycology, protistology and virology. Have initiated a collection of hybridoma cell lines, under contract to National Institute of Allergy and Infectious Diseases. Official depository for strains involved in patentable processes recognized by the US, European Patent Office and other countries.

Aqua-Tek GmbH

Kaiser-Friedrich-Strasse 54, Berlin 12, D-1000, Germany
Telephone: 30 324 7006
Fax: 30 324 0885

Supply cultures for the production of cells, cell fractions and monoclonal antibodies.

Australian National Reference Laboratory in Medical Mycology

The Royal North Shore Hospital of Sydney, Pacific Highway, St Leonards, Sydney, NSW 2065, Australia

Hold culture collection containing algae, fungi and yeasts with a special collection of aerobic actinomycetes for medical mycology research and teaching. Offer identification service.

The Australian Wine Research Institute

Waite Road, Urrbrae, S Australia, 5064, Australia
Telephone: 8 799 681
Fax: 8 795 200

Hold culture collection containing bacteria, fungi and yeasts for general, agricultural, industrial and medical microbiology research and teaching.

Bacillus Collection

Institute of Hygiene, c/o University of Arhus, Universitaetsparken, Arhus, DK-8000, Denmark

Bacterial Culture Collection

Institute Tropical Medicine,
Nationalestraat 155, Antwerpen,
B-2000, Belgium
Telephone: 2 375 24 00

Hold culture collection of bacteria for medical microbiology research and teaching. Offer identification services.

Bayerische Landesanstalt fur Weinbau und Gartenbau

Residenplatz 3, Wurzburg, Germany

Hold culture collection of yeasts for agricultural microbiology research.

BCZ-Friesland BV

Baljeestraat 12, Postbus 396,
BD Leeuwarden, NL-8901,
The Netherlands
Telephone: 58 449444
Fax: 58 449449

Supply bacterial starter cultures for biodegradation of organic components.

BEL Industries

103 boulevard MacDonald, Paris,
F-75019, France
Telephone: 1 42 00 33 11

Produce yeast and bacterial preparations with lactic acid metabolism for use in the agricultural and food industries. Products include animal feed additives based on live organisms (probiotics).

Bercan Environmental Resources Inc

6028 Mount View Rd, PO Box 238
Lantzville, British Colombia,
V0R 2HO, Canada
Telephone: 604 390 3113
Fax: 604 390 2312

Manufacturers of bio-augmentation/bio-remediation products for enhanced decomposition of organic wastes and residues by aerobic or anaerobic processes.

Bicarbolyte Corp

809 Princess St, Alexandria, Virginia,
VA-22314-2223, USA
Telephone: 703 739 0427

Produce microbial cultures for treating waste organics, including ammonia and chlorinated organics, as well as Bacillus thuringiensis strains.

Biologische Bundesanstalt fur Land- und Forstwiwrtschaft - BBA

Institut fur Pflanzenschutz im Forst,
Messeweg 11-12, Braunschweig,
D-3300, Germany
Telephone: 531 399-1
Fax: 531 399 239

Hold culture collection containing fungi for forest mycology research. Offer services in fungal identification.

Bioprox

6 rue Barbes - BP 177,
Levallois-Perret, F-92305, France
Telephone: 1 47 57 74 00

Produce cultures of microorganisms and lactic acid bacterial cultures.

Bioscience Inc

1530 Valley Center Parkway,
Suite 120, Bethlehem, PA-18017, USA
Telephone: 215 974 9693/800 627 3069

Environmental biotechnology firm which develop, manufacture and market specialized biological products, systems and services for pollution control applications. Products contain specialized microorganisms and selected synergistic ingredients for enhancing their action.

Biotal Ltd

5 Chiltern Close, Cardiff, Wales,
CF4 5DL, UK
Telephone: 222 766716
Fax: 222 766945

Exploit naturally occurring microorganisms to develop microbial technology, products and services for the agricultural and environmental biotechnology sectors. Supply microbial inoculants for silage, hay and grain as well as microbial compost accelerator.

Canadian Collection of Fungus Cultures

Room 1015, D W Neatby Building
CEF, Ottawa, Ontario, K1A 0C6,
Canada
Telephone: 613 996 1665
Fax: 613 943 0953

Hold culture collection of fungi for agricultural and forest microbiology research. Offer identification services.

CAYLA Sarl

avenue Larrieu, Toulouse, F-31400,
France
Telephone: 61 44 24 40
Fax: 61 40 20 67

Produce enzymes, microbial cultures and cloning vectors for use in the food and agricultural industries. Products include silage additives and enzymes for use in the preparation of protoplasts. Carry out research on microbial enzyme production as well as strain selection and genetic improvement of microorganisms producing novel antibiotics, herbicides and insecticides.

The CB *Rhizobium* Collection

CSIRO Division of Tropical Crops and,
Pastures, 306 Carmody Road, St Lucia,
Brisbane, Queensland, SA 5064,
Australia
Telephone: 8 799 681
Fax: 8 267 7342

Hold culture collection containing bacteria including a special collection of Rhizobium of tropical crop and pasture legumes for agricultural microbiology research.

CCMCU

Department of Microbiology, Ciudad
Universitaria Madrid, Madrid, Spain

Culture Collection of Microorganisms of Ciudad Universitaria Madrid. Hold culture collection containing bacteria, fungi and yeasts, with special collection of *Rhizobium* for general microbiology research and teaching.

CCTM - Centre de Collection de Types Microbiens

Rue du Bugnon 44, Lausanne,
CH-1011, Switzerland
Telephone: 21 3 14 23 39
Fax: 21 3 14 31 81

Hold culture collection containing mycobacteria, *Shigella* and other medical and non-medical bacteria for general microbiology use in research and teaching.

CECT - Coleccion Espanola de Cultivos Tipo

Valencia University, Microbiology
Department, Burjasot Valencia, Spain
Telephone: 96 3 63 00 11

Hold culture collection containing bacteria and fungi for general and industrial microbiology research and teaching.

Central Bureau for Cultures of Fungi

Oosterstraat 1, Postbus 273, AG Baarn,
NL-3740, The Netherlands
Telephone: 2154 81211
Fax: 2154 16142

Centraalbureau voor Schimmelcultures (CBS). Major collection of bacteria, fungi and yeast and one of the four reference collections of actinomycetes for agricultural, industrial, medical, veterinary and insect microbiology research and teaching. Carry out both microbiological research and collection work. Offer supply of cultures as well as services for identification, contract research and patent deposit.

Central Laboratories

Jaffa Street 86, Jerusalem, Israel

Hold culture collection of bacteria and bacterioviruses for medical microbiology research. Offer identification services.

Central Laboratory Collection

PO Box 9000 Terminal A, Toronto, Ontario, M5W 1R5, Canada
Telephone: 416 235 5705
Fax: 416 235 5951

Hold culture collection of bacteria for medical microbiology research and teaching. Offer identification service.

Centro di Studio dei Microorganismi

Pzale Cascine 27, Firenze, I-50144, Italy

Hold culture collection of algae, bacteria, fungi and yeasts including special collections of phototrophic microorganisms and cyanobacteria for general, agricultural, industrial and forest microbiology research and teaching. Offer identification services.

Centro di Virologia

Circonvallazione Gianicolense 85, Roma, Italy

Hold culture collection of tissue cultures for medical microbiology research.

CERS

Instituto Agronomio de Campinas, Av Barao de Itapura 148, Campinas Sao Paulo, CEP 13 100, Brazil

Hold culture collection of bacteria for agricultural microbiology research. Offer identification service.

Chiba University

1-8-1 Inohana, Chiba-shi, Chiba-ken 280, Japan

Hold culture collection as part of Japanese Federation of Culture Collections containing bacterial and fungal cultures for medical microbiological research.

Collection de Champignons et d'Actinomycetes Pathogenes

Institut Pasteur, 25 rue du Docteur Roux, Paris Cedex 15, F-75724, France

Hold culture collection of bacteria for general, medical and veterinary microbiology research and act as a patent depository. Offer identification services.

Collection Laboratorium voor Microbiologie

Fakulteit Wetenschappen, Rijksuniversiteit, Ledeganckstraat 35, Gent, B-9000, Belgium
Telephone: 91 22 78 21

Hold culture collection of bacteria containing a special collection of aerobic heterotrophs for general, agricultural, microbiology and biotechnology research, especially on plant pathogens.

Collection Nationale de Cultures de Microorganisms

Institut Pasteur, 25 rue du Docteur Roux, Paris Cedex 15, F-75724, France

Hold culture collection of bacteria for general, medical and veterinary microbiology research and act as a patent depository. Offer identification services.

Collezione Instituto di Microbiologia

Spedali Civili, P le Spedali Civili 1, Brescia, I-25100, Italy

Hold culture collection of bacteria, animal viruses and tissue cultures including a special collection of *Pseudomonas* and *Staphylococcus* for general and medical microbiology research. Offer identification services.

Connaught Laboratories Inc

Rte 611, PO Box 187, Stillwater, PA-18370, USA
Telephone: 717 839 7187/800 822 2463

Produce human vaccines and diagnostics distributed through E R Squibb & Sons Inc, US, as well as monkey blood, tissues and cell lines available from Connaught Laboratories Ltd, Canada.

Connecticut Valley Biological Supply Co Inc

82 Valley Rd, Southampton, MA-01073, USA
Telephone: 413 527 4030/800 628 7748

Supply cultures of microorganisms including bacteria, yeasts, fungi and both marine and fresh water algae.

CSIRO Division of Plant Industry

PO Box 1600, Canberra, ACT 2601, Australia
Telephone: 62 46 4911
Fax: 62 47 3785

Hold culture collection of bacteria including a special *Rhizobium* collection for agricultural microbiology research.

CSIRO Marine Laboratories

Castray Esplanade, Hobart, Tasmania 7001, Australia
Telephone: 02 20 6222
Fax: 02 24 0530

Hold culture collection containing algae mainly for marine micro-algae research of mariculture.

Culture Collection of Algae and Protozoa (CCAP)

Institute of Freshwater Ecology, The Windermere Laboratory, Far Sawrey, Ambleside, Cumbria, LA22 0LP, UK
Telephone: 5394 42468
Fax: 5395 46914

Supply cultures of freshwater and terrestrial microalgae and cyanobacteria, free-living protozoa and parasitic protozoa not pathogenic to man or domestic animals etc, which can be maintained *in vitro*. An international depository authority under article 7 of the Budapest Treaty. Also provide confidential safe depository for cultures of commercial value and carry out contract research, including screening and consultancy.

Culture Collection of Algae and Protozoa

Dunstaffnage Marine Laboratory, PO Box 3, Oben, Argyll, PA34 4AD, UK
Telephone: 631 622 44
Fax: 631 655 18

Hold culture collection of marine algae, protozoa and *Rotifera* for general microbiology and cryobiology research. The marine section of the CCAP, offering equivalent identification, depository and other services and facilities as provided by the Institute of Freshwater Ecology for freshwater algae, etc.

Culture Collection University of Goteborg

Department of Clinical Bacteriology, Guldhesgatan 10, Goteborg, S-413 46, Sweden
Telephone: 31 60 46 25
Fax: 31 82 54 84

Hold culture collection of bacteria and yeasts including special collections of *Movaxella, Campylobacter, Haemophilus* and non-fermenters for general and medical microbiology research and teaching. Offer identification service.

Department of Plant Defence Culture Collection

Cattedra di Micologia, P le M Kolbc 4, Udine, Italy

Hold culture collection of bacteria for general microbiology research with particular reference to actinomycetes and related organisms. Offer identification services.

Diamond V Mills Inc

436 G Ave Nw, PO Box 74570, Cedar Rapids, IA-52407, USA
Telephone: 319 366 0745/800 373 7234
Fax: 319 366 6333

Leading suppliers of yeast cultures including probiotics with digestive stimulatory properties to improve feed conversion and animal performance.

DSM - Deutsche Sammlung von Mikroorganismen un Zellkultlureen GmbH

Mascheroder Weg 1b, Braunschweig, D-3300, Germany
Telephone: 531 6187-0
Fax: 531 6187-18

The national service collection of bacteria of most genera, yeasts, fungi, plasmids, phages, plant viruses, plant cell cultures, human and animal cell lines and hybridomas relevant to applied microbiology, biotechnology, education and those of general interest. Cultures are supplied for a fee within and outside Germany and used in research on bacterial systematics. Information is published on the collection. Offer scientific services as follows: deposit of biological material including safe deposit and deposit of strains related to patent procedures under the Budapest Treaty; identification of microorganisms; consultation service on isolation, application, and preservation of microorganisms; long-term maintenance of biological material; patent deposit; on-line databank; individual training facilities.

Electro-Nucleonics Inc

7101 Riverwood Drive, Colombia, MD-21046, USA
Telephone: 301 381 4800/800 346 4364

Supply virus cultures.

European Collection of Animal Cell Cultures

PHLS Centre for Applied Microbiology, Porton Down, Salisbury, Wiltshire, SP4 0JG, UK
Telephone: 980 610391
Fax: 980 611315

Offer the following services to users: a patent depository; a safe deposit back-up storage facility; a collection of readily available cell lines including some plant cell lines; storage for individual collections of specialized cell types; a facility for the distribution and cataloguing of cells; an information service; preservation, characterization and quality control service; a research programme to improve procedures of cell culture and storage. Recognized under the Budapest Treaty as a culture depository for patent procedure.

Eurozyme

75 quai d'Orsay, Paris, F-75007, France
Telephone: 1 45 55 44 30

Produce and market lactic acid fermenting organisms.

Fermentation Research Institute

1-3 Higashi-1-chome, Yatabe-machi Tsukuba-gun, Ibaraki-ken 305, Japan
Telephone: 73 0022

Carrying out research into aspects of fermentation and bioreactor design. Act as a patent depository for fungi, bacteria, yeasts and actinomycetes except those which may be regarded as dangerous or those which need the physical containment of levels of P2, P3 or P4 required for experimentation.

Forintek Canada Corp

2665 East Mall, Vancouver, British Columbia, V6T 1W5, Canada
Telephone: 604 224 3221
Fax: 604 222 5690

Non-profit corporation carrying out research and development studies related to Canada's wood products industries. Interests include control of fungal degradation of wood and the maintenance of a collection of wood-destroying fungi. Developing a biological control approach to sap stain and decay of lumber.

Franklin Products International BV

Ramgatseweg 46, Postbus 136, AC Raamsdonkerveer, NL-4940, The Netherlands
Telephone: 1621 15351
Fax: 1621 16890

Produce live yeast cultures and lactic acid bacteria for use in the feed industry.

Fungal Collection, Belgium

Mycotheque, Lab de Mycologie Systematique et Applique, c/o Universite Catholique de Louvain, 3 Place Croix Sud, B-1348, Belgium
Telephone: 1 41 81 81 37 42

Offers identification, conservation, screening and isolation service.

Gentest Corporation

6 Henshaw Street, Wobnurn, MA-01801, USA
Telephone: 617 935 5115
Fax: 617 938 8644

Carrying out research on effects of mutagens on specific cell lines and developing improved methods for the detection of mutagen effects of industrial chemicals. Marketing human cell lines engineered to express human cytochrome P450 enzymes as well as cytochrome P450 enzyme preparations.

Ghent University

Dept of Microbiology & Microbial Genetics, K L Ledeganckstraat 35, Gent, B-9000, Belgium
Telephone: 91 22 78 21

Supply bacterial cultures.

Guinness (Dublin) Culture Collection

A Guinness Son & Co (Dublin) Ltd, St James Gate, Dublin 8, Ireland

Hold culture collection of bacteria and yeasts for industrial microbiology research.

Human Genetic Mutant Cell Repository

Coriell Institute for Medical Research, Copewood & Davis Sts, Camden, NJ-08103, USA
Telephone: 609 966 7377

Distribute cell cultures from individuals with genetic abnormalities. Sponsored by the National Institutes of Health, US.

IAL - Colecao de Culturas Adolfo Lutz

Instituto Adolfo Lutz, Av Dr Arnaldo 355 CP 01246, Sao Paulo SP, Brazil

Hold culture collection containing bacteria and other cultures including fungi and yeasts (*Leveduras*) for general, industrial, medical and veterinary microbiology research and teaching. Offer identification services.

IBSBF

Instituto Biologico, Rodovia Heitor Penteado, km 4, CP 70 Campinas 13 100 Sao Paolo, Brazil

Hold culture collection of bacteria for agricultural microbiology research. Offer identification service.

In Vitro International Inc

611 P Hammonds Ferry Rd, Linthicum, MD-21090, USA
Telephone: 301 789 3636

A type culture depository. Joint venture between Intra Gene International and University Micro Reference Laboratories, US. Accept only hosts containing plasmids that can be worked on in a P1 or P2 facility.

Industrial Microbiology Dublin

University College, Stillorgan Road, Dublin 4, Ireland

Hold culture collection of bacteria, fungi and yeasts for general microbiology research and teaching.

INRA - Beaucouze

Rue Georges Morel, Beaucouze, F-49070, France
Telephone: 41 73 51 61
Fax: 41 73 51 01

French Collection of Phytopathogenic Bacteria. Collection currently consists of 2,000 strains belonging to the genera *Agrobacterium, Bacillus, Clavibacter, Curtobacterium, Erwinia, Pseudomonas, Rhodococcus, Xanthomonas* and *Xylophilus*, isolated from samples of plants grown in France or in the French Overseas Territories or supplied from collections in other countries, representing taxonomic type strains, saprophytic strains and rhizobacteria. Supplied free to INRA and collaborative laboratories or sold to private companies and government agencies on request.

Institute for Medical Research

Human Genetic Mutant Cell Repository, Copewood & Davis Streets, Camden, NJ-08103, USA
Telephone: 609 966 7377

Distribute cell cultures frrom individuals with genetic abnormalities. Sponsored by the National Institutes of Health, US.

Institute for Microbial Resources

Kudankita 1-9-5-1209, Chiyoda-ku, Tokyo 102, Japan

Hold culture collection of bacteria, yeasts and fungi including *Aspergillus, Monascus, Mucor, Rhizopus* and *Penicillium*, as well as special collections of hydrocarbon utilizers, radio-resistants and brewing yeasts for general, agricultural and industrial microbiology research and teaching.

Institute of Physical and Chemical Research

Hirosawa, Wako-shi, Saitama 351-01, Japan

Hold the Japan Collection of Microorganisms including bacteria, fungi and yeasts for general, agricultural and industrial microbiology research.

International Mycological Institute - IMI

Ferry Lane, Kew Richmond, Surrey, TW9 3AF, UK
Telephone: 81 940 4086
Fax: 81 332 1171

Formerly Commonwealth Mycological Institute formerly CAB International Mycological Institute (CMI). Hold a collection of over 14,700 strains of fungi including species of industrial importance. Incorporate the UK National Collection of Fungus Cultures supported by 30 national governments. Recognized as an international depository authority under article 7 of the Budapest Treaty. The industrial laboratory is NAMAS accredited. Provides consultancy and contract research and development services in biodeterioration and industrial mycology, produces 8 serial publications and books and provides identifications and sells fungal cultures. Training can be provided on any area of work done by the IMI and individual training programmes can be arranged through the training officer.

ITAL Food Technology Institute

Banco de Fermentos Lacticos, Avenue Brasil 2880, Campinas, Sao Paulo, Brazil

Hold culture collection containing bacteria and fungi for industrial microbiology research. Offer identification services.

John Innes Institute Plant Virus Collection

Colney Lane, Norwich, Norfolk, NR4 7UH, UK
Telephone: 603 52571

Hold culture collection of plant viruses for research and teaching.

Kaken Pharmaceutical Co Ltd

6-42 Jujodai 1-chome, Kita-ku, Tokyo, Japan

Hold culture collection of actinomycetes for research in general, agricultural and industrial microbiology.

Laboratoires Unicet

92 rue Baudin, Levallois-Perret, F-92307, France
Telephone: 1 47 39 94 80

Produce monoclonal antibodies including gamma-interferon monoclonal antibody, clones of human lymphocytes and cell cultures.

Laboratory of Microbiology and Pathology

Department of Health, PO Box 495, Brisbane, Queensland 4000, Australia

Hold culture collection of bacteria including special collection of *Leptospira* for medical microbiology research. Offer identification service.

Laboratory Sales (UK) Ltd

Whitworth Rd, Rochdale, Lancashire, OL12 0JL, UK
Telephone: 706 356 444

Supply disposable bacteriological loops containing viable microorganisms in various combination in sets of 25 loops.

Lacto Labo

Zone d'Activite de Buxieres, BP 10, Dange Saint Romain, F-86220, France
Telephone: 49 86 40 82
Fax: 49 86 34 49

Carry out lactic acid fermentations and supply starter cultures for cheese making and yoghurt, surface flowering yeasts and moulds and ripeners for the meat industry, silage inoculants, bacteria for nutritional regulation and balance, as well as malo-lactic bacteria for oenology. Part of Rhone Poulenc, France.

Lausanne University - Microbial Collection Centre

Institute of Microbiology, 44 rue du Bugnon, Lausanne, CH-1011, Switzerland
Telephone: 21 22 33 91

Supply bacterial cultures.

Louvain Catholic University

Mycology Laboratory, Place croix du Sud 3, Louvain-la-Neuve, B-1348, Belgium
Telephone: 10 47 36 55

Offer services in identification, screening and isolation of fungi as well as patent and safe deposits and facilities for preservation, biodeterioration testing and contract work.

MAFF Agricultural Science Services

Plant Pathology Laboratory, Hatching Green, Harpenden, Herts, AL5 2BD, UK
Telephone: 58 27 5241

Hold national culture collection of plant pathogens. Carry out testing of new agrochemicals and is concerned with development of methods for detection of contamination of agricultural produce and safety aspects of genetic engineering in agriculture.

Materials Research Laboratories Collection of Fungi

Commonwealth of Australia Dept of Defence, PO Box 50, Ascot Vale, Victoria 3032, Australia

Hold culture collection of fungi including a special collection of curvularia for biodeterioration research.

Medical Mycology, Belgium

Institut d'Hygiene et d'Epidemiologie, Section de Mycologie, 14 rue Juliette Wytsman, Bruxelles, B-1050, Belgium
Telephone: 2 642 56 30

Covers both non-pathogenic and pathogenic species, offering an identification service.

Microbiologie des Sols INRA

17 Rue Sully BV 1540, Dijon Cedex, F-21034, France

Hold culture collection of bacteria for agricultural microbiology research.

Museum of Natural History

Cryptogram Laboratory, 12 rue Buffon, Paris, F-75005, France
Telephone: 1 33 13 521

Supply fungal cultures.

National Collection of Food Bacteria

AFRC Institute of Food Research,
Reading Laboratory, Early Gate,
Whiteknights Road, Reading,
RG6 2EF, UK
Telephone: 734 357000

Holds a colelction of over 3000 bacterial cultures with special emphasis on food, dairy, beverage and related products. A catalogue of the holdings is available on request.

The National Collection of Industrial and Marine Bacteria Ltd

23 St Machar Drive, Aberdeen, UK
Telephone: 224 273 332
Fax: 224 487 658

Collects and conserves cultures of bacteria of industrial interest which are made available to laboratories worldwide, holding over 5000 strains, including genetically modified organisms and strains manufactured using recombinant DNA technology, as well as plasmids. Recognized as an international depository authority under the Budapest Treaty for patent strains. Offer services in biocide testing, bacterial identification, etc.

National Collection of Pathogenic Fungi

PHLS Mycological Reference Laboratory, 61 Colindale Avenue,
London, NW9 5HT, UK
Telephone: 81 200 4400
Fax: 81 205 6528

Provides identification and safe deposits for fungi and yeasts pathogenic to man and other animals.

National Collection of Plant Pathogenic Bacteria

MAFF Central science Laboratory,
Hatching Green, Harpenden, Herts,
AL5 2BD, UK
Telephone: 582 715 241
Fax: 582 762 178

Maintains over 3500 strains of plant pathogenic bacteria and offers an identification service for a wide range of bacteria.

National Collection of Type Cultures

Central Public Health Laboratory,
61 Colindale Avenue, London,
NW9 5HT, UK
Telephone: 81 200 4400
Fax: 81 200 7874

Maintain reference culture collections of pathogenic bacteria as well as mycoplasmas and plasmids. Also an FAO/WHO Collaborating Centre for Mycoplasmas. Recognized as an international depository authority for patent strains under the terms of the Budapest Treaty.

National Collection of Wood Rotting Fungi

Building Research Establishment,
Bucknalls Lane, Garston, Watford,
Herts, WD2 7JR, UK
Telephone: 923 894 040
Fax: 923 664 010

Maintains cultures of wood rotting basidiomycete fungi. Carries out research and development on decay resistance of wood and wood-based materials as well as the effectiveness of wood-preserving chemicals. Provides an identification service for wood-rotting fungi, as well as acting as a depository for reference material.

National Collection of Yeast Cultures

AFRC Institute of Food Research,
Colney Lane, Norwich, NR4 7UA, UK
Telephone: 603 56122
Fax: 603 507723

Maintains yeast cultures, other than known pathogens. Provides information on yeasts (computer searches and printouts of strain data), identification and preservation services. Also offers a confidential safe deposit service and acts as an international depository under the terms of the Budapest Treaty. Holds over 2000 cultures, with a special emphasis on brewing.

National Culture Collection of Microorganisms - CNCM

Institut Pasteur, 28 rue du Dr Roux,
Paris Cedex 15, F-75724, France
Telephone: 306 19 19

Culture collection contains bacteria (including actinomycetes), bacteria containing plasmids, filamentous fungi and yeasts, and viruses. Supply bacterial cultures.

National Health Institute

Kenepuru Drive, PO Box 50-348
Porirua, Wellington, New Zealand
Telephone: 4 370 149
Fax: 4 378 983

Hold culture collection containing bacteria, fungi, bacterial viruses and yeasts including a special collection of propagating strains for phage typing for medical microbiology research and teaching. Offer identification service.

National Institute of Agrobiological Resources

2-1-1 Kannondai Yatabe-machi,
Tsukuba-gun, Ibaraki-ken 305, Japan
Telephone: 29 756 8155

Hold culture collection containing bacteria and fungi for agricultural microbiology research.

National Institute of Allergy and Infectious Diseases - NIAID

Room 7 A-03 Wetwood Bldg,
Bethesda, MD-20205, USA
Telephone: 301 496 7985

Supply source material for culture of virus and mycoplasma as well as allergenic extracts and antisera against human immunoglobulins, viruses and mycoplasma.

National University of Athens

Department of Microbiology,
109 Leoforos Kifissias, Athens,
GR-11524, Greece

Hold culture collection of bacteria for medical microbiology research and teaching. Offer identification service.

Nationales Referenzzentrum fur Adenoviren

Postfach 610180, Hannover 61,
D-3000, Germany
Telephone: 511 532 4311
Fax: 511 532 5732

Institut fuer Virologie und Seuchenhygiene Medizinische Hochschule Hannover. Hold culture collection containing animal viruses for medical microbiology research. Offer identification service.

NCS Diagnostics Inc

130 Matheson Blvd East, Suite 2
Mississauga, Ontario, L4Z 1Y6,
Canada
Telephone: 416 890 0177/800 268 2857

Agent for American Type Culture Collection.

New Zealand Reference Culture Collection (Plant & Soil)

DSIR Plant Diseases Division,
120 Mt Albert Road (Private Bag),
Auckland, New Zealand
Telephone: 893 660
Fax: 863 330

Hold culture collection containing bacteria for agricultural microbiology research especially in relation to plant disease.

North East Pacific Culture Collection

University of British Columbia,
6270 University Boulevard, Vancouver,
BC V6T 1W5, Canada
Telephone: 604 822 4378
Fax: 604 822 6091

Hold culture collection of marine phytoplankton representing all major groups including a special collection of Dinophyceae. Species list and ordering information available on request.

Norwegian Institute for Water Research - NIVE

PO Box 333, Brekkaveien 19 Blindern,
Oslo 3, N-0314, Norway
Telephone: 2 23 52 80

Hold culture collection of algae including special collections of cyanobacteria, Oscillatoriea and Selenastrum for general, industrial and phycological research as well as aquaculture and bioassays. Offer identification services.

NRRL/ARS Culture Collection

1815 North University Strreet, Peoria,
IL-61604, USA
Telephone: 309 685 4011
Fax: 309 671 7814

United States Department of Agriculture National Center for Agricultural Utilization Research. Culture collection and patent depository for strains of agriculturally and industrially important bacteria, yeasts, moulds and actinomycetes (except for certain pathogens and moulds which cannot be preserved by lyophilization).

Osaka Institute for Fermentation

17-85 Juso-Honmachi 2-chome,
Yodogawa-ku, Osaka, Japan

Hold culture collection containing bacteria, fungi, bacterial viruses, yeasts and cell lines for research and commercial use in general, agriculture and industrial microbiology.

Otto Madsen Dairy Research Laboratory

19 Hercules Street, Hamilton,
Queensland 4007, Australia

Dairy Starter Collection. Hold culture collection containing bacteria, fungi, bacterial viruses and yeasts with special collections of *Streptococcus cremoris* and *S lactis* strains used for cheese manufacture and their bacteriophages for dairy microbiology research.

Pasteur Culture Collection of Cyanobacterial Strains

Institut Pasteur, 24 rue du Docteur
Roux, Paris Cedex 15, F-75724,
France

Hold culture collection of Cyanobacteria for general microbiology research.

Plant Science Ltd

University of Sheffield, Firth Court,
Western Bank, Sheffield,
S10 2TN, UK
Telephone: 742 722922
Fax: 742 727949

Maintain a large Plant Cell Culture Bank and offers this for screening by the pharmaceutical and agrochemical industry (for lead chemicals, biotransformations). Conduct enzyme screens and clones plant genes into bacterial and yeast hosts.

Presque Isle Cultures

PO Box 8191, Presque Isle,
PA-16505, USA
Telephone: 814 833 6262

Supply bacterial and fungal cultures, as well as media for their growth.

Queensland University

Department of Microbiology, St Lucia, Brisbane, Queensland 4067, Australia
Telephone: 7 377 3639
Fax: 7 371 5896

Maintain culture collection of nitrogen-fixing bacterium *Rhizobium*. Carrying out research on fermentation, genetic engineering and microbial diagnostics.

R & A Plant/Soil Inc

24 Pasco/Kahlotus Rd, Pasco,
WA-99301, USA
Telephone: 509 545 6867

Supply cultures of bacteria, algae and cyanobacteria to farmers for use as soil conditioners and fertilizers.

Research Centre Agriculture Canada

1400 Western Rd, London, Ontario,
N6G 2V4, Canada

Hold culture collection of bacteria for agricultural microbiology research.

Research Institute for Plant Protection

Binnenhaven 12, Wageningen,
The Netherlands

Hold culture collection of plant viruses for research in breeding crops for resistance.

Research Station Agriculture Canada

8801 East Saanich Road, Sidney,
British Columbia, V8L 1H3, Canada

Hold culture collection of bacteria and fungi for agricultural microbiology.

Rosell Institute Inc

8480 boulevarde St-Laurent, Montreal,
Quebec, H2P 2M6, Canada
Telephone: 514 381 5631
Fax: 514 383 4493

Produce bacteria, yeast and fungi for the milk and food industries, agriculture, human and animal nutrition, prophylaxis and therapy. Products include lactic ferments, food supplements, probiotics for human and veterinary products, and legume inoculants.

Rothamsted *Rhizobium* Collection

Rothamsted Experimental Station,
Harpenden, Hertfordshire,
AL5 2JQ, UK
Telephone: 582 763133

Hold culture collection of bacteria for agricultural microbiology research.

Sammlung von Cionjugaten-Kulturen

Institut fur Allgemeine Botanik,
Ohnhorststrasse 18, Hamburg 52,
D-2000, Germany

Hold culture collection containing algae for phycology research and teaching.

Servico de Microbiologia e Immunologia

Faculdade de Ciencias Medicas - UERJ, Rua Prof Manuel de Abreu 48, Rio de Janiero, 20550 RJ, Brazil

Hold culture collection containing bacteria including a special collection of Enterobacteriaceae for medical microbiology research and teaching.

Sheffield University

Dept of Molecular Biology and Biotechnology, Western Bank, Sheffield, S10 2TN, UK
Telephone: 742 768555
Fax: 742 728697

Hold culture collection of bacteria for general microbiology research and teaching.

Spanish Culture Collection

Dept de Microbiologia Facultad de Ciencias, Biologicas, Universidad de Valencia, Burjasot, Valenci, Spain
Telephone: 96 3630011

Supply bacterial, fungal and yeast cultures.

Statens Seruminstitut

Amager Boulevard 80, Copenhagen S, DK-2300, Denmark
Telephone: 31 95 28 17
Fax: 31 95 58 22

Supply antibodies, antisera and biological standards. Recognized by the World Health Organization (WHO). Hold culture collection containing bacteria for general, medical and veterinary microbiology research. Offer identification service. Collaborating Centre for Reference and Research on *Escherichia* and *Klebsiella*.

Swedish University of Agricultural Sciences

Department of Microbiology, Uppsala, S-750 07, Sweden

Hold culture collection of bacteria, fungi and yeasts including special collections of *Rhizobium* and other soil organisms for general and agricultural microbiology research and teaching.

Sybron Chemicals Inc - Biochemical Division

Birmingham Rd, PO Box 66, Birmingham, NJ-08011, USA
Telephone: 609 893 1100
Fax: 609 894 8641

Supply selectively-adapted bacterial cultures for treatment of industrial and municipal waste water and contaminated groundwater. Subsidiary of Sybron Corp.

Sydney University - Dept of Microbiology

65 Market Street, Sydney, NSW 2006, Australia
Telephone: 02 692 2536
Fax: 02 692 4202

Hold culture collection containing bacteria, tissue cultures, animal viruses, bacterial viruses and yeasts including a special collection of *Corynebacterium pseudotuberculosis*, veterinary non-sporing anaerobes and *Pasteurella multocida*.

Sylvan Spawn Inc

West Hills Industrial park, Kittanning, PA-16201, USA
Telephone: 412 352 1521
Fax: 412 297 3079

Supply mushroom spawn and mycorrhizal inoculants.

UFPEDA - UFPE Department of Antibiotics

Cidade Universitaria, Recife, Pernambuco, Brazil

Hold culture collection of bacteria, fungi, tissue cultures, yeasts and others including actinomycetes for general, agricultural, industrial, and medical microbiology and antibiotic research and teaching.

University of Alberta - Microfungus Collection and Herbarium

Devonian Botanic Garden, Edmonton, Alberta, T6G 2E1, Canada
Telephone: 403 987 3054
Fax: 403 987 4141

Collect and preserve fungi.

University Micro Reference Laboratory Inc

611 P Hammonds Ferry Rd, Linthicum, MD-21090, USA
Telephone: 301 789 3636/800 468 4876
Fax: 301 789 3638

Hold microbial culture collections including bacterial reference strains. Carry out batch processing of microbial strains.

Valio Research and Development Centre

Kalevankatu 56, Helsinki, Finland
Fax: 0 694 8225

Starter Collection of the Valio Research and Development Centre.
Hold culture collection of bacteria for research into dairy starter cultures.

Volcani Centre *Rhizobium* Collection

Agriculture Research Organization, PO Box 6, Bet Dagan 50250, Israel

Hold culture collection of bacteria for agricultural microbiology research.

VTT Collection of Industrial Microorganisms

Biotech Lab Tech Research Centre of Finland, Tietotie 2, Espoo, SF-02150, Finland
Telephone: 0 4561

Hold culture collection of bacteria, fungi and yeasts for industrial microbiology and biotechnology research. Offer limited services in identification of beer-spoiling bacteria and filamentous fungi of grain.

Wageningen Agricultural University

Laboratory of Microbiology, Hesselink van Suchtelenweg 4, CT Wageningen, NL-6703, The Netherlands
Telephone: 8370 83110
Fax: 8370 83829

Hold culture collection of bacteria, fungi, bacterial viruses and yeasts as well as special collections of *Rhizobium*, *Frankia* and coryneform bacteria for general and agricultural microbiology research and teaching.

Waite Agricultural Research Institute

Private Bag, Glen Osmonde, Adelaide, South Australia, 5064, Australia
Telephone: 8 372 2444

Insect Pathology Pathogen Collection.
Hold culture collection containing bacteria, fungi, protozoa and insect viruses for general, agricultural and insect microbiology research and teaching. Offer identification service.

Western Australian Culture Collection

State Health Laboratory Service, PO Box F312, Perth, Western Australia, Australia

Hold culture collection containing bacteria, fungi and yeasts for general, medical and veterinary microbiology research and teaching. Offer identification service.

Wine Yeast Collection

Dipt di Biologia Vegetale, University of Perugia, Via 20 Giugno 74, Perugia, Italy

Hold culture collection of bacteria, fungi and yeasts for industrial microbiology research and teaching. Offer identification services.

Wood Technology and Forest Research Division

Forestry Commission of New South Wales, PO Box 100, Beecroft, NSW 2119, Australia

Hold culture collection of fungi for forest microbiology and timber mycology research.

HOSTS AND CONTRACT ADDRESSES

BioCommerce Data Ltd

95 High Street, Slough, SL1 1DH, UK
Telephone: 753 51177
Fax: 753 512239

BIOSIS (UK)

Garforth House, 54 Micklegate, York, North Yorkshire, Y01 1LF, UK
Telephone: 904 642816

Telecom Gold: 81:B100002, Mercury Link: 19016585.

BIOSIS (USA)

2100 Arch Street, Philadelphia, PA-19103, USA
Telephone: 215 587 4800

BLAISE-LINE

British Library Bibliographic Service, Division, Marketing and Support Group, 2 Sheraton Street, London, W1V 4BH, UK
Telephone: 71 636 1544

BRS - Bibliographic Retrieval Services Inc

Bibliographic Retrieval Services Inc, 1200 Route 7, Latham, New York, NY-12100, USA
Telephone: 518 783 1161

Host organization for access to various databases including BIOSIS, CA Search and CABI abstracts.

BRS Information Technologies (Europe)

Achilles House, Western Avenue, London, W3 0UA, UK
Telephone: 81 992 3456
Fax: 81 993 7335

BRS Information Technologies (North America)

8000 Westpark Drive, McLean, VA-22102, USA
Telephone: 703 442 0900
Fax: 703 893 4632

CABI

CAB International, Wallingford, Oxon, OX10 8DE, UK
Telephone: 491 32111
Fax: 491 33508

Collects, collates and disseminates information through publications and operation of databases available through various hosts including CAN/OLE, Dialog, DIMDI and ESA-IRS. Activities of most direct relevance to biosciences are centred on the Commonwealth Mycological Institute. Telecom Gold: 84:CAU001

CAN/OLE

Canadian Online Enquiry, National Research Council of Canada, Montreal Road, Ottawa, Ontario, K1A 0S2, Canada
Telephone: 613 993 3791

Host organization for access top various databases including BIOSIS, CA Search and CABI database on-line.

CAN/SND

Canadian Institute for Scientific and Technical Information, NRC, Montreal Road, Ottawa, K1A OS2, Canada
Telephone: 613 993 1210

Central Statistical Office

Databank Manager, Great George Street, London, SW1P 3AQ, UK
Telephone: 71 270 6387

Chemical Databank Service - CDS

Science and Engineering Research Council, Polaris House, North Star Avenue, Swindon, SN2 1ET, UK
Telephone: 793 26222
Fax: 793 511181

The CDS is a central, networked computer system providing interactive access to about 11 chemical databanks. These include: Crystal Data Identification File; The Cambridge Structural Databank; Electrolyte Solutions database; Fine Chemicals Directory; Inorganic Crystal Structure Data file; Expert System for organic synthesis; Index of 20,000 organic reaction transformations.

CIS

Chemical Information System Ltd, London House, London Road South, South Poyton, Cheshire, SK12 1YP, UK
Telephone: 749 86358

CIS Inc

7215 York Road, Baltimore, MD-21212, USA
Telephone: 301 321 8440

Commission of the European Communities

Batimont Jean Monnet, Plateau du Kirchberg, BP 1907, Luxembourg
Telephone: 352 243011

Current Awareness in Biological Sciences

132 New Walk, Leicester, UK
Telephone: 533 548707

DATA-STAR (Switzerland)

Radio-Schweiz AG, Data Star, Laupenstrasse 18a, Berne, CH-3000, Switzerland
Telephone: 31 509 630

DATA-STAR is a Swiss operation, offering access to numerous business and biomedical databases. This host may be the cost effective choice for European users.

DATA-STAR (UK)

Data-Star Marketing Ltd, Plaza Suite, 114 Jermyn Street, London, SW1Y 6HJ, UK
Telephone: 71 930 5503
Fax: 71 930 2581

DATA-STAR (USA)

DATA-STAR Marketing Inc, Suite 110, 485 Devon Park Drive, Wayne, PA-19087, USA
Telephone: 215 687 6777/800 221 7754
Fax: 215 687 0984

DC-Datacentralen

Retoretvej 6-8, Valby, DK-2500, Denmark
Telephone: 451 468122

DIALOG (UK)

Dialog Information Services, Woodside, Badger Lane, Hinksey Hill, Oxford, OX1 5AU, UK
Telephone: 865 730275
Fax: 865 736354

Dialmail: mailbox 9124

DIALOG (USA)

Dialog Information Retrieval Service, 3460 Hillview Avenue, Palo Alto, CA-94304, USA
Telephone: 415 858 3810/800 334 2564
Fax: 415 858 7069

Host organization for access to various databases including BIOSIS, CA Search, CABI and Telegenline. A major US information provider, offering access to hundreds of databases (including BIOSIS, CA Search, CABI and Telegenline), 24 hours a day, from all over the world.

DIMDI - Deutsches Institut fur Medizinisches Dokumentation und Information

Dokumentation und Information, Weisshausstrasse 27, Postfach 420560, Cologne 41, D-5000, Germany
Telephone: 221 47241

Host organization for access to various databases including BIOSIS, CA Search and Telegenline.

DRI (USA)

Data Resources Inc, 24 Hartwell Avenue, Lexington, MA-02173, USA
Telephone: 617 863 5100

ECHO-HOST

Commission of the European Communities, ECHO Customer Service, 177 Route d'Esch, L-1023, Luxembourg
Telephone: 352 48 80 41
Fax: 352 48 80 40

Gives access to databases provided by the Commission of the European Communities, many of which are free of charge.

ESA/IRS - European Space Agency Information

Retrieval Service, Via Galileo Galilei, Fracati, I-00044, Italy

Host organization for access to various databases including those of the Royal Chemical Society (UK), CABS Database Online and Pascal Biotechnologies.

Euronet-DIANE

Commission of the European Communities, DG XIII, BP 777, L-1471, Luxembourg
Telephone: 352 48 80 41

A telecommunications network giving access to various database hosts including ESA-IRS, Data-Star and DIMDI.

EUROSTAT

Statistical Office of the European Community, Batiment Jean Monet, BP 1907, L-2920, Luxembourg
Telephone: 43 01 32 20

Fachinformationszentrum Chemie GmbH

Postfach 12 60 50, Steinplatz 2, Berlin 12, D-1000, Germany
Telephone: 30 31 90 030 0

Serves Austria, Switzerland and the Federal Republic of Germany for CAS Online.

FAO - Food and Agricultural Organisation of the United Nations

Information, Data and Statistics Service, Via delle Terme di Caracalla, Rome, I-00100, Italy
Telephone: 6 57 97 64 14

GNS-Dialplus - Global Network Services

British Telecom, GNS Front Office, BT, Network House, Brindley Way, Hemel Hempsted, HP3 9RR, UK
Telephone: 800 282444

Provides users with simple access to remote computer applications and information databases at reasonable cost through most PCs, workstations or computer terminals using an appropriate communications software package and a modem operating at between 300 and 2,400 bits per second. Generally charged as a local call.

ICGEBnet

Padriciano 99, Trieste, I-34012, Italy
Telephone: 40 226556
Fax: 40 226556

The International Centre for Genetic Engineering and Biotechnology Computer Resource for Molecular Biology. Provides access to the Intelligence Suite, an integrated package of molecular biology programs, supplying all the major functions for sequence analysis including sequence entry and editing, DNA gel fragment assembly, sequencing project management, protein and nucleic acid sequence analyses, similarity searching, pattern recognition searching, restriction mapping and cloning simulation. All users have access to the FASTA program of William Pearson for searching the GenBank daily database updates.

IFIS (FRG)

International Food Information Service, Lyoner Strasse 44-88, Arabella Center, Frankfurt/Main 71, D-6000, Germany
Telephone: 69 6687 338

IFIS (UK)

International Food Information Service, Lane End House, Shinfield, Reading, Berks, RG2 9BB, UK
Telephone: 734 883895

International Data Media (UK) Ltd

Plantation House, 4th Floor, Mincing Lane, London, EC3M 3DX, UK
Telephone: 71 283 4445
Fax: 81 626 0170

ISPRA

Commission of the European Communities, ECDIN, Joint Research Centre, Ispra, I-21020 (Varese), Italy
Telephone: 332 78 97 20

JANET

Joint Academic Network, University of London Computer Centre, 20 Guildford Street, London, WC1N 1DT, UK
Telephone: 71 405 8400

JICST - Japan Information Center of Science and Technology

5-2 Nagatacho 2 Chome, Chiyoda-ku, Tokyo, Japan
Telephone: 3 581 6411

Laboratory of the Government Chemist

Biotechnology Group, Queens Road, Teddington, Middlesex, TW11 OLY, UK
Telephone: 81 943 7612

Leatherhead Food RA

Randells Road, Leatherhead, Surrey, KT22 7RY, UK
Telephone: 372 376761

Act as host for their database FROSTI (Food RA Online Scientific and Technical Information Database).

MEAD Data Central (USA)

Mead Data Central, PO Box 933, Dayton, OH-45401, USA
Telephone: 513 865 6800

Host organization for access to various databases including BIOSIS Previews.

MEAD Data Centre (UK)

International House, 1 St Katherines Way, London, E1 9UN, UK
Telephone: 71 488 9187

Microbial Strain Data Network - MSDN

Cambridge University, Institute of Biology, 307 Huntingdon Road, Girton, Cambridge, CB3 0JX, UK
Telephone: 223 27622

Telephone help in the US is available from: Lyn Vass, ATCC
Telephone: 410 881 2600
Fax: 410 770 2587
An international network accessible through BT Gold and other gateways. Provides a variety of biotechnology-related databases and gives information on the location of microorganisms or cultured cell lines with specific properties. Sponsored by CODATA, WFCC and IUMS.

National Technical Information Service (NTIS)

Dept of Commerce, 14th between E and Constitution NW, Washington, DC-20230, USA

Consolidates the information services of US government departments providing a centre for the distribution of specialized, scientific and technical documentation and information to the public. Publish searches from their own database which is available through Dialog. Publications include full texts of research carried out under contract to government agencies.

NCYC-on-line Services

NCYC, Food Research Institute, Colney Lane, Norwich, Norfolk, NR4 7UA, UK
Telephone: 603 56 122

National Collection of Yeast Cultures. Offers a number of on-line computer services for registered users.

ORBIT Search Service

8000 Westpark Drive, McLean, VA-22102, USA
Telephone: 703 442 0900

RSC - The Royal Society of Chemistry

Thomas Graham House, Science Park, Milton Road, Cambridge, CB4 4WF, UK
Telephone: 602 507411

Serves UK, Denmark, Finland, Ireland, Netherlands, Norway and Sweden for all CAS services.

SCAR - Standing Committee of Agricultural Research

Commission of the EC, DG-VI, rue de la Loi 200, Brussels, B-1049, Belgium
Telephone: 2 235 1111

SDC - System Development Corporation

2500 Colorado Avenue, Santa Monica, CA-90406, USA

Host organization for access to various databases including Derwent publications on biotechnology and patents, BIOSIS Previews and CA Search.

SEQNET (Sequence Network)

EMBNET, Meyerhofstrasse 1, Heidelberg, D-6900, Germany

Contains data on biological macromolecules, providing an up-to-date nucleic acid and protein sequence and structure database service. Provides on-line access to major nucleic acid and protein databases. Receives daily sequence updates from the European Molecular Biology Laboratories (EMBL) for the EMBL database and the GenBank database. Some software is also available. Software includes packages for: data retrieval and analysis packages; sequence similarity searching; selecting conditions for the experimental isolation and purification of molecules; structure prediction; phylogenetic analysis. Fast query languages are also provided.

STN International - Germany

STN, PO Box 2465, Karlsruhe 1,
D-7500, Germany
Telephone: 7247 808 555
Fax: 7248 808 666

The Scientific and Technical Information Network.
European host for large number of databases including BIOSIS, DECHEMA and DEQUIP.

STN International - Japan

Gakkai Center Building, 2-4-16 Uaupo
Bunkyo-ku, Tokyo 113, Japan
Telephone: 3 816 3462

Provides access to a large number of databases including BIOSIS, DECHEMA and DEQUIP.

STN International - US

2540 Olentangy River road,
PO Box 02228, Columbus,
OH-43202, USA
Telephone: 614 421 3600
Fax: 7247 82 46 39

The Scientific and Technical Information Network. Provides access to a large number of databases including BIOSIS, DECHEMA and DEQUIP.

SUNIST

Parc d'Affairs Saint Hubert, 64 rue de Creuzat, BP 112, L'Isle d'Abeucedex,
F-3808, France
Telephone: 1 74 27 28 10
Fax: 1 74 27 03 33

French host of MINE (see database section)

Telesystems-Questel

83-85 boulevard Vincent, Paris,
F-75013, France
Telephone: 1 45 82 64 64

Host organization for access to various databases including Pascal Biotechnology and CA Search.

VCH Verlags GmbH

CA Vertrieb, Postfach 1260/1280,
Weinham, D-6940, Germany
Telephone: 6201 602 231

Serves Austria, Switzerland and Germany for CAS printed and microform services.

PART V

INDEXES

ALPHABETICAL INDEX OF ENTRIES

CO **Consultants**
CU **Culture Collections and Suppliers**
DA **Databases and Abstracts**
DI **Directories and Reference Books**
GO **Government Organizations**
HO **Hosts and Contact Addresses**
IO **International Organizations**
JU **Journals**
NL **Newsletters**
NO **National Organizations**
SO **Computer Software**

Agriculture, Ecosystems & Environment **JU**
Agrifutur Srl **CU**
Agrisearch Services Pty Ltd **CO**
Agroforestry Abstracts **DA**
Agroforestry Today **JU**
Agro-Industry Hi-Tech **JU**
Agronomy Journal **JU**
Agropur Cooperative Agro-Alimentaire **CU**
Agrow **NL**
Agrow OnScreen **DA**
AIDS Research and Human Retroviruses **JU**
AIDS Research Reviews **JU**
Alberta Environmental Centre (Canada) **GO**
Alberta Research Council (Canada) **GO**
Alcohol Update **NL**
Alcohol Week **NL**
Algensammling am Institut fur Botanik **CU**
Alko Ltd **CU**
Allied Colloids Manufacturing GmbH **CO**
Amdel **CO**
AMEC Design and Construction Ltd **CO**
American Association for the Advancement of Science (USA) **NO**
American Association for Clinical Chemistry (USA) **NO**
American Association of Immunologists (USA) **NO**
American Chemical Society (USA) **NO**
American Journal of Human Biology **JU**
American Journal of Veterinary Research **JU**
American Medical Association (USA) **NO**
American Pharmaceutical Association (USA) **NO**
American Phytopathological Society (USA) **NO**
American Society of Agricultural Engineers (USA) **NO**
American Society of Biological Chemistry (USA) **NO**
American Society for Biotechnology (USA) **NO**
American Society for Cell Biology (USA) **NO**
American Society of Clinical Pathologists (USA) **NO**
American Society for Horticultural Science (USA) **NO**
American Society for Horticultural Science Journal **JU**
American Society for Microbiology (USA) **NO**
American Society of Plant Physiologists (USA) **NO**
American Synthesis **SO**
American Type Culture Collection - ATCC **CU, DA**
Analytical Abstracts **DA**
Analytical Biochemistry **JU**
Analytical Imaging Concepts Inc **SO**
Analytical Technologies Inc **CO**
Angewandte Chemie **JU**
Animal Breeding Abstracts **DA**
Animal Genetics **JU**
Animal Pharm **NL**
Animal Reproduction Science **JU**
Annals of Applied Biology **JU**
Annals of Clinical Biochemistry **JU**
Annals of Human Biology **JU**
Annual Reviews **JU**
Annual Reviews in Biodiversity **JU**
Annual Reviews in Cell Biology **JU**
Annual Reviews in Entomology **JU**
Annual Reviews in Genetics **JU**
Annual Reviews in Immunology **JU**
Annual Reviews in Microbiology **JU**
Annual Reviews in Neuroscience **JU**
Annual Reviews in Plant Physiology and Plant Microbial Biology **JU**
ANRED (France) **GO**
Antibiotics and Chemotherapy **JU**
Antimicrobial Agents and Chemotherapy **JU**
Antiviral Research **JU**
ANVAR (France) **GO**
Anville Instruments **SO**
Assemble Permanante Chambres d'Agriculture - APCA (France) **NO**
Aplibio SA **CO**
Applied Agricultural Research **JU**
Applied Biochemistry and Biotechnology **JU**
Applied Biochemistry and Microbiology **JU**
Applied Bioscience International **CO**
Applied and Environmental Microbiology **JU**

Australian Government Analytical Laboratories (Australasia) **GO**
Australian Journal of Biological Sciences **JU**
Australian Journal of Biotechnology **JU**
Australian National Reference Laboratory in Medical Mycology **CU**
Australian and New Zealand Biotechnology Directory **DI**
Australian Pharmaceutical Manufacturers Association (Australasia) **NO**
Australian Scientific Industries Association (Australasia) **NO**
Australian Society for Microbiology (Australasia) **NO**
Australian Water and Wastewater Association (Australasia) **NO**
The Australian Wine Research Institute **CU**
Austrian Biochemical Society (Austria) **NO**
Austrian Biophysical Society (Austria) **NO**
Austrian Bioprocessing Association (Austria) **NO**
Austrian Biotechnology Association (Austria) **NO**
Austrian Brewing Society (Austria) **NO**
Austrian Chemical Association (Austria) **NO**
Austrian Pharmacists Association (Austria) **NO**
Austrian Scientific Research Fund (Austria) **GO**
Austrian Society for Chemical Industry (Austria) **NO**
Austrian Society for Immunology (Austria) **NO**
Autoimmunity **JU**
Autoscribe **SO**
Bachmann Favereau Glover & Associates **CO**
Bacillus Collection **CU**
Bacterial Culture Collection **CU**
Baird Europe BV **SO**
Balzers High Vacuum Ltd **SO**
Balzers Inc **SO**
BASE International Biotechnology Ltd **CO**
Battelle Institute Ltd **CO**
Bayerische Landesanstalt fur Weinbau und Gartenbau **CU**
BBN Software Products Corp **SO**
BCZ-Friesland BV **CU**
BDT/FTPT **DA**
Beak Consultants Ltd **CO**
Beckman Instruments **SO**
Becton Dickinson **SO**
Beilstein **DA**
Beilstein Handbook **DI**
BEL Industries **CU**
Belgian Society of Biochemistry (Belgium) **GO**
Belgium Bioindustries Association (Belgium) **NO**
Belgium Pharmaceutical Association (Belgium) **NO**
BellBiotech Consulting **CO**
Benezech-Simpson & Co **CO**
Bennett & Company (Europe) Ltd **CO**
Bercan Environmental Resources Inc **CU**
Bernard Wolnak & Associates **CO**
BEST - British Expertise in Science & Technology **DA**
Betagen Corp **SO**
Bethyl Laboratories Inc **CO**
BIA Bulletin **NL**
Bibliography of Agriculture **DA**
Bicarbolyte Corp **CU**
Bideco AG **CO**
BIKE - Biotechnology Information Knot for Europe GBF mbH **DA**
Bio Techno Systems Inc **CO**
Biobulletin **NL**
Biocatalysis **JU**
Biocatalysts Ltd **CO**
Biochemical and Biophysical Research Communications **JU**
The Biochemical Engineering Journal **JU**
Biochemical Genetics **JU**
Biochemical Journal **JU**
Biochemical Medicine **JU**
Biochemical Society (Belgium) **NO**
Biochemical Society (Italy) **NO**
The Biochemical Society (UK) **NO**
Biochemical Society, Irish Area Section (Ireland) **NO**
Biochemical Society Transactions **JU**
Biochemical Systematics and Ecology **JU**
Biochemische Synthesetechnik GmbH & Co **CO**
Biochemistry **JU**
Biochemistry Abstracts **DA**
Biochemistry and Cell Biology **JU**

Japan Bioindustries Association (Japan) **NO**
The Japan Biotech Letter **NL**
Japan Biotech and Medical News **NL**
Japan Chemical Week **NL**
Japan Information Centre of Science and Technology (Japan) **GO**
Japan Report: Biotechnology **NL**
Japan Technology **DA**
Japanese Biotechnology Industry **NL**
Japanscan **NL**
Jeol Ltd **SO**
JICST - Japan Information Center of Science and Technology **HO**
JICST-E **DA**
John Brown Engineers & Constructors Ltd **CO**
John Innes Institute Plant Virus Collection **CU**
Journal of Agricultural Engineering Research **JU**
Journal of Agricultural Science **JU**
Journal of the American Medical Association **JU**
Journal of the American Veterinary Medical Association **JU**
Journal of Animal Ecology **JU**
Journal of Antibiotics **JU**
Journal of Antimicrobial Chemotherapy **JU**
Journal of Applied Bacteriology **JU**
Journal of Applied Toxicology **JU**
Journal of Bacteriology **JU**
Journal of Biochemical and Biophysical Methods **JU**
Journal of Biochemistry **JU**
Journal of Biological Chemistry **JU**
Journal of Biological Response Modifiers **JU**
Journal of Bioluminescence and Chemiluminescence **JU**
Journal of Biotechnology **JU**
Journal of Cell Biology **JU**
Journal of Cell Science **JU**
Journal of Cellular Biochemistry **JU**
Journal of Cellular Physiology **JU**
Journal of Chemical Technology and Biotechnology **JU**
Journal of Chromatography **JU**
Journal of Clinical Immunology **JU**
Journal of Clinical Microbiology **JU**
Journal of Economic Entomology **JU**
Journal of Environmental Management **JU**
Journal of Environmental Pathology, Toxicology, and Oncology **JU**
Journal of Enzyme Inhibition **JU**
Journal of Experimental Biology **JU**
Journal of Experimental Botany **JU**
Journal of Experimental Marine Biology and Ecology **JU**
Journal of Experimental Pathology **JU**
Journal of Fermentation Technology **JU**
Journal of General and Applied Microbiology **JU**
Journal of General Microbiology **JU**
Journal of General Virology **JU**
Journal of High Resolution Chromatography **JU**
Journal of Horticultural Science **JU**
Journal of Immunoassay **JU**
Journal of Immunogenetics **JU**
Journal of Immunological Methods **JU**
Journal of Immunology **JU**
Journal of Immunopharmacology **JU**
Journal of Industrial Microbiology **JU**
Journal of Interferon Research **JU**
Journal of Labelled Compounds and Radiopharmaceuticals **JU**
Journal of Leukocyte Biology **JU**
Journal of Medical Microbiology **JU**
Journal of Medical Virology **JU**
Journal of Membrane Science **JU**
Journal of Microbial Biotechnology **JU**
Journal of Microbiological Methods **JU**
Journal of Molecular and Applied Genetics **JU**
Journal of Molecular Biology **JU**
Journal of Molecular Catalysis **JU**
Journal of Molecular and Cellular Immunology **JU**
Journal of Molecular Graphics **JU**
Journal of Molecular Recognition **JU**
Journal of Neurocytology **JU**
The Journal of Nutritional Biochemistry **JU**
Journal of Planar Chromatography - Modern TLC **JU**
Journal of Plant Physiology **JU**
Journal of the Science of Food and Agriculture **JU**
Journal of Soil Contamination **JU**
Journal of Soil Science **JU**
The Journal of Steroid Biochemistry and Molecular Biology **JU**
Journal of Structural Biology **JU**
Journal of Tissue Culture Methods **JU**

National Association for the Food Industry (France) **NO**
National Association for Plant Protection (France) **NO**
National Biological Standards Laboratory (Australasia) **GO**
National Biotechnology Committee (Italy) **GO**
National Board of Health (Finland) **GO**
National Board of Patents and Registration (Finland) **GO**
National Board for Science and Technology (Ireland) **GO**
National Board of Water (Finland) **GO**
National Centre of Agricultural Mechanisation for Forestry & Water (France) **GO**
National Centre of Scientific Research (France) **GO**
National Collection of Food Bacteria **CU**
The National Collection of Industrial and Marine Bacteria Ltd **CU**
National Collection of Pathogenic Fungi **CU**
National Collection of Plant Pathogenic Bacteria **CU**
National Collection of Type Cultures **CU**
National Collection of Wood Rotting Fungi **CU**
National Collection of Yeast Cultures **CU**
National Commission on Agricultural Engineering (Norway) **NO**
National Consultative Committee on Biotechnology (Italy) **GO**
National Council for Research & Development (Israel) **GO**
National Council for Research & Development (Spain) **GO**
National Culture Collection of Microorganisms - CNCM **CU**
National Energy Council (Greece) **GO**
National Environment Protection Board (Sweden) **GO**
National Federation of Farmers and Foresters (Italy) **NO**
National Health Institute **CU**
National Institute of Agrobiological Resources **CU**
National Institute of Allergy and Infectious Diseases - NIAID **CU**
National Institutes of Health (USA) **GO**
National Library of Medicine (USA) **GO**
National Pharmaceutical Association (USA) **NO**
National Research Council (Italy) **GO**
National Research Council - NRC (Canada) **GO**
National Research Council of Canada (Canada) **GO**
National Science Foundation (USA) **GO**
National Standards Bureau (Italy) **GO**
National Swedish Board for Energy Source Development (Sweden) **GO**
National Swedish Board of Occupational Health and Safety (Sweden) **GO**
National Swedish Board for Technical Development (Sweden) **GO**
National Swedish Industrial Board (Sweden) **GO**
National Technical Information Service (NTIS) **HO**
The National Toxicology Program's Chemical Database **DA**
National University of Athens **CU**
Nationales Referenzzentrum fur Adenoviren **CU**
Natural Product Reports **JU**
Natural Sciences and Engineering Research Council - NSERC (Canada) **GO**
Nature **JU**
Nature Genetics **JU**
NCFB Catalogue **DA**
NCS Diagnostics Inc **CU**
NCYC-on-line Services **HO**
Nelson Analytical Inc **SO**
Nematological Abstracts **DA**
Netherlands Biochemical Society (The Netherlands) **NO**
Netherlands Biophysics Society (The Netherlands) **NO**
Netherlands Biotechnology Society (The Netherlands) **NO**
The Netherlands Foreign Investment Agency (The Netherlands) **GO**
Netherlands Pharmaceutical Association (The Netherlands) **NO**
Netherlands Society of Plant Pathology (The Netherlands) **NO**
Network News **NL**
The New Biologist **JU**
New Bioresources **NL**

PESTDOC - Pest Control Literature Documentation **DA**
Pesticide Biochemistry and Physiology **JU**
Pesticide Outlook **JU**
Pesticide Science **JU**
PESTMAN - Pesticide Databank **DA**
PG-TRANS **DA**
Pharmaceased Discontinued Drugs File **DA**
Pharmaceutical Business News **NL**
Pharmaceutical Group of the European Community **IO**
Pharmaceutical Manufacturers Association (USA) **NO**
Pharmaceutical Society of Australia (Australasia) **NO**
Pharmacia LKB Biotechnology **SO**
Pharmacogenetics **JU**
Pharmacontacts Directories File **DA**
Pharmacontacts OnScreen **DA**
Pharmaprojects **NL**
Phase Separations Sales **SO**
Phero Tech Inc **CO**
Philosophical Transactions of the Royal Society: Series B **JU**
PHIND - Pharmaceutical and Healthcare Industries News Database **DA**
Photosynthesis Research **JU**
Physiologia Plantarum **JU**
Physiological and Molecular Plant Pathology **JU**
Phytochemical Analysis **JU**
Phytochemistry **JU**
PHYTOMED **DA**
Phytopathology **JU**
Pias Co Ltd **SO**
PJB Publications Ltd **CO**
Plan National de I+D (Spain) **GO**
Plant Biotechnology **DA**
Plant Biotechnology (UCC) Ltd **CO**
Plant Breeding (Austria) **GO**
Plant Breeding Abstracts **DA**
Plant Cell and Environment **JU**
Plant, Cell, Tissue and Organ Culture **JU**
Plant Disease **JU**
Plant Genetic Resources Abstracts **DA**
Plant Growth Regulation **JU**
Plant Growth Regulator Abstracts **DA**
The Plant Journal **JU**
Plant Molecular Biology **JU**
Plant Pathology **JU**
Plant Physiology **JU**
Plant Physiology and Biochemistry **JU**
Plant Physiology Society (France) **NO**
Plant Protection and Quarantine Import Data **DA**
Plant Science **JU**
Plant Science Association (Austria) **NO**
Plant Science Letters **JU**
Plant Science Ltd **CU**
Planta **JU**
Plants Today **JU**
Plasmid **JU**
Pollutech Environmental Ltd **CO**
Pollution Prevention **JU**
Polmark Directory **DI**
Porton Instruments **SO**
Portuguese Biochemistry Society (Portugal) **NO**
Portuguese Biomass Association (Portugal) **NO**
Portuguese Biotechnology Association (Portugal) **NO**
Portuguese Microbiology Society (Portugal) **NO**
Potato Abstracts **DA**
Poultry Abstracts **DA**
PPR - Plasmid Prefix Registry **DA**
Practical Biotechnology **NL**
Presque Isle Cultures **CU**
PRF/LITDB **DA**
PRF/SEQDB **DA**
Prince Edward Island Department of Industry (Canada) **GO**
PROBIOTECH (France) **NO**
Proceedings of the National Academy of Sciences **JU**
Proceedings of the Society for Experimental Biology and Medicine **JU**
Process Biochemistry **JU**
Producers of Cereals (France) **NO**
Producers of Maize (France) **NO**
Programme Bureau Biotechnology (The Netherlands) **GO**
Progress in Biophysics and Molecular Biology **JU**
PROMOCET (Brazil) **GO**
PROMOTECH (France) **NO**
Protein Databases Inc **SO**
Protein Expression and Purification **JU**
Proteins: Structure, Function, and Genetics **JU**
Proteus Molecular Design Ltd **SO**
Protoplasma **JU**

SECTION INDEXES

INTERNATIONAL ORGANIZATIONS

Biotechnology
Energy
Environmentat and health
Food and agriculture
Funding
Information and education
Regulation and patents
Research and development
Science and technology
Trade and industry

BIOTECHNOLOGY

Association of Biotechnology Companies - ABC
Commission on Biotechnology of the IUPAC
EUROBIOTA - COMETT
European Biotechnology Coordination Group - EBCG
European Federation of Biotechnology
European Molecular Biology Laboratory - EMBL
European Molecular Biology Organization - EMBO
European Society for Animal Cell Technology
Federation of European Biochemical Societies
International Association of Biological Standardization
International Biosciences Networks - IBN
International Biotechnology Suppliers Association
International Cell Research Organization
International Centre for Genetic Engineering and Biotechnology
International Centre of Insect Physiology and Ecology and Entomology
International Committee on Economic and Applied Microbiology
International Federation for Cell Biology
International Federation of Societies for Electron Microscopy - IFSEM
International Life Sciences Institute
International Organization for Biotechnology and Bioengineering - IOBB
International Union of Biochemistry and Molecular Biology - IUBMB
International Union of Biological Sciences - IUBS
International Union of Immunological Societies - IUIS
International Union of Microbiological Societies - IUMS
Microbiological Resource Centre - MIRCEN
MIRCEN Biotechnology and Agriculture Unit
MIRCEN Biotechnology Unit - Commonwealth Mycological Institute
Regional Programme of Biotechnology for Latin America and the Caribbean
Scientific Committee on Biotechnology - COBIOTECH
Scientific Committee on Genetic Experimentation - COGENE
Senior Advisory Group on Biotechnology - SAGB
United Nations Industrial Development Organization - UNIDO
World Federation for Culture Collections

ENERGY

Association Europeenne pour la Biomasse - AEBIOM
Biomass Users Network - BUN
European Petrochemical Association
The European Solar Energy Association - EUROSOLAR
Food and Agricultural Organization - FAO
International Association for Hydrogen Energy
International Atomic Energy Agency
International Crop Research Institute for Semiarid Tropics - ICRISAT
International Development Research Centre
International Energy Agency
International Research Centre on Environment and Development - CIRED
MIRCEN Fermentation Technology Unit
United Nations Institute for Training and Research - UNITAR
World Bank - International Bank for Reconstruction and Development

ENVIRONMENT AND HEALTH

Biomass Users Network - BUN
CIBA Foundation
European Federation of Pharmaceutical Industries Associations - EFPIA
International Association on Water Pollution Research and Control - IAWPRC
International Centre for Genetic Engineering and Biotechnology
International Centre of Insect Physiology and Ecology and Entomology

International Council of Scientific Unions - ICSU
International Federation of Organic Agricultural Movements - IFOAM
International Federation of Pharmaceutical Manufacturers' Associations
International Institute for Applied Systems Analysis - IIASA
International Research Centre on Environment and Development - CIRED
International Union of Immunological Societies - IUIS
International Union of Pharmacology - IUPHAR
Microbiological Resource Centre - MIRCEN
MIRCEN Biotechnology Unit - Commonwealth Mycological Institute
MIRCEN Fermentation Technology Unit
Pharmaceutical Group of the European Community
Scientific Committee on Genetic Experimentation - COGENE
Scientific Committee on Problems of the Environment - SCOPE
United Nations Environment Programme - UNEP
World Health Organization - WHO

FOOD AND AGRICULTURE

Association of the Cheese Processing Industries - ASSIFONTE
Association of the Cider and Fruit Wine Industries of the EC
Association of the Dairy Industries of the EC - ASSILEC
Biomass Users Network - BUN
Centre for International Cooperation in Agricultural Research and Development
Committee of Agricultural Organizations of the European Community
Committee of the Bread Yeast Manufacturers of the EC - COFALEC
Consultative Group on International Agricultural Research - CGIAR
EUROBIOTA - COMETT
European Association for Research on Plant Breeding
European Committee of Starch Products - CECPA
European Federation of Manufacturers of Additives for Animal Nutrition
European Union for Alcohol Brandies and Spirits
Food and Agricultural Organization - FAO
International Agricultural Development Service - IADS
International Centre for Genetic Engineering and Biotechnology
International Crop Research Institute for Semiarid Tropics - ICRISAT
International Development Research Centre
International Federation of Organic Agricultural Movements - IFOAM
International Food Information Service - IFIS
International Food Policy Research Institute
International Institute for Environmental Technology and Management
International Society of Soil Science
International Union of Food Science and Technology - IUFoST
International Union of Forestry Research Organizations - IUFRO
International Union of Nutritional Sciences - IUNS
International Union of Pharmacology - IUPHAR
Liaison Committee of the Glucose Producers in the European Community
MIRCEN Biotechnology and Agriculture Unit
MIRCEN Biotechnology Unit - Commonwealth Mycological Institute
MIRCEN Fermentation Technology Unit
MIRCEN Rhizobium Unit
Permanent International Committee of the Vinegar Industry of the EC
Scandinavian Society of Plant Physiology
World Bank - International Bank for Reconstruction and Development

FUNDING

Consultative Group on International Agricultural Research - CGIAR
International Agricultural Development Service - IADS
International Biosciences Networks - IBN
International Energy Agency
World Bank - International Bank for Reconstruction and Development

INFORMATION AND EDUCATION

Biomass Users Network - BUN
CAB International
CAB International Institute of Biological Control
CAB International Institute of Entomology
CAB International Institute of Parasitology
CAB International Mycological Institute
Commission on Biotechnology of the IUPAC
Committee on Data for Science and Technology - CODATA
Committee on Science and Technology in Developing Countries - COSTED
EUROBIOTA - COMETT
European Society for Animal Cell Technology
European Society for Opinion and Marketing Research
Federation of European Biochemical Societies

REGULATIONS AND PATENTS

RESEARCH AND DEVELOPMENT

International Union of Pharmacology - IUPHAR
International Union of Physiological Sciences
Scientific Committee on Biotechnology - COBIOTECH
United Nations Institute for Training and Research - UNITAR
World Bank - International Bank for Reconstruction and Development
World Health Organization - WHO

SCIENCE AND TECHNOLOGY

CAB International
CAB International Institute of Biological Control
CAB International Mycological Institute
Commission of the European Communities - CEC
Committee on Data for Science and Technology - CODATA
Committee on Science and Technology in Developing Countries - COSTED
European Council of Chemical Industry Federations - CEFIC
European Federation of Chemical Trade
European Federation of National Associations of Engineers - FEANI
European Federation of Pharmaceutical Industries Associations - EFPIA
Food and Agricultural Organization - FAO
International Council for Scientific and Technical Information - ICSTI
International Council of Scientific Unions - ICSU
International Development Association
International Federation of Automatic Control - IFAC
International Federation for Information and Documentation - FID
International Federation for Information Processing - IFIP
International Federation of Institutes for Advanced Study
International Foundation for Science - IFS
International Institute for Applied Systems Analysis - IIASA
International Research Centre on Environment and Development - CIRED
International Union of Food Science and Technology - IUFoST
International Union of Physiological Sciences
International Union for Pure and Applied Biophysics - IUPAB
International Union of Pure and Applied Chemistry - IUPAC
Organization for Economic Cooperation and Development - OECD
UNEP/UNESCO/ICRO Panel on Microbiology
Union of International Associations
United Nations Conference on Trade and Development - UNCTAD
United Nations Industrial Development Organization - UNIDO
World Association of Industrial and Technical Research Organizations
World Intellectual Property Organization

TRADE AND INDUSTRY

Commission of the European Communities - CEC
Confederation of Industries of the European Community
Economic and Social Council of the United Nations
European Council of Chemical Industry Federations - CEFIC
European Federation of Biotechnology
European Federation of Chemical Trade
European Federation of National Associations of Engineers - FEANI
International Bank for Reconstruction and Development
International Biotechnology Suppliers Association
International Committee on Economic and Applied Microbiology
International European Construction Federation
International Federation of Pharmaceutical Manufacturers' Associations
International Research Centre on Environment and Development - CIRED
International Union of Forestry Research Organizations - IUFRO
Organization for Economic Cooperation and Development - OECD
The Packaging Group Inc - TPG
Union of International Association
United Nations Conference on Trade and Development - UNCTAD
United Nations Industrial Development Organization - UNIDO
World Association of Industrial and Technical Research Organizations

GOVERNMENT ORGANIZATIONS

Biotechnology
Energy
Environment and health
Food and agriculture
Funding
Information and education
Research and development
Regulation and patents
Science and technology
Trade and industry

BIOTECHNOLOGY

ENERGY

National Energy Council (Greece)
National Research Council of Canada (Canada)
National Swedish Board for Energy Source Development (Sweden)
Ontario Ministry of Energy (Canada)
Royal Ministry of Petroleum and Energy (Norway)
Stichting Voorlilchting Energisbesparing Nederland (The Netherlands)

ENVIRONMENT AND HEALTH

Agriculture Canada (Canada)
Alberta Environmental Centre (Canada)
ANRED (France)
California State Water Resources Control Board (USA)
Canadian Centre for Occupational Health and Safety (Canada)
Central Veterinary Laboratory (UK)
Commission for Environmental Protection (Sweden)
CSIRO - Division of Animal Health (Australasia)
CSIRO - Division of Animal Production (Australasia)
CSIRO - Division of Biotechnology (Australasia)
CSIRO - Division of Forestry (Australasia)
CSIRO - Division of Plant Industry (Australasia)
CSIRO - Division of Soils (Australasia)
Department of Energy (USA)
Department of the Environment - DoE (UK)
Department of Health and Social Security (UK)
Department of Health and Human Services (USA)
Department of Health Services (USA)
Department of the Interior (USA)
Department of Pollution Control (France)
DSIR - Crop Research (Australasia)
DSIR - Plant Diseases (Australasia)
DSIR - Water Quality Centre (Australasia)
Energy Mines and Resources Canada (Canada)
Environment Canada (Canada)
Environmental Protection Agency - EPA (USA)
Federal Environmental Agency (Germany)
Federal Institute for Agriculture and Forestry (Germany)
Federal Ministry for Health and Environment (Austria)
Federal Office of Environment, Forest and Landscape (Switzerland)
Federal Office for Health (Switzerland)
Health and Safety Commission (UK)
Health and Welfare Canada (Canada)
INCQS (Brazil)
Institute of Health (Italy)
MAFF - Agricultural Science Services (UK)
MAFF - Environmental Prortection Division (UK)
MAFF - Pesticides and Pest Infestation Control (UK)
MAFF - Veterinary Services (UK)
Medical Research Council - MRC (UK)
Ministry for Culture and Environmental Affairs (Italy)
Ministry of Energy and Natural Resources (Greece)
Ministry of Environment (Denmark)
Ministry of Environment (Finland)
Ministry for the Environment (France)
Ministry of Environment (Israel)
Ministry of Health and Welfare (Greece)
Ministry of Health and Welfare (Japan)
Ministry of Health (Belgium)
Ministry of Health (Denmark)
Ministry of Health (France)
Ministry of Health (Israel)
Ministry of Health (Italy)
Ministry of Health (Spain)
Ministry of the Interior (Finland)
Ministry of Planning and Environment (Greece)
Ministry of Public Health and Environment (Belgium)
Ministry of Public Health and Environment Hygiene (The Netherlands)
Ministry of Social Affairs and Health (Finland)
National Biological Standards Laboratory (Australasia)
National Board of Health (Finland)
National Board of Water (Finland)
National Centre of Agricultural
Mechanisation for Forestry & Water (France)
National Environment Protection Board (Sweden)
National Institutes of Health (USA)
National Library of Medicine (USA)
National Research Council - NRC (Canada)
National Swedish Board of Occupational Health and Safety (Sweden)
New Zealand Department of Health (Australasia)
Parliamentary Committee for Environmental Protection (Sweden)
Royal Ministry of Environment (Norway)

FOOD AND AGRICULTURE

Agricultural and Food Research Council - AFRC (UK)
AFRC Institute for Animal Health - IAH (UK)
AFRC Institute of Animal Physiology and Genetics Research - IAPGR (UK)
AFRC Institute of Arable Crops Research - IACR (UK)

FUNDING

INFORMATION AND EDUCATION

European Communities (UK)
European Community Information Service (USA)
Federal Bureau of Statistics (Germany)
Federal Ministry for Food, Agriculture and Forestry (Germany)
Federal Office for Education and Science (Switzerland)
General Secretariat for Press and Information (Greece)
Government Printing Offices (The Netherlands)
Government Publications (France)
Government Publications (Italy)
Government Publications Stationery Office (Ireland)
H M Stationery Office (UK)
IBICT (Brazil)
Information and Documentation Institute (Belgium)
Japan Information Centre of Science and Technology (Japan)
Ministry of Education (Finland)
Ministry of Education (France)
Ministry of Education (Italy)
Ministry of Education (Flemish) (Belgium)
Ministry of Education (French) (Belgium)
Ministry of Education and Religion (Greece)
Ministry of Education and Culture (Israel)
Ministry of Education and Research (Denmark)
Ministry of Education and Science (The Netherlands)
Ministry of Education and Science (Spain)
Ministry of Education Science and Culture (Japan)
Ministry of Information and Infrastructure (Israel)
Office of Research and Statistical Studies (France)
The Patent Office (UK)
Royal Ministry of Education and Research (Norway)
Scientific and Technical Documentation Centre (France)
Supply and Services Canada (Canada)
Technology and Industry Development Authority (Australasia)

RESEARCH AND DEVELOPMENT

Agricultural and Food Research Council - AFRC (UK)
AFRC Institute for Animal Health - IAH (UK)
AFRC Institute of Animal Physiology and Genetics Research - IAPGR (UK)
AFRC Institute of Arable Crops Research - IACR (UK)
AFRC Institute of Engineering Research - IER (UK)
AFRC Institute of Food Research - IFR (UK)
AFRC Institute for Grassland and Animal Production - IGAP (UK)
AFRC Institute of Horticultural Research - IHR (UK)
AFRC Institute of Plant Science Research (UK)
Alberta Research Council (Canada)
Agency for Environment and Energy Control - ADEME (France)
Agricultural Research Centre Library (Finland)
Agriculture Canada (Canada)
Alberta Research Council (Canada)
ANVAR (France)
Australian Government Analytical Laboratories (Australasia)
Austrian Scientific Research Fund (Austria)
CNPq (Brazil)
CSIRO - Division of Animal Production (Australasia)
CSIRO - Division of Biotechnology (Australasia)
CSIRO - Division of Entomology (Australasia)
CSIRO - Division of Food Processing (Australasia)
CSIRO - Division of Forestry (Australasia)
CSIRO - Division of Horticulture (Australasia)
CSIRO - Division of Plant Industry (Australasia)
CSIRO - Division of Soils (Australasia)
Department of Agriculture and Fisheries for Scotland (UK)
Department for Development - DANIDA (Denmark)
Directorate of Research and Development (Italy)
Directorate for Scientific and Technical Research (France)
DSIR - Crop Research (Australasia)
DSIR - Entomology (Australasia)
DSIR - Horticulture (Australasia)
DSIR - Plant Diseases (Australasia)
DSIR - Plant Physiology (Australasia)
DSIR - Water Quality Centre (Australasia)
Economic Diversification and Trade (Canada)
EMBRAPA (Brazil)
Energy Mines and Resources Canada (Canada)
Energy Technology Support Unit - ETSU (UK)
Federal Ministry for Science and Research (Austria)
Fisheries and Oceans Canada (Canada)
Forestry Commission (UK)
Industry, Science and Technology Canada (Canada)

REGULATIONS AND PATENTS

SCIENCE AND TECHNOLOGY

Alberta Research Council (Canada)
ANVAR (France)
Central Commission for Biology (Germany)
Central Veterinary Laboratory (UK)
Committee on Science & Technology (USA)
CSIRO - Division of Animal Production (Australasia)
CSIRO - Division of Biotechnology (Australasia)
CSIRO - Division of Entomology (Australasia)
CSIRO - Division of Food Processing (Australasia)
CSIRO - Division of Forestry (Australasia)
CSIRO - Division of Horticulture (Australasia)
CSIRO - Division of Plant Industry (Australasia)
CSIRO - Division of Soils (Australasia)
Danish Council for Scientific Policy and Planning (Denmark)
Department of Agriculture and Fisheries for Scotland (UK)
Department of Education and Science (UK)
Department of the Interior (USA)
Department of Science and Technology (Australasia)
Directorate for Scientific and Technical Research (France)
DSIR - Crop Research (Australasia)
DSIR - Entomology (Australasia)
DSIR - Horticulture (Australasia)
DSIR - Plant Diseases (Australasia)
DSIR - Plant Physiology (Australasia)
DSIR - Water Quality Centre (Australasia)
Federal Ministry for Food, Agriculture and Forestry (Germany)
Federal Ministry for Science and Research (Austria)
Finnish Council for Scientific and Technical Information - TINFO (Finland)
Industry, Science and Technology Canada (Canada)
Institute pour l'Encouragement de la Recherche Scientifique dans l'Industrie (Belgium)
Interministerial Committee of Scientific and Technical Research (France)
MAFF - Ministry of Agriculture, Fisheries and Food (UK)
MAFF - Agricultural Science Services (UK)
MAFF - Pesticides and Pest Infestation Control (UK)
MAFF - Plant Variety Rights (UK)
MAFF - Veterinary Services (UK)
Manitoba Energy and Mines (Canada)
Manitoba Industry Trade and Technology (Canada)
Ministere de l'Enseignement Superieur et de la Science (Canada)
Ministere des Technologies Nouvelles et des Relations Exterieures (Belgium)
Ministry of Culture and Sciences (Greece)
Ministry of Education and Science (Germany)
Ministry of Education and Science (Spain)
Ministry of Research & Technology (Greece)
Ministry of Science and Technology (Israel)
National Biotechnology Committee (Italy)
National Board for Science and Technology (Ireland)
National Centre of Scientific Research (France)
National Institutes of Health (USA)
National Library of Medicine (USA)
National Research Council - NRC (Canada)
National Science Foundation (USA)
National Swedish Board for Energy Source Development (Sweden)
National Swedish Board for Technical Development (Sweden)
Natural Sciences and Engineering Research Council - NSERC (Canada)
Office of the Chief Scientist (Israel)
Plan National de I+D (Spain)
PROMOCET (Brazil)
Royal Norwegian Council for Scientific and Industrial Research - NTNF (Norway)
Science and Technology Agency (Japan)

TRADE AND INDUSTRY

Administration Bureau (Japan)
British Standards Institution - BSI (UK)
Canadian Commercial Corp (Canada)
Centro para el Desarrollo Tecnologico Industrial - CDTI (Spain)
Department of Commerce (USA)
Department of Foreign Affairs (Ireland)
Department of Industry Commerce and Energy (Ireland)
Department of Industry Trade Commerce and Tourism (Ireland)
Department of Industry (Belgium)
Department of Industry (Sweden)
Department of Regional Industrial Expansion - DRIE (Canada)
Department of Treasury (USA)
Directorate General for International Cooperation (The Netherlands)
Federal Ministry for Trade and Industry (Austria)
Federal Office of Economics (Germany)
Federal Office for Industry and Labour (Switzerland)
Federation of Netherlands' Industry (The Netherlands)
Foreign and Commonwealth Office (UK)
Industrial Development Authority (Ireland)
Manitoba Industry Trade and Technology (Canada)

NATIONAL ORGANIZATIONS

BIOTECHNOLOGY

ENERGY

Biomass And Biofuels Association - BABA (UK)
Center for Renewable Energy Sources (Greece)
Danish Organization for Renewable Energy - OVE (Denmark)
French Biomass Association (France)
Italian Biomass Association - ITABIA (Italy)
Portugese Biomass Association (Portugal)
Spanish Biomass Association - ADABE (Spain)

ENVIRONMENT AND HEALTH

American Association for Clinical Chemistry (USA)
American Association of Immunologists (USA)
American Medical Association (USA)
American Pharmaceutical Association (USA)
American Society of Clinical Pathologists (USA)
American Society for Microbiology (USA)
Association of Applied Biologists (UK)
Association of British Health Care Industries (UK)
Association of the British Pharmaceutical Industry (UK)
Association of the Drug Industries (Belgium)
Association of Finnish Pharmacies (Finland)
Association of Finnish Water Supply (Finland)
Association of Pharmaceutical Industries (Austria)
Association of Pharmaceutical Distributors (Finland)
Association of Pharmaceutical Industries (Finland)
Association of Swiss Speciality
Pharmaceutical Manufacturers (Switzerland)
Association of Water Interests (Austria)
Association of the Water Treatment Industry (Germany)
Australian Pharmaceutical Manufacturers Association (Australasia)
Australian Water and Wastewater Association (Australasia)
Austrian Pharmacists Association (Austria)
Austrian Society for Immunology (Austria)
Belgium Pharmaceutical Association (Belgium)
BioIndustry Association (UK)
British Association of Pharmaceutical Physicians (UK)
British Effluent and Water Association (UK)
British Medical Association (UK)
British Society for Immunology (UK)
British Society of Toxicological Pathologists - BSTP (UK)
Danish Microbiology Society (Denmark)
Danish Pharmaceutical Society (Denmark)
Danish Water Supply Association (Denmark)
Dutch Microbiology Society (The Netherlands)
French Microbiology Society (France)
German Microbiology Society (Germany)
German Pharmaceutical Society (Germany)
Institute of Biology (UK)
Institute of Clinical Molecular Biology (UK)
Institute for Life Sciences (UK)
Italian Association of Clinical Microbiology (Italy)
Italian Society for Microbiology (Italy)
Japan Antibiotics Research Association (Japan)
Japan Association for Tissue Culture (Japan)
Microbiology Research Association (France)
National Association for Plant Protection (France)
National Pharmaceutical Association (USA)
Netherlands Pharmaceutical Association (The Netherlands)
Norwegian Society for Microbiology (Norway)
Pharmaceutical Manufacturers Association (USA)
Portuguese Microbiology Society (Portugal)
Royal Society (UK)
Society of Chemical Industry (UK)
Society for Experimental Biology and Medicine (USA)
Society for Food Hygiene (France)
Society for General Microbiology (UK)
Society for Industrial Microbiology (USA)
Spanish Microbiology Society (Spain)
Swedish Society for Microbiology (Sweden)
Swiss Pharmaceutical Society (Switzerland)
Swiss Society for Microbiology (Switzerland)
Swiss Water Pollution Control Association (Switzerland)
Wastes Trade Association (Belgium)

FOOD AND AGRICULTURE

Agricultural Association (Portugal)
Agricultural Engineering Association (Belgium)
Agricultural Engineers (Spain)
Agricultural Science (Denmark)
Agricultural Society of France (France)
Agricultural Society of Finland (Finland)
Agricultural and Veterinary Chemicals
Association of Australia (Australasia)
American Phytopathological Society (USA)
American Society of Agricultural Engineers (USA)

American Society for Horticultural Science (USA)
American Society of Plant Physiologists (USA)
Assemble Permanante Chambres d'Agriculture - APCA (France)
Association of Agricultural Engineers (Switzerland)
Association of Agricultural Engineers (Germany)
Association of Agricultural Engineers (France)
Association of Agricultural Engineers (Sweden)
Association of Applied Biologists (UK)
Association of Cheese Processors (UK)
Association of Engineers in Food and Agricultural Industries (France)
Association for the Promotion of Agroindustry (France)
Austrian Brewing Society (Austria)
Biomass Center for Energy - CBE (Portugal)
Biomass And Biofuels Association - BABA (UK)
Brewing Association (Germany)
Spanish Brewing Association (Spain)
British Agrochemicals Association (UK)
British Crop Protection Council (UK)
British Pest Control Association (UK)
British Society for Plant Growth Regulation (UK)
British Veterinary Association (UK)
Center for Renewable Energy Sources (Greece)
Central Union of Agricultural Producers (Finland)
Council of Australian Food Technology Associations (Australasia)
Crop Science Society of America (USA)
Federation of Swedish Farmers (Sweden)
Finnish Agricultural Association (Finland)
French Biomass Association (France)
German Agricultural Society (Germany)
German Botanical Society (Germany)
Institute of Biology (UK)
Institute of Food Science and Technology (UK)
Institute for Life Sciences (UK)
Italian Association of Agricultural Engineers (Italy)
Italian Brewing Association (Italy)
National Association for Plant Protection (France)
National Association for the Food Industry (France)
National Commission on Agricultural Engineering (Norway)
National Federation of Farmers and Foresters (Italy)
Netherlands Society of Plant Pathology (The Netherlands)
Norwegian Biochemistry Society (Norway)
Norwegian Farmers Union (Norway)
Pharmaceutical Society of Australia (Australasia)
Plant Physiology Society (France)
Portugese Biomass Association (Portugal)
Producers of Cereals (France)
Producers of Maize (France)
Royal Agricultural Society of Denmark (Denmark)
Royal Netherlands Society of Agricultural Science (The Netherlands)
Royal Society (UK)
Royal Swedish Academy of Agriculture and Forestry (Sweden)
Society of Agricultural Engineering (The Netherlands)
Society of Chemical Industry (UK)
Society for Food Hygiene (France)
Swedish Brewmasters Association (Sweden)
Swedish Society for Microbiology (Sweden)
Swiss Association of Agricultural Engineers - SVIAL (Switzerland)

FUNDING

Deutsche Forschungsgemeinschaft - DFG (Germany)
Israel Academy of Sciences and Humanities (Israel)
National Academy of Sciences - NAS (USA)
Royal Society (UK)
Royal Swedish Academy of Sciences (Sweden)
Swiss Academy of Sciences (Switzerland)

INFORMATION AND EDUCATION

Academie des Sciences (France)
Academy of Sciences of Lisbon (Portugal)
Agricultural Society of France (France)
Association of Agricultural Engineers (Germany)
Association of Consulting Engineers (UK)
Association of Consulting Scientists (UK)
Association for the Promotion of Agroindustry (France)
Australian Academy of Sciences (Australasia)
Biomass And Biofuels Association - BABA (UK)
Center for Renewable Energy Sources (Greece)
Delegation of Finnish Academies of Science and Letters (Finland)
Institute of Clinical Molecular Biology (UK)
Israel Academy of Sciences and Humanities (Israel)
National Academy of Sciences - NAS (USA)
Portugese Biomass Association (Portugal)
PROBIOTECH (France)

PROMOTECH (France)
Royal Danish Academy of Sciences and Letters (Denmark)
Royal Irish Academy (Ireland)
Royal Netherlands Academy of Arts and Sciences (The Netherlands)
Royal Society of New Zealand (Australasia)
Royal Society (UK)
Royal Swedish Academy of Sciences (Sweden)
Science Council of Japan (Japan)
Swiss Academy of Sciences (Switzerland)

REGULATIONS AND PATENTS

Danish Standards Institute (Denmark)
Institute of Patent Attorneys of Australia (Australasia)
Standards Institute (Belgium)

RESEARCH AND DEVELOPMENT

Association of Biotechnology Companies - ABC (USA)
Australian Academy of Sciences (Australasia)
Biomass Center for Energy - CBE (Portugal)
Delegation of Finnish Academies of Science and Letters (Finland)
Deutsche Forschungsgemeinschaft - DFG (Germany)
Federation of American Societies for Experimental Biology (USA)
Institute of Clinical Molecular Biology (UK)
Institute for Life Sciences (UK)
Israel Academy of Sciences and Humanities (Israel)
Israel Biochemical Society (Israel)
Japan Antibiotics Research Association (Japan)
Royal Netherlands Academy of Arts and Sciences (The Netherlands)
Royal Society of New Zealand (Australasia)
Royal Society (UK)
Royal Swedish Academy of Sciences (Sweden)
Society for Experimental Biology (UK)
Society for Experimental Biology and Medicine (USA)
Swiss Union of the Societies for Experimental Biology (Switzerland)

SCIENCE AND TECHNOLOGY

Academie des Sciences (France)
Academy of Athens (Greece)
Academy of Sciences of Lisbon (Portugal)
Academy of Sciences (Austria)
American Association for Clinical Chemistry (USA)
American Association for the Advancement of Science (USA)
American Chemical Society (USA)
American Pharmaceutical Association (USA)
American Society of Biological Chemistry (USA)
American Society for Cell Biology (USA)
American Society for Microbiology (USA)
Association of the British Pharmaceutical Industry (UK)
Association of Chemical Trades (Austria)
Association of Chemical Engineers (Japan)
Association of the Chemical Industry (Japan)
Association of Clinical Biochemists (UK)
Association of Consulting Engineers (UK)
Association of Consulting Scientists (UK)
Association of Finnish Chemical Societies (Finland)
Association of Finnish Pharmacies (Finland)
Asssociation of German Engineers (Germany)
Association of Pharmaceutical Distributors (Finland)
Association of Pharmaceutical Industries (Finland)
Association of Swiss Speciality Pharmaceutical Manufacturers (Switzerland)
Australian Academy of Sciences (Australasia)
Australian Academy of Technological Science (Australasia)
Australian Scientific Industries Association (Australasia)
Australian Society for Microbiology (Australasia)
Austrian Biochemical Society (Austria)
Austrian Biophysical Society (Austria)
Austrian Pharmacists Association (Austria)
Biochemical Society (Ireland)
Biochemical Society (Belgium)
Biochemical Society (Italy)
The Biochemical Society (UK)
Biological Engineering Society (UK)
Biology Society (France)
Biology Society (Italy)
Botanical Society (Portugal)
British Agrochemicals Association (UK)
British Association for the Advancement of Science (UK)
British Association of Pharmaceutical Physicians (UK)
British Biophysical Association (UK)
British Chemical Distributors and Trade Association Ltd (UK)
British Chemical Engineering Contractors Association (UK)
Canadian Chemical Producers Association (Canada)
Center for Renewable Energy Sources (Greece)

Danish Biochemical Society (Biokimisk Forening) (Denmark)
Danish Biology Society (Denmark)
Danish Botanical Society (Denmark)
Danish Microbiology Society (Denmark)
Danish Pharmaceutical Society (Denmark)
Delegation of Finnish Academies of Science and Letters (Finland)
Deutsche Forschungsgemeinschaft - DFG (Germany)
Dutch Microbiology Society (Netherlands)
Federation of American Societies for Experimental Biology (USA)
Federation of Belgium Chemical Industries (Belgium)
French Botanical Society (France)
French Microbiology Society (France)
French Society for Biochemistry and Molecular Biology (France)
German Biochemical Society (Germany)
German Botanical Society (Germany)
German Microbiology Society (Germany)
German Pharmaceutical Society (Germany)
Hellenic Biochemical and Biophysical Society (Greece)
Icelandic Biochemical Society (Iceland)
Institute of Biology (UK)
Israel Academy of Sciences and Humanities (Israel)
Israel Biochemical Society (Israel)
Italian Association of Clinical Microbiology (Italy)
Italian Botanical Society (Italy)
Italian Chemical Society (Italy)
Italian Society for Microbiology (Italy)
Japan Association for Tissue Culture (Japan)
Microbiology Research Association (France)
National Academy of Sciences - NAS (USA)
National Pharmaceutical Association (USA)
Netherlands Biochemical Society (The Netherlands)
Netherlands Biophysics Society (The Netherlands)
Netherlands Pharmaceutical Association (The Netherlands)
Netherlands Society of Plant Pathology (The Netherlands)
Norwegian Academy of Sciences and Letters (Norway)
Norwegian Biochemistry Society (Norway)
Norwegian Botanical Association (Norway)
Norwegian Society for Microbiology (Norway)
Portuguese Biochemistry Society (Portugal)
Portuguese Microbiology Society (Portugal)
PROBIOTECH (France)
PROMOTECH (France)
Royal Academy of Belgium (Belgium)
Royal Australian Chemical Institute (Australasia)
Royal Botanical Society of the Netherlands (The Netherlands)
Royal Danish Academy of Sciences and Letters (Denmark)
Royal Irish Academy (Ireland)
Royal Netherlands Academy of Arts and Sciences (The Netherlands)
Royal Netherlands Society of Agricultural Science (The Netherlands)
Royal Society of New Zealand (Australasia)
Royal Society (UK)
Royal Swedish Academy of Sciences (Sweden)
Science Council of Japan (Japan)
Societas Biochemica, Biophysica et Microbiologica Fenniae (Finland)
Society of Chemical Industry (UK)
Society for the Chemical Industry (France)
Society for Experimental Biology (UK)
Society for Experimental Biology and Medicine (USA)
Society of General Microbiology and Biotechnology (Italy)
Society for General Microbiology (UK)
Society for Industrial Microbiology (USA)
Spanish Biochemical Society (Spain)
Spanish Microbiology Society (Spain)
Swedish Biochemical Society (Sweden)
Swedish Botanical Society (Sweden)
Swedish Society for Microbiology (Sweden)
Swiss Academy of Sciences (Switzerland)
Swiss Pharmaceutical Society (Switzerland)
Swiss Society for Biochemistry (Switzerland)
Swiss Society for Microbiology (Switzerland)
Swiss Union of the Societies for Experimental Biology (Switzerland)

TRADE AND INDUSTRY

Association of Belgium Industries (Belgium)
Association of Biotechnology Companies - ABC (USA)
Association of British Health Care Industries (UK)
Association of the British Pharmaceutical Industry (UK)
Association of the Chemical Industry (Japan)
Association of Engineers in Food and Agricultural Industries (France)
Association of Finnish Pharmacies (Finland)
Association of Finnish Water Supply (Finland)
Association of Pharmaceutical Industries (Austria)
Association of Pharmaceutical Distributors (Finland)
Association of Pharmaceutical Industries (Finland)
Association for the Promotion of Agroindustry (France)

Association for the Promotion of International Trade (Japan)
Association of Swiss Speciality Pharmaceutical Manufacturers (Switzerland)
Association of the Water Treatment Industry (Germany)
Associazione Nazionale per lo Svuluppo delle Biotechnologie - ASSOBIOTEC (Italy)
Australian Scientific Industries Association (Australasia)
Austrian Society for Chemical Industry (Austria)
Belgium Bioindustries Association (Belgium)
BioIndustry Association (UK)
Biomass And Biofuels Association - BABA (UK)
British Chemical Distributors and Trade Association Ltd (UK)
Canadian Chemical Producers Association (Canada)
Central Union of Agricultural Producers (Finland)
Danish Water Supply Association (Denmark)
Federation of Belgium Chemical Industries (Belgium)
Federation of Belgium Industries (Belgium)
Industrial Biotechnology Association (USA)
Institute of Clinical Molecular Biology (UK)
Japan Bioindustries Association (Japan)
National Association for the Food Industry (France)
Organisation Nationale Interprofessionnelle des Bioindustries - ORGANIBIO (France)
Pharmaceutical Manufacturers Association (USA)
Producers of Cereals (France)
Producers of Maize (France)
Society of Chemical Industry (UK)
Society for the Chemical Industry (France)
Society for Industrial Microbiology (USA)

JOURNALS

Agriculture
Analysis
Animal health and nutrition
Aquaculture and marine biology
Biochemistry
Biology
Biophysics
Biotechnology
Brewing/distilling
Chemistry and chemicals
Computer products and applications
Education
Energy
Engineering
Entomology
Environment and ecology
Enzymology
Fermentation
Food and nutrition
Forestry
Genetics
Human biology
Human health care
Immunology
Industry
Legislation and regulatory issues
Microbiology
Modelling
Patents
Pesticides
Plant science
Virology
Waste management

AGRICULTURE

AgBiotech
Agriculture, Ecosystems & Environment
Agroforestry Today
Agro-Industry Hi-Tech
Agronomy Journal
American Society for Horticultural Science Journal
Annual Reviews
Applied Agricultural Research
Arboricultural Journal
Biofouling
Biological Agriculture and Horticulture
Biology and Fertility of Soils
Biomass & Bioenergy
Bioresource Technology
Carlsberg Research Communications
Cellular and Molecular Biology
Communications in Soil Science
Critical Reviews in Biotechnology
Crop Protection
Crop Science
Farm Chemicals International
Fertilizer Research
Geoderma
IMPACT AgBioIndustry
Indian Journal of Microbiology
Industrial Crops and Products
Journal of Agricultural Engineering Research
Journal of Agricultural Science
Journal of Horticultural Science
Journal of Molecular and Applied Genetics
Journal of the Science of Food and Agriculture
Journal of Soil Contamination
Journal of Soil Science
Land Management and Environmental Law Report
Mycopathologia
Pesticide Biochemistry and Physiology
Pesticide Outlook
Pesticide Science

Photosynthesis Research
Phytochemical Analysis
Phytopathology
Plant Cell and Environment
Plant Disease
Plant Pathology
Research in Veterinary Science
Review of Agricultural Entomology
Revue d'Ecologie et de Biologie du Sol
RIA (Revue de l'Industrie Agro-Alimentaire)
Soil Biology and Biochemistry
Soil Science
Soviet Biotechnology (Biotekhnologiya)
Tropical Animal Health and Production
Veterinary Immunology and Immunopathology
Veterinary Microbiology
Veterinary Parasitology
Veterinary Record
Zeitschrift fur Pflanzenernahrung und Bodenkunde
Zentralblatt fur Mikrobiology

ANALYSIS

Analytical Biochemistry
Annual Reviews
Applied and Theoretical Electrophoresis
Biological Mass Spectrometry
Biomedical Chromatography
BioPharm
Bioscience, Biotechnology and Biochemistry
CRC Critical Reviews
Cytometry
Diagnostic Immunology
Diagnostic Microbiology and Infectious Disease
DNA Probes
DNA Sequence
Electroanalysis
Electrophoresis
Environmental Pollution
Environmental Toxicology and Water Quality
Gene
Hybridoma
Journal of Bioluminescence and Chemiluminescence
Journal of Chemical Technology and Biotechnology
Journal of Chromatography
Journal of High Resolution Chromatography
Journal of Immunoassay
Journal of Immunological Methods
Journal of Planar Chromatography - Modern TLC
Journal of Soil Contamination
Journal of Structural Biology
Macromolecular Structures
Micron and Microscopica Acta
Molecular and Cellular Probes
Molecular Immunology
Phytochemical Analysis
Protoplasma
Receptors and Channels
Spectra Biologie
Spectroscopy Europe
Supramolecular Chemistry
Topics in Fluorescence Spectroscopy
Viral Immunology

ANIMAL HEALTH AND NUTRITION

American Journal of Veterinary Research
Animal Genetics
Animal Reproduction Science
Antiviral Research
Comptes Rendus de l'Academie des Sciences
Indian Journal of Microbiology
Interferon Y Biotecnologia
Journal of the American Veterinary Medical Association
Journal of Economic Entomology
Journal of Environmental Pathology, Toxicology, and Oncology
Journal of Immunogenetics
Journal of Medical Microbiology
The Journal of Nutritional Biochemistry
The Journal of Steroid Biochemistry and Molecular Biology
Mycopathologia
Pesticide Outlook
Pharmacogenetics
Reproduction Nutrition Development
Research in Veterinary Science
Review of Medical and Veterinary Entomology
Review of Medical and Veterinary Mycology
Toxicology
Toxicology and Applied Pharmacology
Toxicology Letters
Tropical Animal Health and Production
Vaccine
Veterinary Immunology and Immunopathology
Veterinary Microbiology
Veterinary Parasitology
Veterinary Record
Viral Immunology

AQUACULTURE AND MARINE BIOLOGY

Aquaculture
Aquaculture Engineering
Aquatic Conservation
Aquatic Toxicology
Journal of Experimental Marine Biology and Ecology
Marine Environmental Research
Reviews in Fish Biology and Fisheries

BIOCHEMISTRY

AIDS Research Reviews
Analytical Biochemistry
Annals of Clinical Biochemistry
Annual Reviews
Annual Reviews in Cell Biology
Applied Biochemistry and Biotechnology
Applied Biochemistry and Microbiology
Archives of Biochemistry and Biophysics
Arzneimittel-Forschung/Drug Research
Australian Journal of Biological Sciences
Biocatalysis
Biochemical and Biophysical Research Communications
The Biochemical Engineering Journal
Biochemical Genetics
Biochemical Journal
Biochemical Medicine
Biochemical Society Transactions
Biochemical Systematics and Ecology
Biochemistry
Biochemistry and Cell Biology
Biochimica et Biophysica Acta
Biochimica et Biophysica Acta - Bioenergetics
Biochimica et Biophysica Acta - Biomembranes
Biochimica et Biophysica Acta - Gene Structure and Expression
Biochimica et Biophysica Acta - General Subjects
Biochimica et Biophysica Acta - Lipids and Lipid Metabolism
Biochimica et Biophysica Acta - Molecular Cell Research
Biochimica et Biophysica Acta - Reviews on Bioenergetics
Biochimica et Biophysica Acta - Reviews on Biomembranes
Biochimie
BioEssays
BioFactors
Biofutur
Biological Membranes
Bioorganic and Medicinal Chemistry Letters
Biopolymers
Bioscience, Biotechnology and Biochemistry
Bioscience Reports
Bio/Technology
Biotechnology Advances
Biotechnology and Applied Biochemistry
Biotechnology and Bioengineering
Biotechnology Letters
Biotechnology Progress
Botanica Acta
Canadian Journal of Biochemistry and Cell Biology
Cancer Research
Carbohydrate Research
Carlsberg Research Communications
Cell Biochemistry and Function
Cell Biology: International Reports
Cell Motility and the Cytoskeleton
Cellular Biochemistry
Cellular and Molecular Biology
Cellular Signalling
Chemico-Biological Interactions
Chinese Journal of Physiological Sciences
Clinical Biochemistry
Clinical Science
Comments on Molecular and Cellular Biophysics
Comparative Biochemistry and Physiology
Comptes Rendus de l'Academie des Sciences
CRC Critical Reviews
Critical Reviews in Biochemistry and Molecular Biology
Current Advances in Biochemistry
Current Advances in Cell and Developmental Biology
Current Methods in Molecular Biology
Current Opinion in Cell Biology
Current Opinion in Neurobiology
Current Opinion in Structural Biology
Cytobios
Cytometry
Cytotechnology
Developmental Genetics
DNA A Journal of Molecular Biology
Electrophoresis
EMBO Journal
Enzyme and Fermentation Biotechnology
European Journal of Biochemistry
European Journal of Cell Biology
Experimental Cell Research
FEBS Letters
Food Chemistry
General and Comparative Endocrinology
Glycoconjugate Journal
Growth Factors
The Histochemical Journal
Histochemistry
Immunogenetics
Insect Biochemistry and Molecular Biology
International Journal of Biochemistry
International Journal of Biological Macromolecules
International Journal of Peptide and Protein Research
International Research Communications Systems
Journal of Bacteriology
Journal of Biochemical and Biophysical Methods
Journal of Biochemistry
Journal of Biological Chemistry
Journal of Biological Response Modifiers
Journal of Bioluminescence and Chemiluminescence
Journal of Biotechnology
Journal of Cell Biology
Journal of Cell Science
Journal of Cellular Biochemistry
Journal of Cellular Physiology

Journal of Chromatography
Journal of Economic Entomology
Journal of Experimental Botany
Journal of Experimental Marine Biology and Ecology
Journal of Experimental Pathology
Journal of Membrane Science
Journal of Molecular and Applied Genetics
Journal of Molecular Biology
Journal of Molecular Catalysis
Journal of Molecular Recognition
Journal of Neurocytology
The Journal of Nutritional Biochemistry
Journal of Planar Chromatography - Modern TLC
Journal of Plant Physiology
The Journal of Steroid Biochemistry and Molecular Biology
Journal of Virology
Life Sciences
Macromolecular Structures
Membrane Biochemistry
Methods in Molecular and Cellular Biology
Molecular Biology Reports
Molecular and Cellular Biochemistry
Molecular and Cellular Biology
Molecular Microbiology
The Mycologist
Natural Product Reports
Nature
The New Biologist
New Phytologist
The New Phytologist
New Scientist
Nucleic Acids Research
Nucleosides and Nucleotides
Oncogene
Peptides
Pesticide Biochemistry and Physiology
Phytochemical Analysis
Phytochemistry
Plant Cell and Environment
Plant, Cell, Tissue and Organ Culture
Plant Growth Regulation
The Plant Journal
Plant Molecular Biology
Plant Physiology
Plant Physiology and Biochemistry
Plant Science
Plant Science Letters
Planta
Proceedings of the National Academy of Sciences
Process Biochemistry
Progress in Biophysics and Molecular Biology
Protein Expression and Purification
Proteins: Structure, Function, and Genetics
Protoplasma
Receptors and Channels
Regulatory Peptides
Science
Scientific American
Seed Science Research
Seed Science and Technology
Soil Biology and Biochemistry
Soviet Biotechnology (Biotekhnologiya)
Starch
Technique
Topics in Fluorescence Spectroscopy
Toxicology
Toxicology and Applied Pharmacology
Toxicology Letters
Transgenic Research
Trees - Structure and Function
Trends in Biochemical Sciences
Trends in Cell Biology
Virology
Wood Science and Technology
Yeast
Zeitschrift fur Pflanzenernahrung und Bodenkunde

BIOLOGY

Annals of Applied Biology
Australian Journal of Biological Sciences
BioEssays
Biologist
Biology Digest
Biology and Fertility of Soils
Biosystems
Biotechnology Advances
Biotechnology and Bioengineering
Cell
Cell Biology: International Reports
Cell Motility and the Cytoskeleton
Cellular and Molecular Biology
Comments on Theoretical Biology
Comptes Rendus de l'Academie des Sciences
Current Advances in Cell and Developmental Biology
Current Biology
Current Opinion in Structural Biology
DNA A Journal of Molecular Biology
Environmental Pollution
The FASEB Journal
Journal of Economic Entomology
Journal of Experimental Biology
Journal of Molecular Recognition
Journal of Structural Biology
Macromolecular Structures
Marine Environmental Research
Micron and Microscopica Acta
Molecular Biology Reports
Molecular and Cellular Biology
Molecular and Cellular Endocrinology
Molecular Phylogenetics and Evolution
M/S Medicine Sciences
Mycopathologia
Nature
The New Biologist
New Scientist
Nucleosides and Nucleotides

Philosophical Transactions of the Royal Society: Series B
Proceedings of the National Academy of Sciences
Protoplasma
Quantitative Structure-Activity Relationships
Reproduction Nutrition Development
Reviews in Fish Biology and Fisheries
Revue d'Ecologie et de Biologie du Sol
Science
Scientific American
Soil Biology and Biochemistry
Supramolecular Chemistry
Theriogenology - An International Journal of Animal Reproduction
Toxicology

BIOPHYSICS

Annual Reviews
Archives of Biochemistry and Biophysics
Biochemical and Biophysical Research Communications
Biochimica et Biophysica Acta
Biochimica et Biophysica Acta - Bioenergetics
Biochimica et Biophysica Acta - Biomembranes
Biochimica et Biophysica Acta - Gene Structure and Expression
Biochimica et Biophysica Acta - General Subjects
Biochimica et Biophysica Acta - Lipids and Lipid Metabolism
Biochimica et Biophysica Acta - Molecular Cell Research
Biochimica et Biophysica Acta - Reviews on Bioenergetics
Biochimica et Biophysica Acta - Reviews on Biomembranes
Biophysical Chemistry
Biotechnology and Bioengineering
Chinese Journal of Physiological Sciences
Comments on Molecular and Cellular Biophysics
Experimental Cell Research
FEBS Letters
International Journal of Radiation Biology
Journal of Biochemical and Biophysical Methods
Journal of Cell Biology
Journal of Cell Science
Journal of Cellular Biochemistry
Journal of Cellular Physiology
Journal of Experimental Botany
Journal of Experimental Pathology
Journal of Membrane Science
Journal of Virology
Marine Environmental Research
New Phytologist
New Scientist
Plant Cell and Environment
Plant Physiology and Biochemistry
Proceedings of the National Academy of Sciences
Progress in Biophysics and Molecular Biology
Wood Science and Technology
Zeitschrift fur Pflanzenernahrung und Bodenkunde

BIOTECHNOLOGY

Acta Biotechnologica
Acta Pathologica et Immunologica Scandinavica
Advances in Neuroimmunology
AgBiotech
AIDS Research and Human Retroviruses
AIDS Research Reviews
Annual Reviews in Biodiversity
Annual Reviews in Cell Biology
Antibiotics and Chemotherapy
Antimicrobial Agents and Chemotherapy
Antiviral Research
Applied Biochemistry and Biotechnology
Applied and Theoretical Electrophoresis
Arzneimittel-Forschung/Drug Research
Australian Journal of Biotechnology
Biodiversity and Conservation
Bioelectrochemistry and Bioenergetics
Bioelectromagnetics
Bioelectronics and Biosensors
BioEngineering - Forschung & Praxis
Biofutur
Biology and Fertility of Soils
Biomedical Chromatography
BioPharm
Biopolymers
Bioprocess Engineering
Bioscience, Biotechnology and Biochemistry
Biosensors and Bioelectronics
Biotech Forum Europe
Biotech Knowledge Sources
Bio/Technology
Biotechnology Advances
Biotechnology and Applied Biochemistry
Biotechnology and Bioengineering
Biotechnology Chemonomics
Biotechnology and Genetic Engineering Reviews
Biotechnology Insight
Biotechnology Letters
Biotechnology Progress
Biotechnology Techniques
Biotechnology Therapeutics
Carlsberg Research Communications
Cell Adhesion and Communication
Cell Biochemistry and Function
Cellular Biochemistry
Cellular Signalling
Chemistry and Industry
Chimica oggi
Chromosoma
Critical Reviews in Biotechnology

BREWING/DISTILLING

CHEMISTRY AND CHEMICALS

Phytochemistry
Phytopathology
Plant Pathology
Pure and Applied Chemistry
Quantitative Structure-Activity Relationships
Science
Scientific American
Supramolecular Chemistry
Toxicology
Toxicology and Applied Pharmacology
Toxicology Letters
Zeitschrift fur Pflanzenernahrung und Bodenkunde

COMPUTER PRODUCTS AND APPLICATIONS

Applied Agricultural Research
Bioprocess Engineering
Clinical Chemistry and Enzymology Communications
Computer Applications in the Biosciences - CABIOS
International Journal of Clinical Monitoring and Computing
Journal of Biotechnology
Journal of Chemical Technology and Biotechnology
Journal of Environmental Management
Journal of Molecular Graphics
Journal of Planar Chromatography - Modern TLC
Quantitative Structure-Activity Relationships

EDUCATION

Biology Digest
Food Science and Technology Today

ENERGY

Annual Reviews
Applied Biochemistry and Biotechnology
The Biochemical Engineering Journal
Biomass - an International Journal
Biomass & Bioenergy
Biorecovery
Bioresource Technology
Bio/Technology
Critical Reviews in Plant Sciences
Environmental Conservation
Geomicrobiology Journal
Industrial Crops and Products
International Biodeterioration
Photosynthesis Research
Renewable Energy
Resources, Conservation and Recycling
Water Services
Wood Science and Technology

ENGINEERING

Aquaculture Engineering
Bioprocess Engineering
Biotechnology Advances
Biotechnology and Bioengineering
Biotechnology Letters
Journal of Agricultural Engineering Research

ENTOMOLOGY

Annals of Applied Biology
Annual Reviews in Entomology
International Journal of Insect Morphology and Embryology
Journal of Economic Entomology
Review of Agricultural Entomology
Review of Medical and Veterinary Entomology

ENVIRONMENT AND ECOLOGY

Agriculture, Ecosystems & Environment
Annals of Human Biology
Annual Reviews in Biodiversity
Applied and Environmental Microbiology
Aquatic Conservation
Aquatic Toxicology
Arboricultural Journal
Biochemical Systematics and Ecology
Biodiversity and Conservation
Biofouling
Biological Conservation
Biological Mass Spectrometry
Biology Digest
Biology and Fertility of Soils
Biomass & Bioenergy
British Phycological Journal
Comptes Rendus de l'Academie des Sciences
CRC Critical Reviews
Critical Reviews in Plant Sciences
Crop Science
Ecotoxicology
Electroanalysis
Environmental Conservation
Environmental and Molecular Mutagenesis
Environmental Pollution
Environmental Science & Technology
Environmental Toxicology and Water Quality
Evolutionary Ecology
FEMS Microbiology Ecology
Genetica
Geomicrobiology Journal
Indian Journal of Microbiology
Journal of Animal Ecology
Journal of Applied Toxicology
Journal of Chemical Technology and Biotechnology

Journal of Economic Entomology
Journal of Environmental Management
Journal of Environmental Pathology, Toxicology, and Oncology
Journal of Experimental Marine Biology and Ecology
Journal of Industrial Microbiology
Journal of Planar Chromatography - Modern TLC
Journal of Soil Contamination
Land Management and Environmental Law Report
Marine Environmental Research
The Mycologist
Pesticide Science
Photosynthesis Research
Plant Cell and Environment
Plant Physiology and Biochemistry
Pollution Prevention
Protoplasma
Quantitative Structure-Activity Relationships
Renewable Energy
Resources, Conservation and Recycling
Reviews in Fish Biology and Fisheries
Revue d'Ecologie et de Biologie du Sol
Theriogenology - An International Journal of Animal Reproduction
Transgenic Research
Trees - Structure and Function
Waste Management Today
Water Air and Soil Pollution
Water Pollution Control Federation Journal
Zeitschrift fur Pflanzenernahrung und Bodenkunde
Zentralblatt fur Mikrobiology

ENZYMOLOGY

Analytical Biochemistry
Applied Biochemistry and Biotechnology
Archives of Biochemistry and Biophysics
Biocatalysis
Biochemistry
Biochimica et Biophysica Acta
Biochimie
Biomass & Bioenergy
Biotechnology and Bioengineering
Biotechnology Letters
Clinical Chemistry and Enzymology Communications
Enzyme
Enzyme and Fermentation Biotechnology
Enzyme and Microbial Technology: Biotechnology Research and Reviews
European Journal of Biochemistry
FEBS Letters
Food Microbiology
Journal of Bacteriology
Journal of Biological Chemistry
Journal of Chemical Technology and Biotechnology
Journal of Enzyme Inhibition
Journal of Microbial Biotechnology
Journal of Molecular Catalysis
Journal of Plant Physiology
Molecular and Cellular Biochemistry
Nature
Nucleic Acids Research
Plant Physiology
Plant Science
Plant Science Letters
Topics in Enzyme and Fermentation Biotechnology
ZFL

FERMENTATION

Applied Biochemistry and Biotechnology
BioEngineering - Forschung & Praxis
Biomass & Bioenergy
Bioprocess Engineering
Bioresource Technology
Bioscience, Biotechnology and Biochemistry
Enzyme and Fermentation Biotechnology
Food Microbiology
Journal of Chemical Technology and Biotechnology
Journal of Fermentation Technology
Journal of Industrial Microbiology
Topics in Enzyme and Fermentation Biotechnology
ZFL

FOOD AND NUTRITION

Agro-Industry Hi-Tech
American Society for Horticultural Science Journal
Applied Agricultural Research
Applied and Environmental Microbiology
Aquaculture
The Biochemical Engineering Journal
Bioresource Technology
Bioscience, Biotechnology and Biochemistry
Bio/Technology
Carbohydrate Polymers
Catering and Health
CRC Critical Reviews
Critical Reviews in Food Science and Nutrition
Electroanalysis
Electrophoresis
Flavour and Fragrance Journal
Food and Chemical Toxicology
Food Chemistry
Food Control
Food Microbiology
Food Research Quarterly
Food Sciences and Nutrition
Food Science and Technology Today
IMPACT AgBioIndustry
Intern Zeitschrift fur Lebensmittle Technologie und Verfahrentechnik
International Journal of Food Microbiology
The Journal of Nutritional Biochemistry

Journal of the Science of Food and Agriculture
Phytochemical Analysis
Process Biochemistry
RIA (Revue de l'Industrie Agro-Alimentaire)
Starch
Zeitschrift fur Pflanzenernahrung und Bodenkunde
ZFL

FORESTRY

Agroforestry Today
Arboricultural Journal
Biology and Fertility of Soils
Trees - Structure and Function

GENETICS

American Society for Horticultural Science Journal
Animal Genetics
Animal Reproduction Science
Annals of Human Biology
Annual Reviews
Annual Reviews in Genetics
Applied Biochemistry and Biotechnology
Archives of Biochemistry and Biophysics
Australian Journal of Biological Sciences
Autoimmunity
Biochemical Genetics
Biochimica et Biophysica Acta
Biochimica et Biophysica Acta - Gene Structure and Expression
Biochimie
BioEngineering - Forschung & Praxis
Bioscience Reports
Biotechnology Advances
Biotechnology and Genetic Engineering Reviews
Biotechnology Letters
Biotechnology Progress
Biotechnology Therapeutics
Carlsberg Research Communications
Cellular Biochemistry
Chromosoma
Comptes Rendus de l'Academie des Sciences
Critical Reviews in Eukaryotic Gene Expression
Critical Reviews in Plant Sciences
Crop Science
Current Genetics
Current Opinion in Cell Biology
Current Opinion in Genetics & Development
Developmental Genetics
DNA A Journal of Molecular Biology
DNA Probes
DNA and Protein Engineering Techniques
DNA Sequence
Electrophoresis
Environmental and Molecular Mutagenesis
European Journal of Biochemistry
European Journal of Immunology
Evolutionary Ecology
Experimental and Clinical Immunogenetics
FEBS Letters
Gene
Genes, Chromosomes and Cancer
Genes and Development
The Genetic Engineer and Biotechnologist
Genetic Engineering
Genetica
Genetical Research
Genetics
Genome
Genomics
Human Genetics
Human Molecular Genetics
Immunogenetics
Indian Journal of Genetics and Plant Breeding
International Journal of Bioethics
Journal of Bacteriology
Journal of Biological Chemistry
Journal of Biotechnology
Journal of Cell Biology
Journal of Experimental Marine Biology and Ecology
Journal of Horticultural Science
Journal of Immunogenetics
Journal of Microbial Biotechnology
Journal of Molecular and Applied Genetics
Journal of Molecular Biology
Journal of Virology
Mammalian Genome
Molecular and General Genetics
Molecular Microbiology
Mutation Research
The Mycologist
Nature
Nature Genetics
New Scientist
Pharmacogenetics
Photosynthesis Research
Plant, Cell, Tissue and Organ Culture
Plant Molecular Biology
Plant Physiology and Biochemistry
Plant Science
Plant Science Letters
Planta
Plasmid
Proceedings of the Society for Experimental Biology and Medicine
Proteins: Structure, Function, and Genetics
Receptors and Channels
Science
Scientific American
Somatic Cell and Molecular Genetics
Soviet Biotechnology (Biotekhnologiya)
Theoretical and Applied Genetics
Transgenic Research
Trends in Genetics

HUMAN BIOLOGY

American Journal of Human Biology
Annals of Human Biology
Applied Agricultural Research
Cellular and Molecular Biology
Current Advances In Neuroscience
Current Opinion in Neurobiology
Environmental Conservation
The European Journal of Neuroscience
Genomics
Human Genetics
Human Molecular Genetics
International Journal of Bioethics
Journal of Environmental Pathology, Toxicology, and Oncology
Journal of Experimental Biology
Journal of Neurocytology
The Journal of Nutritional Biochemistry
M/S Medicine Sciences
Nature Genetics
Proceedings of the National Academy of Sciences
Regulatory Peptides
Reproduction Nutrition Development
Toxicology
Toxicology and Applied Pharmacology
Toxicology Letters

HUMAN HEALTH CARE

Acta Pathologica et Immunologica Scandinavica
Advances in Neuroimmunology
AIDS Research and Human Retroviruses
AIDS Research Reviews
Annals of Clinical Biochemistry
Annual Reviews
Annual Reviews in Neuroscience
Antibiotics and Chemotherapy
Antimicrobial Agents and Chemotherapy
Antiviral Research
Arzneimittel-Forschung/Drug Research
Autoimmunity
The Biochemical Engineering Journal
Biochemical Genetics
Biochemical Medicine
Biochemistry
Bioelectromagnetics
Biological Mass Spectrometry
Biological Membranes
Biomedical Chromatography
Bioorganic and Medicinal Chemistry Letters
BioPharm
Biopharmaceutics and Drug Disposition
Bioscience Reports
Bio/Technology
Biotechnology Progress
Biotechnology Therapeutics
British Medical Journal
Cancer Drug Delivery
Cancer Research
Carbohydrate Polymers
Catering and Health
Cell Biochemistry and Function
Cellular Biochemistry
Cellular Immunology
Cellular and Molecular Biology
Chemico-Biological Interactions
Chinese Journal of Physiological Sciences
Clinical Biochemistry
Clinical Chemistry
Clinical Chemistry and Enzymology Communications
Clinical and Experimental Immunology
Clinical Microbiology Reviews
Clinical Science
CRC Critical Reviews
Critical Reviews in Biotechnology
Critical Reviews in Immunology
Current Advances In Neuroscience
Current Microbiology
Current Opinion in Genetics & Development
Current Opinion in Immunology
Current Opinion in Neurobiology
Cytokine
Diagnostic Immunology
Diagnostic Microbiology and Infectious Disease
Electroanalysis
Electrophoresis
Environmental and Molecular Mutagenesis
European Journal of Immunology
The European Journal of Neuroscience
Experimental and Clinical Immunogenetics
Genes, Chromosomes and Cancer
Immunogenetics
Immunology
Indian Journal of Microbiology
Infection and Immunity
Interferon Y Biotecnologia
International Journal of Biochemistry
International Journal of Clinical Monitoring and Computing
International Journal of Radiation Biology
International Research Communications Systems
Journal of the American Medical Association
Journal of Antimicrobial Chemotherapy
Journal of Biological Response Modifiers
Journal of Clinical Immunology
Journal of Clinical Microbiology
Journal of Experimental Pathology
Journal of Immunogenetics
Journal of Immunology
Journal of Immunopharmacology
Journal of Interferon Research
Journal of Labelled Compounds and Radiopharmaceuticals
Journal of Leukocyte Biology
Journal of Medical Microbiology
Journal of Medical Virology
Journal of Membrane Science
Journal of Molecular and Applied Genetics

Journal of Molecular Recognition
Journal of Neurocytology
Journal of Planar Chromatography - Modern TLC
The Journal of Steroid Biochemistry and Molecular Biology
The Journal of Trace Elements in Experimental Medicine
La Recherche
Life Sciences
Microbial Pathogenesis
Micron and Microscopica Acta
Molecular Biotherapy
Molecular and Cellular Probes
M/S Medicine Sciences
Mycopathologia
Nature Genetics
Oncogene
Pesticide Outlook
Pharmacogenetics
Proceedings of the National Academy of Sciences
Proceedings of the Society for Experimental Biology and Medicine
Quantitative Structure-Activity Relationships
Receptors and Channels
Review of Medical and Veterinary Entomology
Review of Medical and Veterinary Mycology
Soviet Biotechnology (Biotekhnologiya)
Spectra Biologie
Vaccine
Viral Immunology

IMMUNOLOGY

Acta Pathologica et Immunologica Scandinavica
Advances in Neuroimmunology
AIDS Research and Human Retroviruses
AIDS Research Reviews
Analytical Biochemistry
Annual Reviews in Immunology
Arzneimittel-Forschung/Drug Research
Autoimmunity
Biochimie
Biological Membranes
Bioscience Reports
Cancer Research
Cellular Immunology
Clinical and Experimental Immunology
Clinical Microbiology Reviews
Clinical Science
Critical Reviews in Immunology
Current Advances in Immunology
Current Opinion in Immunology
Diagnostic Immunology
Electrophoresis
European Journal of Biochemistry
European Journal of Immunology
Experimental and Clinical Immunogenetics
FEBS Letters
The Histochemical Journal
Immunogenetics
Immunology
Immunology Letters
Immunology Today
Infection and Immunity
International Immunology
International Research Communications Systems
International Reviews of Immunology
Journal of Cell Biology
Journal of Clinical Immunology
Journal of Immunogenetics
Journal of Immunological Methods
Journal of Immunology
Journal of Immunopharmacology
Journal of Medical Microbiology
Journal of Molecular and Cellular Immunology
Journal of Virology
Medical Microbiology and Immunology
Microbiology and Immunology
Molecular Immunology
Nature
Proceedings of the National Academy of Sciences
Proceedings of the Society for Experimental Biology and Medicine
Receptors and Channels
Scandinavian Journal of Immunology
Soviet Biotechnology (Biotekhnologiya)
Veterinary Immunology and Immunopathology
Viral Immunology

INDUSTRY

Agro-Industry Hi-Tech
Australian Journal of Biotechnology
The Biochemical Engineering Journal
Biofouling
Biology and Fertility of Soils
BIOS-Conditionnement Brasserie-Malterie
Biotechnology Chemonomics
Carbohydrate Polymers
Chemistry and Industry
Chimica oggi
Critical Reviews in Biotechnology
Electroanalysis
Food Control
Food Science and Technology Today
The Genetic Engineer and Biotechnologist
IMPACT AgBioIndustry
Industrial Crops and Products
Intern Zeitschrift fur Lebensmittle Technologie und Verfahrentechnik
Journal of Applied Toxicology
Journal of Industrial Microbiology
Mycopathologia
Process Biochemistry
RIA (Revue de l'Industrie Agro-Alimentaire)

Water Pollution Control Federation Journal
Water Services
Wood Science and Technology
ZFL

LEGISLATION AND REGULATORY ISSUES

BioEngineering - Forschung & Praxis
Environmental Toxicology and Water Quality
Food Control
IMPACT AgBioIndustry
Journal of Soil Contamination
Land Management and Environmental Law Report
Waste Management Today

MICROBIOLOGY

Acta Pathologica et Immunologica Scandinavica
Annual Reviews in Microbiology
Annual Reviews in Plant Physiology and Plant Microbial Biology
Applied Biochemistry and Biotechnology
Applied Biochemistry and Microbiology
Applied and Environmental Microbiology
Archives of Microbiology
Australian Journal of Biological Sciences
Biochimie
Biodegradation
BioEngineering - Forschung & Praxis
Biofouling
Biomass & Bioenergy
Biorecovery
Bioscience, Biotechnology and Biochemistry
Biotechnology Letters
Carlsberg Research Communications
Clinical Microbiology Reviews
CRC Critical Reviews
Critical Reviews in Microbiology
Current Advances in Microbiology
Current Microbiology
Diagnostic Microbiology and Infectious Disease
Electrophoresis
Enzyme and Microbial Technology: Biotechnology Research and Reviews
FEMS Microbiology Ecology
FEMS Microbiology Letters
Folia Microbiologica
Food Microbiology
Geomicrobiology Journal
Indian Journal of Microbiology
International Journal of Food Microbiology
Journal of Antibiotics
Journal of Antimicrobial Chemotherapy
Journal of Applied Bacteriology
Journal of Bacteriology
Journal of Bioluminescence and Chemiluminescence
Journal of Biotechnology
Journal of Cell Science
Journal of Chemical Technology and Biotechnology
Journal of Clinical Microbiology
Journal of Economic Entomology
Journal of General and Applied Microbiology
Journal of General Microbiology
Journal of Industrial Microbiology
Journal of Medical Microbiology
Journal of Microbial Biotechnology
Journal of Microbiological Methods
Medical Microbiology and Immunology
Microbial Pathogenesis
Microbial Releases
Microbiological Reviews
Microbiology
Microbiology and Immunology
MIRCEN
Molecular and Cellular Biology
Molecular Microbiology
Physiological and Molecular Plant Pathology
Proceedings of the National Academy of Sciences
Proceedings of the Society for Experimental Biology and Medicine
Review of Medical and Veterinary Mycology
Soviet Biotechnology (Biotekhnologiya)
Systema Ascomycetum
Veterinary Microbiology
Wood Science and Technology
Yeast
Zentralblatt fur Mikrobiology

MODELLING

Biological Membranes
Biophysical Chemistry
Biopolymers
Bioprocess Engineering
Biotechnology Progress
Journal of Environmental Management
Journal of Molecular Graphics
Journal of Soil Contamination
Quantitative Structure-Activity Relationships
Receptors and Channels
Waste Management Today

PATENTS

Current Opinion in Biotechnology
Enzyme and Microbial Technology: Biotechnology Research and Reviews
The Genetic Engineer and Biotechnologist

PESTICIDES

American Society for Horticultural Science Journal
Annals of Applied Biology

Biology and Fertility of Soils
Bioscience, Biotechnology and Biochemistry
Crop Protection
Farm Chemicals International
Journal of Economic Entomology
Pesticide Biochemistry and Physiology
Pesticide Outlook
Pesticide Science
Phytopathology
Plant Disease
Plant Pathology
Plant Physiology and Biochemistry

PLANT SCIENCE

Agronomy Journal
American Society for Horticultural Science Journal
Annals of Applied Biology
Annual Reviews
Annual Reviews in Plant Physiology and Plant Microbial Biology
Antiviral Research
Archives of Biochemistry and Biophysics
Biological Agriculture and Horticulture
Botanica Acta
Canadian Journal of Plant Pathology
Canadian Journal of Plant Science
Carlsberg Research Communications
Cellular and Molecular Biology
Communications in Soil Science
Comptes Rendus de l'Academie des Sciences
CRC Critical Reviews
Critical Reviews in Plant Sciences
Crop Protection
Crop Science
Euphytica
Indian Journal of Genetics and Plant Breeding
Journal of Biotechnology
Journal of Experimental Botany
Journal of Horticultural Science
Journal of Molecular Catalysis
Journal of Plant Physiology
Mycopathologia
New Phytologist
The New Phytologist
Photosynthesis Research
Physiologia Plantarum
Physiological and Molecular Plant Pathology
Phytochemical Analysis
Phytochemistry
Phytopathology
Plant Cell and Environment
Plant, Cell, Tissue and Organ Culture
Plant Disease
Plant Growth Regulation
The Plant Journal
Plant Molecular Biology
Plant Pathology
Plant Physiology
Plant Physiology and Biochemistry
Plant Science
Plant Science Letters
Planta
Plants Today
Protoplasma
Seed Science Research
Seed Science and Technology
Trees - Structure and Function
Zeitschrift fur Pflanzenernahrung und Bodenkunde

VIROLOGY

AIDS Research and Human Retroviruses
Antiviral Research
Archives of Biochemistry and Biophysics Biochemistry
Cancer Research
Critical Reviews in Microbiology
Current Opinion in Genetics & Development
Journal of Biotechnology
Journal of Cell Science
Journal of General Microbiology
Journal of General Virology
Journal of Medical Microbiology
Journal of Medical Virology
Journal of Molecular Biology
Journal of Virological Methods
Journal of Virology
Medical Microbiology and Immunology
Microbial Pathogenesis
Microbiology and Immunology
Physiological and Molecular Plant Pathology
Plant Science Letters
Veterinary Microbiology
Viral Immunology
Virology
Zentralblatt fur Mikrobiology

WASTE MANAGEMENT

The Biochemical Engineering Journal
Biocycle
Biodegradation
Biology and Fertility of Soils
Biomass & Bioenergy
Biorecovery
Bioresource Technology
Bio/Technology
Environmental Conservation
International Biodeterioration
Journal of Industrial Microbiology
Process Biochemistry
Resources, Conservation and Recycling
Soviet Biotechnology (Biotekhnologiya)
Waste Management Today
Water Research
Water Science Technology
Water Services
ZFL

NEWSLETTERS

AGRICULTURE

AFRC News
AgBiotech News and Information
Agribusiness Intelligence
Agricell Report
Agricultural Genetics Report
Agrow
Alcohol Update
Alcohol Week
Animal Pharm
Applied Genetics News
Biocontrol News and Information
Bioenergi
Biofutur FLASH
Biojoule
Biotech News
Brewing and Distilling International
The Digest
ICRAF Newsletter
IEA Bioenergy Newsletter
LINK - Crops for Industrial Use
Network News
REVIEW
Weekly Bulletin
Weekly Digest

ANALYSIS, ANALYTICAL SYSTEMS AND DIAGNOSTICS

Biologics and Biotech Regulatory Report
BioPharm
Biotech Market News and Strategies
The Biotechnology Instrument Market
Biotechnology Software
Clinica
Frontiers in Immunoassay and Biotechnology
New Biotech
Radioassay-Ligand Assay News

ANIMAL HEALTH AND NUTRITION

Animal Pharm

BIOCHEMISTRY

Biobulletin
Bio Sciences
Biotechnology Software

BIOLOGY

Bulletin of the Society for Experimental Biology

BIOTECHNOLOGY

ABA Bulletin
AgBiotech News and Information
Agribusiness Intelligence
BIA Bulletin
Biobulletin
BioEngineering News
Biofutur FLASH
BioGen
Bio: La Lettre Des Biotechnologies
Biologics and Biotech Regulatory Report
BioPharm
Bioscan
Bio Sciences
Biotech Knowledge Sources
Biotech Market News and Strategies
Biotech News
Biotechnologia
BioTechnologie
Biotechnology Bulletin
Biotechnology Business News
Biotechnology Information News
Biotechnology Insight
The Biotechnology Instrument Market
Biotechnology in Japan Newservice
Biotechnology Law Report
Biotechnology News
Biotechnology Newswatch
Biotechnology Software
Biotechnology Techniques
BioVenture View

BioWorld Today
Canadian Biotech News
Changing Medical Markets
Details
DNA and Protein Engineering Techniques
EBIS newsletter
EFB Newsletter
European Biotechnology Newsletter
Frontiers in Immunoassay and Biotechnology
Genetic Engineering and Biotechnology Monitor
IBSA BioNews
Irish Biotechnology News
The Japan Biotech Letter
Japan Biotech and Medical News
Japan Report: Biotechnology
Japanese Biotechnology Industry
Japanscan
Monoclonal Antibody News
New Bioresources
New Biotech
New Biotech Busine$$
Nostrum
Pharmaprojects
Practical Biotechnology
Recombinant DNA Technical Bulletin
Swiss Biotech
UK Monoclonal Bulletin

BREWING/DISTILLING

Biotech News
Brewing and Distilling International

CHEMISTRY AND CHEMICALS

Agrow
Animal Pharm
Bioprocessing Technology
Biotech News
The Government Chemist
Hydro Oakland Research Bulletins
Japan Chemical Week
Pharmaceutical Business News
Pharmaprojects
US Regulatory Reporter

COMPANIES

ABA Bulletin
Agrow
BIA Bulletin
BioGen
Bioscan
Biotechnologia
BioTechnologie
Biotechnology Bulletin
Biotechnology Business News
Biotechnology Information News
Biotechnology News
Biotechnology Newswatch
BioWorld Today
Brewing and Distilling International
Details
The Digest
ENDS
European Clinical Laboratory
Irish Biotechnology News
Japan Report: Biotechnology
Japanese Biotechnology Industry
New Biotech Busine$$
Nostrum
Pharmaprojects
UK Monoclonal Bulletin

COMPUTER PRODUCTS

Biotech Knowledge Sources
Biotechnology Software

ECONOMICS AND FINANCE

Agrow
Alcohol Update
Alcohol Week
Applied Genetics News
BioEngineering News
Biotechnology Business News
Biotechnology Insight
BioWorld Today
European Biotechnology Newsletter
Microbial Clean-up
New Biotech Busine$$
Pharmaceutical Business News

ENERGY

Agribusiness Intelligence
Alcohol Update
Alcohol Week
Bioenergi
Biofutur FLASH
Biojoule
Biomass Bulletin
Bioprocessing Technology
The Digest
IEA Bioenergy Newsletter
Network News
Renewable Energy Bulletin
Renewable Energy News
REVIEW
Soft Energy Worldwide
UNITAR/UNDP Newsletter
Warmer Bulletin

ENGINEERING

BioEngineering News
Bio Sciences
Hydro Oakland Research Bulletins

ENTOMOLOGY

Biocontrol News and Information

ENVIRONMENT AND ECOLOGY

ENZYMOLOGY

FERMENTATION

FOOD AND NUTRITION

FORESTRY

GENETICS

HUMAN HEALTH CARE

IMMUNOLOGY

INDUSTRY

INFORMATION SOURCES AND SERVICES

LEGISLATION AND REGULATORY ISSUES

Agrow
Alcohol Update
Alcohol Week
At the CENTRE
BIA Bulletin
BioEngineering News
Biologics and Biotech Regulatory Report
Biotechnology Business News
Biotechnology Law Report
BioWorld Today
EBIS newsletter
ENDS
Environmental Impact
Genetic Engineering and Biotechnology Monitor
Japanscan
Pharmaceutical Business News
Resource Recovery Report
REVIEW
Soft Energy Worldwide
US Regulatory Reporter
Water Quality International

MARKETS AND MARKETING

Agrow
BioGen
Biotech Knowledge Sources
Biotech Market News and Strategies
BioTechnologie
Biotechnology Bulletin
Biotechnology Business News
Biotechnology Insight
BioWorld Today
European Biotechnology Newsletter
Japan Chemical Week
New Biotech Busine$$

MICROBIOLOGY

Biobulletin
Bio Sciences
Biotechnology Software
Microbial Clean-up
Microbial Genetics Bulletin
Society for General Microbiology Quarterly

MODELLING

Biotechnology Software

PATENTS

Agricell Report
BioEngineering News
Bioinvention
Biotechnologia
Biotechnology Law Report
Biotechnology News
D-J-M Enzyme Report
Patents Information News

PESTICIDES

Agrow
Biocontrol News and Information

PLANT SCIENCE

Biobulletin
FESPP Newsletter

PRODUCTS

ABA Bulletin
Alcohol Update
Alcohol Week
BioGen
Bioscan
Biotech Market News and Strategies
Biotechnologia
BioTechnologie
Biotechnology Business News
The Biotechnology Instrument Market
BioWorld Today
Brewing and Distilling International
European Clinical Laboratory
Japanese Biotechnology Industry
Japanscan
New Biotech
Pharmaceutical Business News
Pharmaprojects
Water Quality International

WASTE MANAGEMENT

Biomass Bulletin
The Digest
Environment Business
Environmental Impact
Hazardous Substances
IEA Bioenergy Newsletter
Industrial Waste Management
Microbial Clean-up
Recycling and Resource Management
Resource Recovery Report
REVIEW
Warmer Bulletin
Water Quality International

DIRECTORIES AND REFERENCE BOOKS

Biotechnology
Chemistry and pharmaceuticals
Culture collections
Energy
Environment and health
Food and agriculture
Funding
Information and education
Regulations and patents

State-by-State Biotechnology Directory, Centres, Companies and Contacts
The UK Biotechnology Handbook

SCIENCE AND TECHNOLOGY

The Academic File - Biosciences
Directory of British Associations - DBA
Directory of EC Industry Information Sources
The UK Biotechnology Handbook

TRADE AND INDUSTRY

Directory of British Associations - DBA
Directory of EC Industry Information Sources
A Guide to the British Food Manufacturing Industry - 3rd Edition
Northern Ireland Trade Directory, Marketing Development Division

WASTE MANAGEMENT

Polmark Directory
The Sitefile Digest 1991 Edition
The UK Biotechnology Handbook

DATABASES AND ABSTRACTS

Abstracts
Agriculture
Analysis, analytical supplies and diagnostics
Animal health and nutrition
Aquaculture
Biochemistry
Biology
Biophysics
Biotechnology
Chemistry
Companies
Computer products
Culture collections
Databases and on-line services
Economics and finance
Energy
Engineering
Environment
Enzymology
Fermentation
Food, feeds and nutrition
Forestry
Genetics
Human biology
Human health care
Immunology
Industry
Legislation and regulatory issues
Markets and marketing
Microbiology
Patents
Pesticides
Plant science
Products and equipment
Research programmes and projects
Science and technology
Virology
Waste management and treatment

ABSTRACTS

Abstracts in BioCommerce - ABC
Abstracts of Mycology
AGDEX
Agricultural Engineering Abstracts
Agroforestry Abstracts
Analytical Abstracts
Animal Breeding Abstracts
Bibliography of Agriculture
Biochemistry Abstracts
BioCommerce Abstracts and Directory Online
Biodeterioration Abstracts
Biological Abstracts
BIOSIS Previews
BIOSIS/CAS Selects
Biotech Knowledge Sources - BKS
Biotechnology Abstracts
Biotechnology Japanese Patent Alert
Biotechnology Patent Fast-Alert
Biotechnology Research Abstracts
The British Library Biotechnology Information Service
The British Library Science Reference and Information Service - SRIS
CA Search
CA Selects
CAB Abstracts
CABS - Current Awareness in Biological Sciences
CABS Database Online
CAS Biotech Updates
CAS ONLINE
CASORF
CBA Database
CEA Database
Chemical Abstracts
Chemical Engineering Abstracts
Chemical Engineering and Biotechnology Abstracts
CHEMTRACTS - Biochemistry and Molecular Biology
CHEMTRACTS - Inorganic Chemistry
CHEMTRACTS - Organic Chemistry
Chromatography Abstracts
CNIC - Centre National de l'Information Chimique
COMPENDEX
CORDIS
Crop Physiology Abstracts
CSA Biochemistry Abstracts
Current Advances in Genetics & Molecular Biology
Current Advances in Microbiology
Current Biotechnology Abstracts

Current Clinical Chemistry
Current Contents
Current Contents on Diskette - Life Sciences
Dairy Science Abstracts
DECHEMA - Dechema Chemical Engineering and Biotechnology Abstracts Database
DRHPL - Database for the Repository of Human Probes and Libraries
EABS
EUREKA
Excerpta Medica
Field Crop Abstracts
Focus On: Global Change
Food Science and Technology Abstracts
Forest Products Abstracts
Forestry Abstracts
GENBANK
Genetics Abstracts
Helminthological Abstracts
Herbage Abstracts
Horticultural Abstracts
Immunology Abstracts
Index Medicus
International Pharmaceutical Abstracts
Irrigation and Drainage Abstracts
Japan Technology
JICST-E
Maize Abstracts
MEDLINE
MiCIS - Microbial Culture Information Service
MiCIS - Information Centre for European Culture Collections
Microbiology Abstracts
Nematological Abstracts
NEWAT
NTIS Bibliographic Database
Nutrition Abstracts and Reviews, Series A - Human and Experimental
Nutrition Abstracts and Reviews, Series B - Livestock Feeds and Feeding
Oncology Abstracts
Packaging Science and Technology Abstracts
PC-ABC
PHYTOMED
Plant Biotechnology
Plant Breeding Abstracts
Plant Genetic Resources Abstracts
Plant Growth Regulator Abstracts
Potato Abstracts
Poultry Abstracts
PRF/LITDB
Protozoological Abstracts
PTG - Protein Translation
Review of Medical and Veterinary Mycology
Review of Plant Pathology
Rice Abstracts
Seed Abstracts
SEQANALREF
SIGPEP
Sorghum and Millets Abstracts
Soyabean Abstracts
Sugar Industry Abstracts
TELEGEN
TOXLINE
TOXNET
TRF - Taxonomic Reference File
TRNAC - tRNA Compilation
VECTOR - Cloning Vector Sequence Database
Virology and AIDS
Weed Abstracts
Wheat, Barley and Triticale Abstracts
World Agricultural Economics and Rural Sociology Abstracts

AGRICULTURE

Abstracts in BioCommerce - ABC
Abstracts of Mycology
AGDEX
AGREP - Agricultural Research Projects
Agricultural Engineering Abstracts
Agroforestry Abstracts
Agrow OnScreen
Analytical Abstracts
Animal Breeding Abstracts
Bibliography of Agriculture
BioCommerce Abstracts and Directory Online
BIOSIS Previews
Biotechnology Citation Index
Biotechnology Research Abstracts
The British Library Science Reference and Information Service - SRIS
CAS Biotech Updates
CAS ONLINE
CFISM
CNIC - Centre National de l'Information Chimique
Crop Physiology Abstracts
Current Contents
Current Contents on Diskette - Life Sciences
Dairy Science Abstracts
The Farmstat System
Field Crop Abstracts
Focus On: Global Change
Herbage Abstracts
Horticultural Abstracts
Irrigation and Drainage Abstracts
Maize Abstracts
Nematological Abstracts
NTIS Bibliographic Database
Nutrition Abstracts and Reviews, Series B - Livestock Feeds and Feeding
PC-ABC
PESTDOC - Pest Control Literature Documentation
PESTMAN - Pesticide Databank
Pharmacontacts Directories File
Plant Growth Regulator Abstracts

Potato Abstracts
Poultry Abstracts
Review of Medical and Veterinary Mycology
Review of Plant Pathology
Rice Abstracts
Seed Abstracts
Sorghum and Millets Abstracts
Soyabean Abstracts
Sugar Industry Abstracts
Sugar Statistical Compendium
Virology and AIDS
Weed Abstracts
Wheat, Barley and Triticale Abstracts
World Agricultural Economics and Rural Sociology Abstracts

ANALYSIS, ANALYTICAL SUPPLIES AND DIAGNOSTICS

Abstracts in BioCommerce - ABC
Analytical Abstracts
BioCommerce Abstracts and Directory Online
CA Selects
Chromatography Abstracts
COMPENDEX
DEQUIP
Diagnostics Business
EPA's Sampling and Analysis Methods Database
Oncology Abstracts
PC-ABC
PESTMAN - Pesticide Databank
QTDGPD-2D Gel Protein Database

ANIMAL HEALTH AND NUTRITION

AGREP - Agricultural Research Projects
Biodeterioration Abstracts
Genetics Abstracts
Helminthological Abstracts
Immunology Abstracts
Microbiological Resource Databank
Nutrition Abstracts and Reviews, Series B - Livestock Feeds and Feeding
Pharmacontacts Directories File
Protozoological Abstracts
Review of Medical and Veterinary Mycology
Virology and AIDS

AQUACULTURE

AGREP - Agricultural Research Projects
Biotechnology Research Abstracts
Review of Medical and Veterinary Mycology

BIOCHEMISTRY

Analytical Abstracts
Biochemistry Abstracts
BIOSIS Previews
BIOSIS/CAS Selects
Biotechnology Research Abstracts
CABS Database Online
CARBBANK
CAS ONLINE
Chemical Abstracts
CHEMTRACTS - Biochemistry and Molecular Biology
COMPENDEX
CSA Biochemistry Abstracts
Current Advances in Genetics & Molecular Biology
Current Awareness in Biological Sciences
Current Biotechnology Abstracts
Dnastar Inc
Dnastar Ltd
European Molecular Biology Laboratory - EMBL
LiMB Database
PESTDOC - Pest Control Literature Documentation
QTDGPD-2D Gel Protein Database

BIOLOGY

Abstracts in BioCommerce - ABC
Biochemistry Abstracts
BioCommerce Abstracts and Directory Online
Biological Abstracts
BIOSIS Previews
CABS Database Online
COMPENDEX
Current Awareness in Biological Sciences
Current Contents
Current Contents on Diskette - Life Sciences
GENBANK
NTIS Bibliographic Database
PC-ABC
TRF - Taxonomic Reference File

BIOPHYSICS

COMPENDEX

BIOTECHNOLOGY

AANSPII
Abstracts in BioCommerce - ABC
BIKE - Biotechnology Information Knot for Europe GBF mbH
BioCommerce Abstracts and Directory Online
BIOREP
BIOSIS/CAS Selects
Biotech Knowledge Sources - BKS
Biotechnology Abstracts
Biotechnology Citation Index
Biotechnology Japanese Patent Alert
Biotechnology Patent Fast-Alert
Biotechnology Research Abstracts

CHEMISTRY

COMPANIES

COMPUTER PRODUCTS

CULTURE COLLECTIONS

MiCIS - Microbial Culture Information Service
MiCIS - Information Centre for European Culture Collections
Microbial Strain Data Network - MSDN
Microbiological Resource Databank
MINE - Microbial Information Network in Europe
MINE - Microbial Information Network in Europe, French Office
NCFB Catalogue
Tropical Data Base
World Data Centre on Microorganisms

DATABASES AMD ON-LINE SERVICES

AANSPII
Abstracts in BioCommerce - ABC
AGDEX
AGREP - Agricultural Research Projects
Agricultural Engineering Abstracts
Agroforestry Abstracts
Agrow OnScreen
American Type Culture Collection - ATCC
The Applied Environmetrics Meteorological Tables
BDT/FTPT
Beilstein
BIKE - Biotechnology Information Knot for Europe GBF mbH
BioCommerce Abstracts and Directory Online
Biodeterioration Abstracts
BIOREP
BIOSIS Previews
Biotech Knowledge Sources - BKS
Biotechnology Citation Index
The British Library Biotechnology Information Service
The British Library Science Reference and Information Service - SRIS
CA Search
CAB Abstracts
CABS Database Online
CAS ONLINE
CBA Database
CCD - Cambridge Structural Database
CEA Database
CFISM
COMPASS - Computer Assisted Yeast Identification System
COMPENDEX
CORDIS
Crop Physiology Abstracts
Current Advances in Microbiology
Current Awareness in Biological Sciences
Current Contents
Current Contents on Diskette - Life Sciences
CUTG
Dairy Science Abstracts
DECHEMA - Dechema Chemical Engineering and Biotechnology Abstracts Database
DEQUIP
DIANEGUIDE
Dnastar Inc
Dnastar Ltd
DOMIS
DRHPL - Database for the Repository of Human Probes and Libraries
EABS
ENREP
ENZIDEX
EPA's Sampling and Analysis Methods Database
EUREKA
European Bank of Computer Programs in Biotechnology
European Molecular Biology Laboratory - EMBL
The Farmstat System
Field Crop Abstracts
Focus On: Global Change
Food Science and Technology Abstracts
Forest Products Abstracts
Forestry Abstracts
FROSTI - Food RA Online Scientific and Technical Information Database
GBSOFT - GenBank Software Clearing House
GENBANK
GeneSeq
Hazardous Chemical Regulations Database
Helminthological Abstracts
Herbage Abstracts
HGIR - Human Genome Information Resource
HGMCR - Human Genetic Mutant Cell Repository
HGML - Human Gene Mapping Library
Horticultural Abstracts
Hybridoma Data Bank
Hybridoma Database
ICDB - Immunoclone Database
ICGEBnet
IES-DC
Immunology Abstracts
In Vitro Conservation Database
Irrigation and Drainage Abstracts
Japan Technology
Lewis Publishers
LIFO-DISK - LIposome inFOrmation Electronic Liposome Library
LiMB Database
Maize Abstracts
MEDLINE
MiCIS - Microbial Culture Information Service
MiCIS - Information Centre for European Culture Collections
Microbial Strain Data Network - MSDN
Microbiological Resource Databank

MINE - Microbial Information Network in Europe
MINE - Microbial Information Network in Europe, French Office
MIRDAB - Microbiological Resource Databank
NADL - Nucleic Acid Data Library
The National Toxicology Program's Chemical Database
NCFB Catalogue
Nematological Abstracts
NEWAT
NTIS Bibliographic Database
Nutrition Abstracts and Reviews, Series A - Human and Experimental
Nutrition Abstracts and Reviews, Series B - Livestock Feeds and Feeding
Packaging Science and Technology Abstracts
PatentImages
PC-ABC
PESTDOC - Pest Control Literature Documentation
PESTMAN - Pesticide Databank
PG-TRANS
Pharmaceased Discontinued Drugs File
Pharmacontacts Directories File
Pharmacontacts OnScreen
PHIND - Pharmaceutical and Healthcare Industries News Database
PHYTOMED
Plant Genetic Resources Abstracts
Plant Growth Regulator Abstracts
Potato Abstracts
Poultry Abstracts
PPR - Plasmid Prefix Registry
PRF/LITDB
PRF/SEQDB
Protozoological Abstracts
PSEQIP
PTG - Protein Translation
QTDGPD-2D Gel Protein Database
RED - Restriction Enzyme Database
Rice Abstracts
RFLPD-RLFP Database
Seed Abstracts
SESAME
SIGPEP
Sorghum and Millets Abstracts
Soyabean Abstracts
SPF - Standard Pesticides File
SRRSD - 16S Ribosomal RNA Sequence Database
SRSRSC - Small Ribosomal Subunit RNA Sequence Compilation
Sugar Industry Abstracts
Sugar Statistical Compendium
SWISS-PROT
TELEGEN
TOXNET
TRF - Taxonomic Reference File
TRNAC - tRNA Compilation
Tropical Data Base
VECTOR - Cloning Vector Sequence Database
Weed Abstracts
Wheat, Barley and Triticale Abstracts
World Agricultural Economics and Rural Sociology Abstracts
World Data Centre on Microorganisms

ECONOMICS AND FINANCE

Agrow OnScreen
Biodeterioration Abstracts
Sugar Industry Abstracts
TELEGEN
World Agricultural Economics and Rural Sociology Abstracts

ENERGY

Biodeterioration Abstracts
CAS Biotech Updates
CNIC - Centre National de l'Information Chimique
EABS
SESAME

ENGINEERING

Agricultural Engineering Abstracts
The British Library Science Reference and Information Service - SRIS
CEA Database
Chemical Abstracts
Chemical Engineering Abstracts
Chemical Engineering and Biotechnology Abstracts
COMPENDEX
Current Contents
DECHEMA - Dechema Chemical Engineering and Biotechnology Abstracts Database
DEQUIP
NTIS Bibliographic Database

ENVIRONMENT

Abstracts of Mycology
Analytical Abstracts
The Applied Environmetrics Meteorological Tables
Biotechnology Citation Index
CABS Database Online
CAS Biotech Updates
CNIC - Centre National de l'Information Chimique
Current Contents
Current Contents on Diskette - Life Sciences
DECHEMA - Dechema Chemical Engineering and Biotechnology Abstracts Database
EABS
ENREP

EPA's Sampling and Analysis Methods Database
Focus On: Global Change
Hazardous Chemical Regulations Database
PHYTOMED
Sugar Industry Abstracts

ENZYMOLOGY

CHEMTRACTS - Biochemistry and Molecular Biology
CHEMTRACTS - Organic Chemistry
Dnastar Inc
ENZIDEX
MiCIS - Microbial Culture Information Service
MiCIS - Information Centre for European Culture Collections
RED - Restriction Enzyme Database

FERMENTATION

CBA Database
Current Biotechnology Abstracts

FOOD, FEEDS AND NUTRITION

Abstracts of Mycology
AGREP - Agricultural Research Projects
Analytical Abstracts
Bibliography of Agriculture
Biodeterioration Abstracts
BIOSIS/CAS Selects
CA Selects
Food Science and Technology Abstracts
FROSTI - Food RA Online Scientific and Technical Information Database
Packaging Science and Technology Abstracts
Plant Protection and Quarantine Import Data
Sugar Industry Abstracts
Sugar Statistical Compendium

FORESTRY

Abstracts of Mycology
AGREP - Agricultural Research Projects
Agroforestry Abstracts
Forest Products Abstracts
Forestry Abstracts
Review of Plant Pathology

GENETICS

Abstracts of Mycology
BIOSIS Previews
BIOSIS/CAS Selects
Biotechnology Citation Index
Biotechnology Research Abstracts
CABS Database Online
CAS Biotech Updates
CBA Database
Current Advances in Genetics & Molecular Biology
Current Biotechnology Abstracts
CUTG
DCT - Drosophila Codon Tables
Dnastar Inc
Dnastar Ltd
DRHPL - Database for the Repository of Human Probes and Libraries
GC - Gene Communications
GENBANK
GeneSeq
Genetics Abstracts
HGIR - Human Genome Information Resource
HGMCR - Human Genetic Mutant Cell Repository
HGML - Human Gene Mapping Library
Immunology Abstracts
Microbiological Resource Databank
MIM - Mendelian Inheritance in Man
Plant Breeding Abstracts
Plant Genetic Resources Abstracts
TELEGEN
VECTOR - Cloning Vector Sequence Database

HUMAN BIOLOGY

Genetics Abstracts
HGIR - Human Genome Information Resource
HGMCR - Human Genetic Mutant Cell Repository
HGML - Human Gene Mapping Library
Immunology Abstracts

HUMAN HEALTH CARE

Abstracts of Mycology
Analytical Abstracts
BIOSIS Previews
BIOSIS/CAS Selects
Biotechnology Citation Index
Biotechnology Research Abstracts
The British Library Science Reference and Information Service - SRIS
CABS Database Online
CBA Database
CHEMTRACTS - Organic Chemistry
CNIC - Centre National de l'Information Chimique
Current Biotechnology Abstracts
Current Clinical Chemistry
Current Contents
Current Contents on Diskette - Life Sciences
Dnastar Inc
Excerpta Medica
FROSTI - Food RA Online Scientific and Technical Information Database
Hybridoma Data Bank
Immunology Abstracts

IMMUNOLOGY

INDUSTRY

LEGISLATION AND REGULATORY ISSUES

MARKETS AND MARKETING

MICROBIOLOGY

PATENTS

GeneSeq
MiCIS - Microbial Culture Information Service
MiCIS - Information Centre for European Culture Collections
PatentImages
TELEGEN

PESTICIDES

Agrow OnScreen
CAS ONLINE
EPA's Sampling and Analysis Methods Database
Lewis Publishers
Nematological Abstracts
PESTDOC - Pest Control Literature Documentation
PESTMAN - Pesticide Databank
Review of Plant Pathology
SPF - Standard Pesticides File
Weed Abstracts

PLANT SCIENCE

BIOSIS/CAS Selects
CABS Database Online
Current Awareness in Biological Sciences
Microbiological Resource Databank
PHYTOMED
Plant Biotechnology
Plant Breeding Abstracts
Plant Genetic Resources Abstracts
Plant Growth Regulator Abstracts
Review of Plant Pathology
Virology and AIDS

PRODUCTS AND EQUIPMENT

The British Library Science Reference and Information Service - SRIS
DEQUIP
DOMIS
LIFO-DISK - LIposome inFOrmation Electronic Liposome Library
Pharmacontacts Directories File
Pharmacontacts OnScreen

RESEARCH PROGRAMMES AND PROJECTS

BDT/FTPT
BIOREP
BIOSIS Previews
Biotechnology Citation Index
Biotechnology Research Abstracts
CORDIS
EABS
ENREP
EUREKA
IES-DC
NTIS Bibliographic Database
Pharmacontacts OnScreen
SESAME
TED - Tenders Electronic Daily
Tropical Data Base

SCIENCE AND TECHNOLOGY

The Applied Environmetrics Meteorological Tables
BEST - British Expertise in Science & Technology
The British Library Science Reference and Information Service - SRIS
CABS - Current Awareness in Biological Sciences
EABS
Food Science and Technology Abstracts
FROSTI - Food RA Online Scientific and Technical Information Database
INIST Diffusion
Japan Technology
JICST-E
NTIS Bibliographic Database
Packaging Science and Technology Abstracts
Pascal Biotechnology

VIROLOGY

Review of Plant Pathology
Virology and AIDS

WASTE MANAGEMENT AND TREATMENT

Biodeterioration Abstracts
CA Selects
DECHEMA - Dechema Chemical Engineering and Biotechnology Abstracts Database
EPA's Sampling and Analysis Methods Database
Hazardous Chemical Regulations Database
Lewis Publishers

COMPUTER SOFTWARE

Analysis data handling
Chemical reaction simulation
Genetic engineering
Image analysis
Process control
Reactor control
Sequence analysis
Simulation and modelling

ANALYSIS DATA HANDLING

Anville Instruments
Astra Scientific Instruments Inc
Baird Europe BV
Balzers High Vacuum Ltd
Balzers Inc
BBN Software Products Corp

CHEMICAL REACTION SIMULATION

GENETIC ENGINEERING

IMAGE ANALYSIS

Imstar
IntelliGenetics Inc
Kubota Computer Inc
Large Scale Biology Corp
Molecular Design Ltd
Molecular Dynamics
Nachet Vision
Pias Co Ltd
Scanalytics/CSPI
Societe Bioinformatique et Biotechnologie - SBB
Tietz Video and Image Processing System GmbH
Vision Systems Ltd

PROCESS CONTROL

Astra Scientific Instruments Inc
Biosystems
Biotechnology Computer Systems Ltd
Digital Equipment Corp
FPS Computing Canada Ltd
HCS Industrial Automation BV
IMA GmbH
Keithley MetraByte/Asyst/DAC
Kratos Analytical
Laboratory Technologies Inc
SLT Labinstruments Austria
Societe Bioinformatique et Biotechnologie - SBB
Spectra Physics Inc

REACTOR CONTROL

Astra Scientific Instruments Inc
Enprotech Corp/BEA
HCS Industrial Automation BV
Research Information Systems Inc

SEQUENCE ANALYSIS

Abel Electronics
American Synthesis
Bio-Rad Laboratories
Biosoft
Clontech Laboratories
CompuGene
Cyborg Corp
Dnastar Inc
Dnastar Ltd
Enprotech Corp/BEA
Genesearch Pty
Genofit
Genomic
Hitachi Software Engineering Co Ltd
Hoefer Scientific Instruments
IntelliGenetics Inc
Pharmacia LKB Biotechnology
Porton Instruments
Protein Databases Inc
Soft Gene Berlin
Vision Systems Ltd
VWR Scientific

SIMULATION AND MODELLING

Asahi Kasei Information Systems Co Ltd
Autoscribe
Biostructure
Chemical Design Ltd
Chemical Design US Ltd
Daikin Industries Ltd
Evans & Sutherland
GeneSystems Computer Software
Genomic
Hypercube Inc
IntelliGenetics Inc
Jandel Scientific
Kubota Computer Inc
Lewis Publishers
Mitsui & Co Ltd
Molecule Co Ltd
NAG Ltd
Oxford Molecular
Oxford University Press
Phase Separations Sales
Proteus Molecular Design Ltd
Sierra Instruments
Sigma Chemical Company
Societe Bioinformatique et Biotechnologie - SBB
Soft Gene Berlin
SPI Synthetic Peptides Inc
Tripos Associates
Tripos Associates Inc

CONSULTANTS

Engineering
Environment and energy
Food and agriculture
Genetic engineering and molecular biology
Information and training
Management and investment
Pharmaceuticals and chemicals
Regulations and licensing
Safety and testing
Technology and processing

ENGINEERING

ABE Trading Co Ltd
AMEC Design and Construction Ltd
APV Baker Pty Ltd
The Austin Co UK Ltd
BellBiotech Consulting
Bideco AG
IMC Netherlands BV
Lansdowne Bodycombe and Partners
Spie Batignolles
Tebodin BV
Triad Technologies Inc

ENVIRONMENT AND ENERGY

Allied Colloids Manufacturing GmbH

FOOD AND AGRICULTURE

GENETIC ENGINEERING AND MOLECULAR BIOLOGY

INFORMATION AND TRAINING

Frost & Sullivan Inc
GB Biotechnology Ltd
GEM
Griffith Hack & Co
J F Coates Inc
Leatherhead Food Research Association
McKnight Laboratories GmbH
Modern Talking Picture Service Inc
NATEC
North Carolina Biotechnology Centre
Novabio
Operations Analysis Centre A/S
PJB Publications Ltd
RauCon GmbH
Ravvan International BV
Scansearch
SEAI Technical Publications
SRI International
Stratagene GmbH
TBZ Hamburg
Technology Marketing Group
Theta Corp
TNO
Venture Information Services

MANAGEMENT AND INVESTMENT

Adams Business Associates
AEM Consultants
Ag-West Biotech Inc
AMEC Design and Construction Ltd
Applied Bioscience International
Aquaculture Nederland BV
Bernard Wolnak & Associates
Bio Techno Systems Inc
Biointelligence Service
BioLabs Inc
Biolink Developments Ltd
Bioquest Resources Corp
Biosurvey SA
Biotechnica
Biotechnology Consultants Ltd
Bogart Delafield Ferrier (Europe) Ltd
Boston Biomedical Consultants
Bromont Economic Development Corporation
Cambio
Chandler-Lorenz International
Consulting Resources Corporation
Cranfield Biotechnology Ltd
Croon Business Support BV
Euro BioMed
H O M Consultancy BV
HGF Hamburgische GmbH
Hill & Knowlton (UK) Ltd
IBRD
IWACO BV
J R Snyder Inc
Korda & Company Ltd
Landell Mills Commodities Studies - LMC
Monkman International Consulting
North Carolina Biotechnology Centre
PA Consultancy Group
PA Consulting Group
Roan Ventures
Strategic Technologies International Inc
Summit Biotechnology Associates
TBZ Hamburg
United Biotechnology Sc

PHARMACEUTICALS AND CHEMICALS

AMEC Design and Construction Ltd
Aplibio SA
Applied Bioscience International
The Austin Co UK Ltd
Battelle Institute Ltd
Bernard Wolnak & Associates
Bethyl Laboratories Inc
Bio Techno Systems Inc
Biochemische Synthesetechnik GmbH & Co
Bio/dynamics Inc
Biolink Developments Ltd
BIOS (Consultancy and Contract Research) Ltd
Bogart Delafield Ferrier (Europe) Ltd
Boston Biomedical Consultants
Business Communications Co Inc
Cambridge Biomedical Consultants
CETEC
Chapel Laboratories
Chathamborough Research Group Inc
Clinical Science Research International Ltd
Corporation BDG
DELPHI Associates SPRL
Drug Development (Scotland) Ltd
Euro BioMed
Euro Laboratories Ltd
Frost & Sullivan Inc
H G Engineering International (1980) Ltd
H O M Consultancy BV
Hazleton Europe
Huntingdon Research Centre Ltd
IBRD
Institute of Clinical Pharmacology
Institute of Drug Technology Australia Ltd
John Brown Engineers & Constructors Ltd
Laboratoire Midac
Landell Mills Commodities Studies - LMC
Lansdowne Bodycombe and Partners
Life Sciences International
Monkman International Consulting
Oxford Biomedical Ltd
PA Consulting Group
Saros Ltd
Strategic Technologies International Inc
Theta Corp
TNO
Triad Technologies Inc
Venture Information Services
Western Chemical Research Corp

REGUALATIONS AND LICENSING

The Austin Co UK Ltd

SAFETY AND TESTING

TECHNOLOGY AND PROCESSING

Euro BioMed
EuroForum
GB Biotechnology Ltd
GEM
H G Engineering International (1980) Ltd
H O M Consultancy BV
Heineken NV
Helix CV
Innovation 128
Intek Pty Ltd
Inveresk Research International Ltd
IWACO BV
J F Coates Inc
J R Snyder Inc
John Brown Engineers & Constructors Ltd
Lansdowne Bodycombe and Partners
A Lassonde & Fils Inc
Life Sciences International
Monkman International Consulting
NATEC
North Carolina Biotechnology Centre
Operations Analysis Centre A/S
Oranjewoud Engineers BV
PA Consultancy Group
PA Technology
Pambec Laboratories
PJB Publications Ltd
Reading Scientific Services Ltd
Roan Ventures
Sequal
Summit Biotechnology Associates
TBZ Hamburg
Technoma GmbH
TNO
Trimach BV
Twose Anthony & Associates
Venture Information Services
Webb Technical Group Inc
Western Chemical Research Corp

CULTURE COLLECTIONS AND SUPPLIERS

Algae
Antibodies
Bacteria
Cell culture
Commercial depository
Food and agriculture
Fungi
Identification service
Industrial applications
Medical applications
Patent depositories
Plankton
Probiotics
Protozoa
Research
Rhizobium
Sera and antisera
Tissue culture
Viruses and viral products
Waste treatment
Yeasts

ALGAE

Algensammling am Institut fur Botanik
Australian National Reference Laboratory in Medical Mycology
Centro di Studio dei Microorganismi
Connecticut Valley Biological Supply Co Inc
CSIRO Marine Laboratories
Culture Collection of Algae and Protozoa (CCAP)
Culture Collection of Algae and Protozoa
Norwegian Institute for Water Research - NIVE
R & A Plant/Soil Inc
Sammlung von Cionjugaten-Kulturen

ANTIBODIES

Aqua-Tek GmbH
Laboratoires Unicet
Statens Seruminstitut

BACTERIA

ABC Research Corp
Advanced Biotechnologies Inc
AFRC Institute of Arable Crops Research
Agrifutur Srl
Agropur Cooperative Agro-Alimentaire
American Type Culture Collection - ATCC
The Australian Wine Research Institute
Bacillus Collection
Bacterial Culture Collection
BCZ-Friesland BV
BEL Industries
Bicarbolyte Corp
Bioprox
The CB *Rhizobium* Collection
CCMCU
CCTM - Centre de Collection de Types Microbiens
CECT - Coleccion Espanola de Cultivos Tipo
Central Bureau for Cultures of Fungi
Central Laboratories
Central Laboratory Collection
Centro di Studio dei Microorganismi
CERS
Chiba University
Collection de Champignons et d'Actinomycetes Pathogenes
Collection Laboratorium voor Microbiologie
Collection Nationale de Cultures de Microorganisms
Collezione Instituto di Microbiologia
Connecticut Valley Biological Supply Co Inc

CELL CULTURE

COMMERCIAL DEPOSITORY

FOOD AND AGRICULTURE

Diamond V Mills Inc
Eurozyme
Fermentation Research Institute
Franklin Products International BV
Guinness (Dublin) Culture Collection
IAL - Colecao de Culturas Adolfo Lutz
IBSBF
Institute for Microbial Resources
Institute of Physical and Chemical Research
Kaken Pharmaceutical Co Ltd
Lacto Labo
MAFF Agricultural Science Services
Microbiologie des Sols INRA
National Collection of Food Bacteria
National Institute of Agrobiological Resources
New Zealand Reference Culture Collection (Plant & Soil)
NRRL/ARS Culture Collection
Osaka Institute for Fermentation
Otto Madsen Dairy Research Laboratory
Plant Science Ltd
R & A Plant/Soil Inc
Research Centre Agriculture Canada
Research Institute for Plant Protection
Research Station Agriculture Canada
Rosell Institute Inc
Rothamsted *Rhizobium* Collection
Statens Seruminstitut
Swedish University of Agricultural Sciences
Sydney University - Dept of Microbiology
UFPEDA - UFPE Department of Antibiotics
Valio Research and Development Centre
Volcani Centre *Rhizobium* Collection
Wageningen Agricultural University
Waite Agricultural Research Institute
Western Australian Culture Collection
Wine Yeast Collection

FUNGI

American Type Culture Collection - ATCC
Australian National Reference Laboratory in Medical Mycology
The Australian Wine Research Institute
Biologische Bundesanstalt fur Land- und Forstwiwrtschaft - BBA
Canadian Collection of Fungus Cultures
CCMCU
CECT - Coleccion Espanola de Cultivos Tipo
Central Bureau for Cultures of Fungi
Centro di Studio dei Microorganismi Chiba University
Connecticut Valley Biological Supply Co Inc
DSM - Deutsche Sammlung von
Mikroorganismen un Zellkultlureen GmbH
Fermentation Research Institute
Forintek Canada Corp
Fungal Collection, Belgium
IAL - Colecao de Culturas Adolfo Lutz
Industrial Microbiology Dublin
Institute for Microbial Resources
Institute of Physical and Chemical Research
International Mycological Institute - IMI
ITAL Food Technology Institute
Louvain Catholic University
Materials Research Laboratories Collection of Fungi
Medical Mycology, Belgium
Museum of Natural History
National Collection of Pathogenic Fungi
National Collection of Wood Rotting Fungi
National Culture Collection of Microorganisms - CNCM
National Health Institute
National Institute of Agrobiological Resources
NCS Diagnostics Inc
NRRL/ARS Culture Collection
Osaka Institute for Fermentation
Otto Madsen Dairy Research Laboratory
Presque Isle Cultures
Research Station Agriculture Canada
Rosell Institute Inc
Spanish Culture Collection
Swedish University of Agricultural Sciences
Sylvan Spawn Inc
UFPEDA - UFPE Department of Antibiotics
University of Alberta - Microfungus Collection and Herbarium
VTT Collection of Industrial Microorganisms
Wageningen Agricultural University
Waite Agricultural Research Institute
Western Australian Culture Collection
Wine Yeast Collection
Wood Technology and Forest Research Division

IDENTIFICATION SERVICE

Australian National Reference Laboratory in Medical Mycology
Bacterial Culture Collection
Biologische Bundesanstalt fur Land- und Forstwiwrtschaft - BBA
Canadian Collection of Fungus Cultures
Central Bureau for Cultures of Fungi
Central Laboratories
Central Laboratory Collection
Centro di Studio dei Microorganismi
CERS
Collection de Champignons et d'Actinomycetes Pathogenes
Collezione Instituto di Microbiologia
Culture Collection of Algae and Protozoa
Culture Collection University of Goteborg
Department of Plant Defence Culture Collection
Fungal Collection, Belgium
IAL - Colecao de Culturas Adolfo Lutz

INDUSTRIAL APPLICATIONS

MEDICAL APPLICATIONS

PATENT DEPOSITORIES

PLANKTON

RESEARCH

Australian National Reference Laboratory in Medical Mycology
The Australian Wine Research Institute
Bacterial Culture Collection
Bayerische Landesanstalt fur Weinbau und Gartenbau
CAYLA Sarl
The CB *Rhizobium* Collection
CCMCU
CCTM - Centre de Collection de Types Microbiens
CECT - Coleccion Espanola de Cultivos Tipo
Central Bureau for Cultures of Fungi
Central Laboratories
Central Laboratory Collection
Centro di Studio dei Microorganismi
Centro di Virologia
Chiba University
Collection Laboratorium voor Microbiologie
Collection Nationale de Cultures de Microorganisms
Collezione Instituto di Microbiologia
CSIRO Division of Plant Industry
Culture Collection of Algae and Protozoa (CCAP)
Culture Collection University of Goteborg
Department of Plant Defence Culture Collection
DSM - Deutsche Sammlung von Mikroorganismen un Zellkultlureen GmbH
European Collection of Animal Cell Cultures
Forintek Canada Corp
Guinness (Dublin) Culture Collection
IAL - Colecao de Culturas Adolfo Lutz
Industrial Microbiology Dublin
Institute for Medical Research
Institute for Microbial Resources
Institute of Physical and Chemical Research
International Mycological Institute - IMI
ITAL Food Technology Institute
John Innes Institute Plant Virus Collection
Kaken Pharmaceutical Co Ltd
Laboratory of Microbiology and Pathology
Microbiologie des Sols INRA
National Collection of Wood Rotting Fungi
National Health Institute
National Institute of Agrobiological Resources
National University of Athens
New Zealand Reference Culture Collection (Plant & Soil)
Osaka Institute for Fermentation
Otto Madsen Dairy Research Laboratory
Pasteur Culture Collection of Cyanobacterial Strains
Queensland University
Research Centre Agriculture Canada
Research Institute for Plant Protection
Research Station Agriculture Canada
Rothamsted *Rhizobium* Collection
Sammlung von Cionjugaten-Kulturen
Servico de Microbiologia e Immunologia
Sheffield University
Statens Seruminstitut
Swedish University of Agricultural Sciences
UFPEDA - UFPE Department of Antibiotics
Valio Research and Development Centre
VTT Collection of Industrial Microorganisms
Wageningen Agricultural University
Waite Agricultural Research Institute
Western Australian Culture Collection
Wine Yeast Collection
Wood Technology and Forest Research Division

RHIZOBIUM

Agrifutur Srl
The CB *Rhizobium* Collection
CCMCU
CSIRO Division of Plant Industry
Queensland University
Rothamsted *Rhizobium* Collection
Swedish University of Agricultural Sciences
Volcani Centre *Rhizobium* Collection
Wageningen Agricultural University

SERA AND ANTISERA

Advanced Biotechnologies Inc
American Type Culture Collection - ATCC
National Institute of Allergy and Infectious Diseases - NIAID
NCS Diagnostics Inc
Statens Seruminstitut

TISSUE CULTURE

Advanced Biotechnologies Inc
American Type Culture Collection - ATCC
Centro di Virologia
Collezione Instituto di Microbiologia
Connaught Laboratories Inc
In Vitro International Inc
NCS Diagnostics Inc
Sydney University - Dept of Microbiology
UFPEDA - UFPE Department of Antibiotics

VIRUSES AND VIRAL PRODUCTS

Advanced Biotechnologies Inc
Agriculture Canada
American Type Culture Collection - ATCC
Central Laboratories
Centro di Virologia
Collezione Instituto di Microbiologia
DSM - Deutsche Sammlung von Mikroorganismen un Zellkultlureen GmbH
Electro-Nucleonics Inc

WASTE TREATMENT

YEASTS